TIDALITES: PROCESSES & PRODUCTS

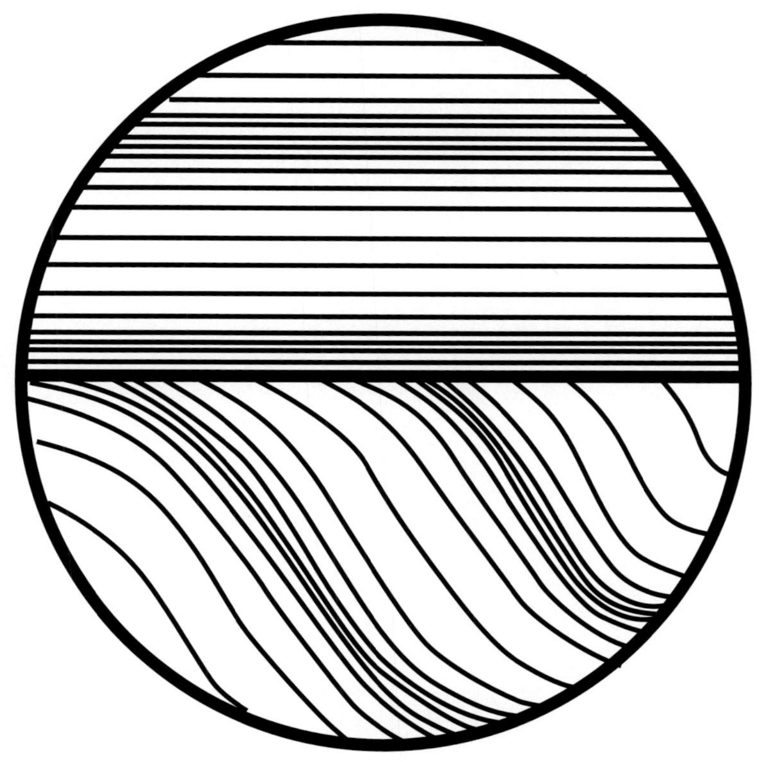

Edited by

Clark R. Alexander, Skidaway Institute of Oceanography, Savannah, GA

Richard A. Davis, University of South Florida, Tampa, FL

and

Vernon J. Henry, Georgia Southern University, Statesboro, GA

Robert W. Dalrymple, Editor of Special Publications
Special Publication No. 61

Tulsa, Oklahoma, U.S.A. *November, 1998*

ISBN 1-56576-059-X

1731 E. 71st Street
Tulsa, OK 74136-5108
1-800-865-9765

Printed in the United States of America

CONTENTS

PREFACE

Approximately every four years, an international meeting is held to disseminate the results of current research in tidal sedimentology. Because there is no formal organization of tidal sedimentologists, these meetings are planned and conducted by volunteer effort. Meetings have typically been sponsored by the International Association of Sedimentologists and SEPM, the Society for Sedimentary Geology. The editors wish to acknowledge these organizations for support of the 1996 meeting, Tidalites '96, which was held in Savannah, Georgia, USA. The meetings generally draw a core group of attendees in addition to those attending for the first time. Students commonly play a prominent role in the program; five of the papers in this volume have a graduate student as senior author.

Research papers were solicited for this volume solely from those who presented a paper or poster at the Savannah meeting. Each manuscript was reviewed by at least two of the editors, and by at least two others, some of whom attended the conference and others who did not. After careful scrutiny the papers presented here were selected for inclusion in this volume because of their quality and because they addressed some of the main themes of tidal sedimentology. Several papers that were reviewed were not included.

The editors would like to thank the following reviewers for their constructive evaluation of the manuscripts: Dennis Anthony, Allen Archer, Jesper Bartholdy, John Boothroyd, George Klein, Poppe de Boer, John-Claude Dionne, Gerhard Els, Kenneth Eriksson, Graham Evans, Berg Flemming, William Frazier, Robert Ginsburg, Stephen Greb, Thomas Gross, Inge Kaas, Diane Kamola, Gail Kineke, Frank Krogel, Erik Kvale, Charles Fletcher, Rudi Meyer, Paul Myrow, Greg Ojakangas, Peter Popenoe, Denise Reed, Stan Riggs, Don Swift, Rod Tillman, and John Wells.

Clark R. Alexander
Richard A. Davis, Jr.
Vernon J. Henry

December 1997

TIDAL SEDIMENTOLOGY: HISTORICAL BACKGROUND AND CURRENT CONTRIBUTIONS

RICHARD A. DAVIS, JR.
Coastal Research Laboratory, Department of Geology, University of South Florida, Tampa, Florida 33620
CLARK R. ALEXANDER
Skidaway Institute of Oceanography, 10 Ocean Science Circle, Savannah, Georgia 31411
AND
VERNON J. HENRY
Applied Coastal Research Laboratory, Georgia Southern University, 10 Ocean Science Circle, Savannah, Georgia 31411

ABSTRACT: Sedimentation in tidal environments covers a broad spectrum of conditions, and produces a wide range of tidal signatures. These sediment sequences that have accumulated as the result of significant tidal influence are called tidalites. Over the past few decades, several models have been developed that are useful in interpreting tidal sequences in the ancient stratigraphic record. The models include tidal bedding, tidal bundles, and the relationships of cycles with various time scales to these tidal deposits. Although the efforts of sedimentologists studying both modern and ancient sequences have provided techniques for detailed analysis of tidalites, there is still much to be learned about tidal sedimentation and the resulting sediment deposits. This volume is a step toward that end. This introductory paper will provide the reader with a brief history of the organized effort in tidal sedimentology along with a synopsis of this volume.

PREVIOUS MEETINGS AND VOLUMES ON TIDAL SEDIMENTOLOGY

In February 1973, R. N. Ginsburg organized the first in what was to become a series of meetings on tidal sedimentology. A group of about 50 scientists from around the world met for a two-day field trip at Sapelo Island, Georgia (USA), to visit modern siliciclastic tidal environments under the leadership of James D. Howard. The group then moved to Fishers Island in Miami, Florida, for technical sessions on modern and ancient siliciclastic and carbonate tidal deposits. Oral sessions were co-chaired by George D. Klein and were supplemented by poster sessions and core workshops. One of the primary group activities at the meeting was to develop definitions for terms in tidal sedimentology and symbols that could be used in describing stratigraphic sequences deposited in tidally influenced environments. A field trip to the Bahamas completed the conference.

Each contributor at the meeting was urged to prepare a short paper for a special volume using the newly developed symbols. This effort produced the landmark publication *Tidal Deposits* (Ginsburg, 1975). Unlike the succeeding works on the topic, the Ginsburg volume is equally divided between siliciclastic and carbonate tidalites. The major focus of the papers is on tidal flat deposits to the exclusion of channel and shelf-tidal environments. Criteria for recognizing tidal flat deposits in the stratigraphic record were summarized. Obviously the criteria are different for recognizing tidal influence in each major category of sediments and most agreed that siliciclastic sequences present the greatest challenge in this regard.

After a long hiatus, the next meeting held specifically on the topic was on the Island of Texel, The Netherlands, in 1985. Because of the outstanding contributions of several Dutch sedimentologists over many years, this was a natural place to have an international conference on tidal sedimentology. This meeting began the regular scheduling of such conferences, and it was agreed that a three- or four-year interval between meetings would be appropriate. A proceedings volume (de Boer et al., 1988) was also published. The contents of this volume stress modern tidal deposits with emphasis on offshore environments, although a spectrum of environments and geologic ages is included. The highlight of the contributions made by the volume to the study of tidalites was the popularization of the concept of the tidal bundle. A field trip to view several modern examples of these features confirmed their interpretation and importance.

The next meeting was held in Calgary, Canada, in 1989. Although Calgary is not a location that is convenient to modern tidal environments, the Cretaceous tidalite sequences of southern Alberta proved excellent venues for field trips. Numerous outstanding papers are presented in the proceedings volume (Smith et al., 1991), which includes a broad spectrum of topics evenly divided between modern tidal environments and their ancient counterparts. The field trips and the papers emphasized tidal rhythmites, especially tidal bedding, with the estuarine environment as the site of accumulation.

The succeeding conference in 1992 was a return to another northern European tidal coast at Wilhelmshaven, Germany, where much of the pioneering work on tidal flats was conducted by scientists at the Senckenberg Institute. Field trips were taken to the classic study locations, including Jade Bay. In addition, H. E. Reineck provided the keynote address. The proceedings volume (Flemming and Bartholoma, 1995) included many excellent papers with an emphasis on the modern nearshore environments along the northern coast of Europe. This volume also included tide-dominated deltas for the first time in the spectrum of depositional environments where tidalites accumulate.

Plans were made to hold the next meeting in Savannah on the coast of Georgia (USA) in May 1996. This region is significant because Reineck and his colleagues, along with Jim Howard, John Hoyt, and Jim Henry, collaborated on research in coastal tidal environments of this region in the 1960s.

The Savannah meeting, from which this volume arose, hosted nearly 100 attendees from 17 countries and featured about 60 papers and a poster session. Field trips were held before and after the meeting, including one to Sapelo Island, the site of much of the original tidal sedimentology research by the groups from Germany and Georgia (USA). The attendees voted to accept the invitation from Y. A. Park and his colleagues to convene the next meeting in Seoul, Korea, during August 2000.

SYNOPSIS OF THE VOLUME

The papers in this volume are divided into three thematic groups: Tidal Sedimentation in the Wadden Sea, Tidal Rhythmites, and Stratigraphy of Tidal Sequences. These follow the retrospective paper by George Klein (*Clastic Tidalites—A Retrospective View*). Klein was a natural choice to present the

Tidalites: Processes and Products, SEPM Special Publication No. 61

keynote paper at the conference. As one of the first sedimentologists in North America to conduct detailed process-response investigations in tide-dominated environments, he coined the term tidalites. His paper provides an historical perspective on research in tidal environments. It is meant to inform the reader of some of the highlights of tidal sedimentology over the years and also to point out the value of understanding this topic when interpreting ancient sedimentary sequences.

TIDAL SEDIMENTATION IN THE WADDEN SEA

A special half-day session on the Wadden Sea was included in the conference program. The Wadden Sea is the tidal flat/ tidal channel complex that separates the barrier island system from the mainland at the south end of the North Sea along the coasts of The Netherlands, Germany and Denmark. It is this extensive tide-dominated coastal area that was, in many respects, the birthplace of tidal sedimentology through the efforts of Van Straaten, Reineck and others. The papers on this topic are found in the first part of the volume. The focus of these papers is on fine-grained sediment processes and products with emphasis on suspended sediment flux and patterns of sediment distribution. Included are innovative ways of looking at the behavior of fine-grained sediments under tide-dominated conditions. This emphasis on muddy sediments has long been neglected in comparison to sand facies because of the difficulty in studying fine-grained material in both the field and the laboratory. These papers represent a step toward increasing both our efforts and our understanding of these problematic sediments.

In an investigation of fine sand-dominated tidal flats behind the island of Spiekroog in the eastern part of the German Wadden Sea, Nyandwi (in "Sediment Distribution Patterns in the Back-barrier Areas of the Wadden Sea, Spiekeroog Island, Germany") is able to demonstrate the existence of two distinct patterns of landward fining of sediment that appear to be in conflict. One follows the orientation of tidal inlets and is restricted to these areas. The other follows a shore-normal trend of landward fining, and is related to the landward decrease in overbank flow velocities. The latter can only be recognized by high-density sampling. The paper shows that great care must be taken in interpreting sediment dispersal pathways from granulometric data.

Hertweck, working in the same general area of the German Wadden Sea, provides important tidal flat facies information based on both physical sediment characteristics and biogenic structures (in "Facies Characteristics of Back-barrier Tidal Flats of the East Frisian Island of Spiekeroog, Southern North Sea"). This comprehensive study of tidal flat facies provides a picture of how these two major aspects of tidal flats interact and how their relative distributions are dictated by the level of physical energy, in much the same way as the biological and physical processes interact across the inner continental shelf.

Krögel and Flemming provide data in their paper that question usual interpretations of fine-grained sediment distribution (in "Evidence for Temperature-adjusted Sediment Distributions in the Back-barrier Tidal Flats of the East Frisian Wadden Sea Southern North Sea"). They find that the seasonal range in water temperature and its effect on viscosity plays a significant role in the transport and accumulation of fine-grained sediments across the tidal-flat environment because of the variation in settling velocity. As a result, a given sediment particle may have various settling velocities over the annual climatic cycle.

In a pair of companion papers, Bartholdy and Anthony consider fine-grained sediment flux in tidal channels, over both seasonal periods and during calm weather conditions (in "Tidal Dynamics and Seasonal Dependent Import and Export of Fine-grained Sediment through a Back-barrier Tidal Channel of the Danish Wadden Sea; and Variability of Fine-grained Suspended Sediment Concentrations in a Back-barrier Tidal Channel during Calm Weather Conditions"). The first of these papers shows that fine-grained sediment is preferentially transported both out to sea and onto the marsh surface during storm periods. It is during periods of windy conditions after long periods of calm (summer) that sediment is imported into the Danish Wadden Sea. During calm weather there is significantly more sediment suspension during the ebb cycle than during the flood portion of the tidal cycle. There is also variation in suspended sediment associated with the spring/neap tidal cycle. The overall result is the net export of suspended sediment to the open North Sea during calm conditions.

This group of papers considers detailed elements of sediment transport and sediment distribution patterns in order to understand how the tidal flat and related tidal channel system functions. Some of these findings are contrary to the general ideas and interpretations of tidal sedimentology, but they illustrate the need for continued examination of how tidal flux influences sediment dynamics.

TIDAL RHYTHMITES

The recognition of various tidal cycles in the stratigraphic record has been a major contribution to the interpretation of ancient tidally influenced environments, and to the understanding of rates of sediment accumulation and preservation. The range of frequencies preserved in these cycles extends from semi-diurnal to at least annual. The recognition of sediment accumulation that has occurred in as little as six hours represents the most detailed time-scale recorded in the rock record from any depositional environment. Although these sequences represent only short periods of geologic time they provide unequaled resolution of rates of sediment accumulation. This group of papers concentrates on the details of these types of rhythmic accumulations, and provides some cautionary points to consider when making temporal interpretations from the ancient record.

Archer provides important insight into the controls that dictate the occurrence and type of rhythmic tidal deposits (in "Hierarchy of Controls on Cyclic Rhythmite Deposition: Carboniferous of Eastern and Mid-continental U.S.A."). Using well-preserved sequences that have been recognized and studied in detail for many years, Archer looks at a variety of scales of these factors. They range from global climate conditions, through the unique setting of large, shallow basins, to special tidal conditions.

Just how much time is represented by the various, thin sedimentation packages that comprise rhythmite sequences? Tessier examines rhythmite deposits from the Mt. Saint Michel area on the northern coast of France (in "Tidal Cycles: Annual versus Semi-lunar Records") to answer this question. In this marsh environment, patterns typically interpreted to be lunar cycles are shown instead to be the result of annual cycles. This observation is an important lesson on the necessity of making detailed, careful, and long-term observations in the modern environment.

An example of some of the conclusions reached in the above paper is provided by Greb and Archer (in "Annual Sedimentation Cycles in Rhythmites of Carboniferous Tidal Channels") from stratigraphic sequences in the Appalachian and Illinois basins. The scale of their stratigraphic analysis extends to the sub-millimeter level in order to make the proper interpretations. The authors find that the periodicity of rhythmite cycles can be easily misinterpreted in the absence of detailed observations.

Another example of a range of tidal periodicities in rhythmites from the Carboniferous is provided by Adkins and Eriksson (in "Rhythmic Sedimentation in a Mid-Pennsylvanian Delta Front Succession, Magoffin Member (Four Corners Formation, Breathitt Group), Eastern Kentucky: A Near-complete Record of Daily, Semi-monthly, and Monthly Tidal Periodicities"). The huge sediment supply to the delta front, coupled with the low degree of bioturbation and the high quality of the preservation of the laminations, provide data that show fortnightly lunar cycles of up to a meter of accumulation, and daily accumulations up to 30 cm thick for short intervals of time.

The common interpretation of tidal rhythmite deposition is associated with marine conditions. The paper by Kvale and Mastalerz (in "Evidence of Ancient Freshwater Tidal Deposits") shows that although within the coastal regime, tidal signatures may extend well into the freshwater environment. They examine rhythmites that accumulated in an environment that experienced tidal conditions but that was not encroached upon by brackish or salt water. The absence of biogenic activity to disturb the tidal rhythmites is a major factor in their preservation, and it is likely that these types of deposits are far more common in the stratigraphic record than has been previously recognized.

The final paper in this group by Lanier and Tessier (in "Climbing-ripple Bedding in the Fluvio-estuarine Transition: A Common Feature Associated with Tidal Dynamics (Modern and Ancient Analogues") examines tidal influences on small-scale bedform preservation. Their work shows that suspension of silt and very-fine sand produces rhythmites in various tidal environments. They document exactly the same detail in the sequences from the modern tidal environment as is preserved in the Carboniferous of Kentucky.

This group of papers illustrates the wide range of appearances of tidal rhythmites and most importantly, it points out some of the important problems when interpreting the periodicity of the cycles that produce these rhythmites. These papers also show that tidal bedding is generally restricted to the landward portions of the tidal environments in which they accumulate. Another critical point brought out by these papers is the need for detailed observations in modern environments, including time-series data collection. Such interpretations must be applied with care when examining tidal sequences from the ancient record.

STRATIGRAPHIC SIGNATURES OF TIDAL PROCESSES

The last group of papers in this volume considers tidalites in a broader stratigraphic context. Each of the papers in this group considers stratigraphic sequences that were deposited in tidal environments associated with a variety of other coastal and nearshore marine environments. These papers demonstrate the application to the bigger geologic picture of process-oriented information, and the various types of rhythmites and their time-scales as discussed in the first two sections of the volume. The papers are presented in geochronologic order starting with the youngest.

Sager and Riggs show that the Holocene valley-fill deposits from the Albemarle Sound area of North Carolina, USA (in "Models for the Holocene Valley-fill History of Albemarle Sound, North Carolina, U.S.A.") contain tidally influenced sediment accumulations from several environments. Tidal flats and channels are the primary tidalite producers within the succession that reflect sea-level change and the formation of barrier islands.

A contrasting estuarine deposit from the Cretaceous of the Western Interior Seaway shows tidalites representing more energetic conditions. Meyer et al. (in "Unconformities within a Progradational Estuarine System: The Upper Santonian Virgelle Member, Milk River Formation. Writing-on-Stone Provincial Park, Alberta, Canada") point out the importance of unconformities in these deposits, which are discussed in a sequence stratigraphic framework. The unconformity in question is termed a tide-cut source diastem. Tidal bars and tidal channels are the primary tide-dominated environments preserved. The prograding estuarine model presented is a contrast to traditional sequence-stratigraphic models of transgressive estuaries.

Myrow shows that tidal sand waves and superimposed subaqueous dunes occupied the primary tidal environments in the Cambrian of Colorado (in "Transgressive Stratigraphy and Depositional Framework of Cambrian Tidal Dune Deposits, Peerless Formation, Central Colorado, U.S.A."). This energetic and sand-dominated depositional environment was probably a tide-dominated open coast in an epicontinental setting, perhaps similar to that on the southern Queensland coast of Australia. These sand waves were quite large and accumulated as a transgressive systems tract above a tidal ravinement surface.

Coarse tidalites were accumulating in the fluvio-estuarine system of the Precambrian in South Africa as reported by Els and Mayer (in "Coarse Clastic Tidal and Fluvial Sedimentation during a Large Late Archean Sea-level Rise: The Turffontein Subgroup in the Vredefort Structure, South Africa"). Tidal bundles and mud drapes characterize the sand-dominated tidalites that accumulated along this high-energy coast.

This group of papers shows that successions of tidalites may represent a wide range of specific conditions within tidally influenced or dominated environments. Whereas one commonly thinks of such sequences as representing transgressive estuarine or open coastal conditions, here are additional models of prograding estuaries and fluvial deltas.

SUMMARY

The papers in this volulme examine elements of tidal sedimentology with differing but complimentary approaches. The processes and sediment distribution characteristics of the Wadden Sea that emphasize fine-grained sediments provide a good introduction to the cyclic patterns described in the papers on rhythmites. Although the origin and periodicity of these rhythmites are generally considered to be well-understood, the authors emphasize important cautionary points that must be considered in interpreting the temporal significance of these deposits. The final group of papers spans much of the geologic history of the Earth and illustrates how various tidalite deposits are incorporated into coastal sedimentary successions, and how they fit into a sequence-stratigraphic framework.

REFERENCES

DE BOER, P. L., VAN GELDER, A., AND NIO, S. D., eds., 1988, Tide-influenced Sedimentary Environments and Facies: Dordrecht, D. Reidel Publishing Company, 530 p.

FLEMMING, B. W., AND BARTHOLOMA, A., eds., 1995, Tidal Signatures in Modern and Ancient Sediments: Oxford, International Association of Sedimentologists, Special Publication 24, 358 p.

GINSBURG, R. N., ed., 1975, Tidal Deposits, A Casebook of Recent Examples and Fossil Counterparts: New York, Springer-Verlag, 428 p.

SMITH, D. G., REINSON, G. E., ZAITLIN, B. A., AND RAHMANI, R. A., eds., 1991, Clastic Tidal Sedimentology: Calgary, Canadian Society of Petroleum Geologists, Special Publication Number 16, 387 p.

CLASTIC TIDALITES—A PARTIAL RETROSPECTIVE VIEW

GEORGE D. KLEIN
SED-STRAT Geoscience Consultants, Inc., P.O. Box 42188, Houston, Texas, 77242, USA. e-mail: GDKlein97@worldnet.att.net

ABSTRACT: Research on tidalites, sediments deposited by tidal currents, evolved through four phases during the last half century:

PHASE I. Facies mapping of Holocene tidalites in Germany, Holland, the United Kingdom, and Canada identified the seaward-coarsening pattern of sediment distribution, a distinct zonation of sedimentary structures, and provided a fining-upward facies model used to recognize ancient counterparts. Mapping in subtidal areas showed that extensive sheets of tidally molded-and-deposited sand accumulations characterized continental shelves that were both wide, and funnel-shaped in plain view. Similarly, extensive work was completed on carbonate tidalites, although it is not discussed herein.

PHASE II. Study of sedimentary structures was followed by a detailed analysis of sediment transport dynamics on intertidal sand bodies in Canada, where time-velocity asymmetry is the major factor controlling sand body geometry, orientation of bedforms, grain size distribution, sediment dispersal, and sand body orientation. Parallel work in tide-dominated continental shelves of the Yellow Sea of Korea and the southern North Sea showed similar patterns. These studies confirmed that tidal sand bodies are likely to be preserved in the rock record and provide a counterpart facies that is likely to dominate ancient cratonic seas.

PHASE III. In tide-dominated estuaries of The Netherlands, cross-bedded units were observed to be organized into discrete bundles that were correlated to neap and spring tides. These observations were replicated in ancient counterparts.

PHASE IV. Detailed analysis of the Schelde Estuary, The Netherlands, demonstrated that parallel-bedded couplets of sand and mud (tidal bedding) could be correlated directly to neap-spring tidal cycles. Recognition of such couplets, particularly in Mississippian and Pennsylvanian sediments of the midcontinent of North America, can be correlated to lunar dynamics and tidal patterns.

All of these studies demonstrated that tidalites accumulated rapidly and were preserved widely. Where preserved in the stratigraphic record, tidalites represent accumulation during very short time intervals. Consequently, in many sequences where such facies are preserved, the time gaps in the stratigraphic record were far longer than previously interpreted.

INTRODUCTION

The term Tidalites was introduced by Klein (1971,1972) to designate (at that time) a new process-sedimentary facies, namely, sediments deposited by tidal currents. I proposed the term after completing a sabbatical leave at Oxford University and integrating work I had completed in the Minas Basin of the Bay of Fundy, Canada, and Precambrian counterparts (Klein 1970a,b). It seemed to me, particularly after completing field work examining some ancient tidalites, that a process facies for these sediments was needed paralleling the Turbidite concept for sediments deposited by turbidity currents. Thus, I formally proposed the term in 1971 (Klein, 1971), and the term was accepted almost immediately, but not without some detractors. At the time I proposed the term, I did not envisage either periodic tidalites conferences or an invitation to present this paper 25 years later.

Many geologists assume that tidal sedimentation research focuses only on intertidal flat environments (see discussion by Klein, 1976, 1977a). This assumption (derived from accessibility) overlooked the significance of tidal sedimentation processes in subtidal shelf and deepwater marine settings. Thus, in subtidal continental shelves, particularly where an increase in shelf width tends to enhance tidal current intensity (Cram, 1979; Klein, 1977a,b; Klein and Ryer, 1978), large tidal current sand ridges are common and they show evidence of tide-dominated deposition (Off, 1963; Houbolt 1968; Boggs, 1974; Stride, 1963, 1982; Belderson, 1964; Klein et al., 1982; Belderson et al., 1972; McCave and Langhorne, 1982; Kenyon et al., 1981). Such subtidal sand bodies also are tidalites.

By way of background, the intertidal zone represents a diverse depositional surface demarcated by the elevation of both high- and low-tide stage. Intertidal flats are low-sloping features within this interval that are exposed at low-tide stage. Intertidal sand bodies represent linear shoals or bars deposited by tidal currents that occur in the lower portions of intertidal flats and the intertidal zone. They are exposed also at low tide, in contrast to tidal current sand ridges, which are always subtidal.

Intertidal environments have fascinated sedimentologists for some time. Within them, a variety of sedimentary features are known, many of which are common to ancient sedimentary rocks. Because of geologists' urge to understand the origin, depositional processes, environment, and predictive trends of sedimentary rocks, they have examined modern systems of sediment transport and deposition to interpret the ancient sedimentary rock record. A uniformitarian approach has underlain many of the interpretations common to practitioners of sedimentology, and the intertidal zone is no exception to such analysis, largely because of its accessibility.

Earliest-studied tidal flats occur on the coasts of northwestern Europe along the North Sea. Much of the earlier and descriptive work done there was summarized in literature inaccessible to most of the English-speaking segment of the geological community. The earlier papers were small-scale studies focusing on specific features. Hantzschel (1939) provided the first English-language synthesis on German work.

Intertidal flat sedimentation studies exploded immediately following World War II. The first major work was by Van Straaten, in The Netherlands, who presented a series of papers focusing first on specific sediment features, especially sedimentary structures, and later expanding to include comparative studies of other areas and a synthesis of depositional processes (Van Straaten, 1952, 1953, 1954, 1959, 1961; Van Straaten and Kuenen, 1957). These studies were followed by an extensive on-going program of research along the northwest coast of Germany by Reineck (1963, 1967, 1972) and Reineck and Wunderlich (1968a,b); these are summarized in a chapter in Reineck and Singh (1973, 1980). Evans' (1965, 1975) work along the coast of eastern England in The Wash and work by Bajard (1996) in France comprise other earlier European work. In North America, intertidal flat sedimentary structures were first described by Kindle (1917) at the Bay of Fundy; these were followed by a regional summary by Klein (1963), a comparative study of Bay of Fundy and western European intertidal flats by Klein and Sanders (1964), and detailed work on intertidal

Tidalites: Processes and Products, SEPM Special Publication No. 61

sand bodies by Klein (1970a) and Knight (1980), including their bedform migration (Klein and Whaley, 1972; Dalrymple et al., 1978). Thompson (1968) provided a most detailed summary of intertidal flats in an arid setting along the Gulf of California, and Pestrong (1972) completed a distributional sedimentological study of tidal flats along San Francisco Bay. In Australia, Gellatly (1970) described sedimentary structures from intertidal sand bodies in King Sound; the regional intertidal flats there were described later by Semeniuk (1981).

These studies provided a baseline for comparison and development of facies models. Facies models on intertidal flats were provided by Klein (1971), whereas facies models for intertidal sand bodies were proposed first by Knight and Dalrymple (1975).

Tidal processes are common also to many other coastal environments including deltas (Coleman, 1980), barrier islands, and tidal marshes other than those associated with intertidal flats and intertidal sand bodies. Tidal sedimentation processes are active also in deepwater settings in water depths ranging from 2000 to 2500 m (Keller et al., 1973; Shepard et al., 1969; Shepard and Marshall, 1973; Lonsdale et al., 1972; Lonsdale and Malfait, 1974), and the sedimentary features so produced show some element of commonality in type to features produced in coastal and relatively shallow subtidal conditions.

Global Distribution

It has long been known that tidal ranges and tidal current velocities along coastlines tend to vary with shelf width and the associated relative intensity of wave action (Davies, 1964; Hayes, 1975, 1979; Cram, 1979; Klein, 1977a,b; Klein, 1985a, Klein and Ryer, 1978). Tidal ranges were used to classify coastlines according to the following scheme of Davies (1964): microtidal (<2 m), mesotidal (2-4 m), and macrotidal (>4 m; see Klein, 1985a, p. 189, his Fig. 3-1). Hayes (1979) discussed variation in coastline morphology and sedimentary systems in each of these settings. His summaries (Hayes, 1975, 1979) demonstrated that extensive intertidal flats occur in macrotidal areas, whereas narrower but significant intertidal flats occur also in mesotidal areas.

In fact, few of these places have been studied in detail. Perhaps the most spectacular intertidal flats are those from the Yellow Sea of Korea, a macrotidal coast. These were studied by Chung and Park (1977), Wells and Huh (1979) and Alexander et al. (1991), who demonstrated a textural sediment distribution there similar to intertidal flats from the North Sea (see below). Both intertidal and subtidal sand bodies also have been observed in the Yellow Sea (Klein et al., 1982). The Bay of Fundy intertidal flats of the Minas Basin are the only intertidal flats and intertidal sand bodies from a macrotidal coast that have been examined in detail (Klein, 1963, 1970a; Knight, 1980; Knight and Dalrymple, 1975; Dalrymple et al., 1978; Lambiase, 1980). The Baie du Mont Saint Michel of France is an area of well-described flats also occurring in a macrotidal domain (Larsonnieur, 1975; Bajard, 1966). The Wash of eastern England (Evans, 1965; McCave and Geiser, 1979) is also well described and it, too, occurs in a macrotidal setting (Davies, 1964).

Intertidal flats have been described from many mesotidal coasts. These include the North Sea coastline of The Netherlands (Van Straaten, 1952, 1959, 1961), and of Germany (Reineck, 1963, 1967, 1972). Mesotidal-range intertidal flats along the Gulf of California were described by Thompson (1968) and those of San Francisco Bay by Pestrong (1972). Similarly, intertidal sand bodies (Gellantly, 1970) and intertidal flats (Semeniuk, 1981) from the mesotidal domain of King Sound, Australia, are noteworthy because of their combined semitropical and local setting. Muddy intertidal flats have been reported and analyzed in detail from mesotidal coasts in New Hampshire (Anderson et al., 1981) and the north coast of Surinam (Wells and Coleman, 1981a,b).

PHASE 1: FACIES MAPPING

The earlier descriptive work mentioned above ultimately lead to mapping of subzones and sedimentary facies of Holocene tidalites that are controlled by zonation of sediment transport along a coast. Sediment transport processes in the intertidal zone are distributed in a contour-parallel fashion from high-tide level to low-tide level (see Klein, 1977a, p. 81, his Fig. 73, Klein, 1985a, p. 191, his Fig. 3-2). These transport process zones include bed-load tidal current sediment transport in combination with late-stage, sheet-like runoff prior to exposure, a second zone where bed-load tidal current processes alternate with suspension deposition (transitional zone), and a third zone dominated by suspension sedimentation. These zones are superimposed by processes and features caused by exposure to air, tidal scour, and bioturbation. These sediment transport zones also control the textural distribution of sediment.

Suspension processes of sediment transport occur over intertidal flats during periods of submergence. Deposition of sediment from suspension occurs only during periods of negligible velocity with slack water periods around the time of peak submergence or high tide. Such sediment, however, is resuspended as tidal current velocities increase, by periodic wave action or when storms occur.

The source of suspended sediment appears to be twofold: offshore continental shelf zones and resuspended material from the intertidal flats themselves (Postma, 1954, 1961; Van Straaten and Kuenen, 1957; Groen, 1967; Anderson et al., 1981; Wells and Coleman, 1981a,b). Suspended sediment from offshore areas is brought landward by tidal currents containing relatively large concentrations of sediment. During peak submergence of the intertidal flat, the slack water stage is characterized by a negligible bottom current velocity and minimal turbulence so that suspended material settles to the intertidal zones' seabed. This slack water period may last as long as 2 hours (Postma, 1961), permitting sufficient time for suspended material to settle to the bottom because the settling velocity of the particulate matter exceeds bottom current velocities capable of maintaining material in suspension. In landward zones of intertidal flats (known as high-tidal flat), water levels begin to fall in the early stages of the ebb tidal cycle and as it does so, suspended sediment that settled to the intertidal flat surface becomes exposed and remains there. Some of this fine-grained sediment is resuspended by ebb tidal currents of slightly increased velocity. Because this velocity is relatively small, the seaward distance of sediment transport off the intertidal zone or downslope on the intertidal flat is less than the combination of resuspension, additional yield of suspended sediment, and increased relative distance of landward transport associated with the succeeding phase of flood-dominated deposition. This process favors sediment accumulation on the high intertidal flats and is termed settling lag (Postma, 1954, 1961).

A second factor that aids in concentrating relatively larger volumes of suspended sediment is the time-velocity asymmetry of tidal currents (Postma, 1961; Groen, 1967). Along the North Sea coast of The Netherlands, tidal currents on some tidal flats and offshore zones are characterized by a flood-dominated, time-velocity asymmetry, which means that current velocities are greater during the flood stage of a tidal cycle than the ebb stage of a tidal cycle (Klein, 1977a, p. 23, his Fig. 7; Klein, 1985a, p. 192, his Fig. 3-3). These greater tidal current velocities erode find-grained sediment and also are the cause of a larger competence of tidal currents resulting in transport of relatively large concentrations of sediment. Because flow directions are landward, such sediment is dispersed landward also and accumulates during the high-water, slack-water stage of the tidal cycle.

Van Straaten and Kuenen (1957) demonstrated also that a scour lag exists under these suspension-dominated processes. This scour lag involves the resuspension of fine-grained sediment from the intertidal flat surface. Net accumulation of fine-grained sediment from suspension on high-tidal flats comes about also because with exposure and subsequent desiccation, material cannot be eroded by tidal currents. However, in the submerged part of intertidal flats, mud can be and is resuspended and thus net mud accumulation diminishes in the middle portion of intertidal flats. Mud is seldom preserved on lower intertidal flats because of this process.

Most of the resuspension of mud from tidal flats occurs during periods of storm activity along a coast or during relatively moderate to large wave energy expended along coasts (Anderson et al., 1981; Wells and Coleman, 1981b). Nevertheless, on many intertidal flats, it is not uncommon to observe water-saturated muds showing a gel-like character (Wells and Coleman, 1981b). This material tends to move downslope as a slurry or a slide with associated small-scale slump scars, but it acts to baffle wave action along an intertidal flat (Wells et al., 1980). This baffling effect damps out wave action along intertidal flats unless wave energy increases. When wave energy increases, resuspension of fine-grained sediment occurs. A seasonal periodicity to resuspension during winter months coupled with net sediment accumulation during fair-weather spring and summer months has been documented along the intertidal flats of New Hampshire and elsewhere along the northeastern coast of North America (Anderson et al., 1981) and southwestern Korea (Wells et al., 1981).

Transport and deposition of sediment by bedload processes occurs by means of tidal current systems. Tidal current systems operate, obviously, in the intertidal zone only during periods of submergence. When such tidal currents flow across the intertidal zone, sand-sized sediment is moved by bed shear when bottom current velocities exceed at least 10 cm/s (Reineck and Wunderlich, 1969a). With increased bottom tidal current velocity, sandy zones are deformed into large and small bedforms, with current ripples moving with an average threshold velocity of 50 cm/s, dunes with an average threshold velocity of 45 cm/s, and sand waves with average threshold velocities of 55 to 60 cm/s (Dalrymple et al., 1978). Minimal threshold bottom-current velocities for migration of dunes and sand waves of 45–47 cm/s, respectively, were observed by Klein (1970a) and Klein and Whaley (1972).

The internal nature and orientation of cross-stratification, cross-strata set boundaries, and other discontinuities are controlled by the time-velocity asymmetry of tidal currents, and the longer term alternation of neap and spring tides (Reineck, 1963; Klein, 1970a). When maximum tidal current velocities tend to be of nearly equal intensity, a vertical stacking of units of herringbone cross-stratification can be developed by accumulation of sediment coupled with opposite migration of dunes and sand waves with each reversing tidal phase (Reineck, 1963). Moreover, the depth of scour and reworking of sediment will change from the neap to the spring stage, and as Klein (1970a) demonstrated, during the change from the spring to the neap stage of a lunar tidal cycle, the depth of scour decreases, the velocity spectrum changes, and vertically stacked herringbone cross-stratification is preserved. When time-velocity asymmetry is a dominant characteristic of tidal currents, truncation surfaces, termed reactivation surfaces (Collinson, 1969; Klein, 1970a; See Klein, 1977a, p. 24–25, his Figs. 8 and 9), are formed during the subordinate-velocity phase. As a consequence of this velocity asymmetry, unimodal orientation of cross-strata is observed, with bundles of such cross-strata bounded by reactivation surfaces dipping in the same direction, but truncating avalanche cross-stratification at a lower angle.

Superimposed on these reversals of flow directions of tidal currents during a tidal cycle are longer term cycles of changes in magnitude of bottom-current velocities during the change from neap to spring tide and back to neap tide. DeRaaf and Boersma (1971), Boothroyd and Hubbard (1975), Allen and Friend (1976), Visser (1980), and Boersma and Terwindt (1981) all observed that the distance of migration of bedforms is greater during spring tides and minimal, approaching zero, during neap tides.

Exposure of the entire tidal flat superimposes its own depositional modifications. As tidal elevation falls during the ebb-phase of a tidal cycle, the intertidal zone becomes progressively exposed. This changing water depth is coupled with changes in flow directions and changing bottom-current velocities. These three combined processes define late-stage emergence runoff.

On both the lower reaches of intertidal flats and intertidal sand bodies, the direction of flow of tidal currents will change as both the crests of dunes and sand waves, and as the crest of intertidal sand bodies, become emergent. Flow directions change from shore-parallel to downslope; some of this flow becomes confined to troughs of large-scale bedforms and moves in a manner similar to open-channel flow. Some of this downslope flow may parallel the main direction of ebb current flow, or flow at right angles or opposite to such flow, depending on the direction of slope locally, and regionally depending on the depositional surface. Because of the continuing reduction in water depths, the height and size of bedforms that are migrating will decrease. Some of the larger bedforms stop migrating completely. Because of changing direction in flow, it is normal that smaller scale bedforms will become superimposed on larger scale bedforms and the orientation of the superimposed bedforms will be oblique, at right angles, or opposite to the larger bedforms developed during the main ebb or flood current flow. These superimposed smaller-scale features involve several sizes of diminishing bedforms. Thus, current ripples may be superimposed on dunes or sand waves, or small current ripples may be superimposed on larger current ripples. As the flow velocity is maintained while water depth decreases, double-crested current ripples will develop as the bed shear is maintained, although depth of migration is reduced on the crest but continues in a slightly deeper trough (McMullen, 1964; Klein, 1970a, 1977a, p. 31–39, 1985a). Locally, scour pits may develop, and these too would show such superimposed features. Reduction of water level during this late-stage emergence runoff may also produce horizontal, step-like hachure marks. If wind-driven small waves move over the water surface, small currents with scouring capability may be generated. These currents partially destroy current ripple geometries and leave behind plane beds that truncate these ripples. These features were termed washout structures by Van Straaten (1954, 1959).

Local air entrapment is also common during both emergence runoff (Emery, 1945; Stewart, 1956) and later submergence. This process forms internal air hole cavities. During submergence, this air is observed bubbling off into the water. The air hole cavities so formed are comparable to birds-eye structure in carbonates (Shinn, 1968).

Within the middle portion of tidal flats, the dominant process of sedimentation is the alternation of both bedload and suspension transport and deposition. As a consequence, sedimentary features there show a preserved record of such processes. Suspension deposition occurs under the conditions discussed above involving both submergence and negligible velocity of transport by tidal currents during high-water slack tide. Bedload transport and deposition typically begin under bottom-current velocities in excess of 10 cm/s (Reineck and Wunderlich, 1968a,b; see Klein, 1985a, p. 203, his Fig. 3-14).

Because the volume of available sand is relatively small, the type of sedimentary features that develop in response to this alternation of bedload and suspension deposition includes small-scale dunes and current ripples, with their internal cross-stratification and thin lamina of clay organized as flaser bedding. A large variety of lenticular, wavy, and flaser bedding is characteristic of this depositional process (Reineck and Wunderlich, 1968a,b; see also Klein, 1985a, p. 204, his Fig. 3-15). The differentiation of these types of structures is controlled, however, by the relative volume of sand and mud, by the relative duration of both the bedload and suspension mode of deposition, and by current velocities. If sand exceeds mud, ripples form with isolated clay drapes in ripple troughs and crests. If mud and sand content is nearly equal, the volume of flaser bedding increases. When the relative volume of mud increases significantly with respect to sand, current ripples become isolated and are preserved as lenticular beds, some with internal flaser bedding. Wavy bedding occurs if the clay layers are draped continuously over both symmetrical and asymmetrical ripples. It should be observed, however, that in subtidal areas, including those whose circulation is tidally subordinate, similar flasers form by storms (McCave, 1970).

This process of alternation of bedload and suspension sedimentation also gives rise to thin, parallel-layered beds of alternating and interbedded sands and muds, termed tidal bedding by Reineck and Wunderlich (1968b), who monitored the duration, volume, and type of sedimentation under this regime more precisely (see Klein, 1985a, p. 203, his Fig. 3-14). They showed that deposition of alternating suspension muds and parallel-layered bedload sands was controlled by critical bottom current velocities. Sand accumulated whenever velocities exceeded 10 cm/s, whereas mud accumulated when velocities were less than 10 cm/s. Using a series of color markers, suspension processes were observed to coincide with slack-water high tide and negligible-velocity low tide stage (Reineck and Wunderlich, 1968b).

Depositional processes in the intertidal flats were integrated by Van Straaten (1952, 1953, 1957), who characterized two integrative styles of deposition. One was vertical sedimentation, in which the combinations of bedload and suspension sedimentation accumulated sediment vertically and by progradation. At the same time, because of the lateral cutting and filing of channels across many tidal flats, Van Straaten also recognized a second mode of deposition, namely lateral sedimentation, which contributed to tidal flat progradation, reworking and stacking of internal units.

Intertidal Flats

Intertidal flats have been examined by sedimentologists since the early 1930s (see Introduction). The earliest studies focused on the intertidal flats of the North Sea and because of integrative and detailed studies by Van Straaten (1959, 1961), Reineck (1963), and Evans (1965), the intertidal flats of The Netherlands (Wadden Sea), northwest Germany (Wadden Sea), and eastern England (The Wash) became type areas from which comparisons were made.

In the intertidal zone of these three areas in the North Sea, sediment processes of transport and deposition are zoned in contour-parallel fashion. Above the mean high water line is a supratidal zone that is dominated by submergence only during periods of elevated spring tide and storms. There, tidal marshes with *Spartina* grass are commonplace. They developed typically on a substrate mud; however, marsh taxa there show little preference for sediment texture. With additional accumulation of sediment, these tidal marshes encroach seaward as the position of high tide also progrades seaward.

This motif of supratidal sedimentation occurs in other areas. *Spartina* marshes have been reported from the intertidal zone of the Bay of Fundy (Klein, 1963, 1970a), San Francisco Bay (Pestrong, 1972), and Boundary Bay (British Columbia) (Kellerhals and Murray, 1969) and the coast of Massachusetts (Hayes, 1969). These all occur in a temperate-humid region. As climates become more tropical, this zone tends to give way to mangrove swamps such as those reported from the tidal flats associated with the tide-dominated Klang-Langat Delta of Malaysia (Coleman et al., 1970) and the Niger Delta, Nigeria (Oomkens, 1974). Along the northwest coast of Australia, Semeniuk (1981) reported that the mangrove supratidal zone graded laterally into a salt-pan evaporitic zone. Evaporite salt flats have also been described from supratidal zones adjoining intertidal flats around the Gulf of California (Thompson, 1968).

The sediment distribution of the main intertidal flat between mean high and mean low water coarsens in a seaward direction. The intertidal flat is zoned into three sections consisting of a high-tidal flat, a mid flat, and a low-tidal flat. These zones are distinguished on the basis of sediment distribution and the dominant process of sedimentation. The high-tidal flats are areas dominated by deposition of fine-grained silts and clays. There, submergence lasts for less than one-third of a tidal cycle associated with the high-water level stage when velocities are negligible. Mud deposition is favored by suspension processes (as reviewed above). Sedimentary structures preserved here include mudcracks, silty current ripples, and bioturbation features. Bioturbation tends to be fairly strong in this setting because the fine-grained sediment appears to serve as a suitable substrate for local infauna, and because of elevation and long exposure, many of the organisms burrow fairly deeply (Rhoads, 1967).

High intertidal flats at most locations contained mud and are, in fact, similar to the general case reviewed above from the North Sea. Thus the high-tidal flats of the Baie du Mont Saint Michel (Bajard, 1966, Larsonnieur, 1975), San Francisco Bay (Pestrong, 1972), the Bay of Fundy (Klein, 1963, 1970a), northwest Australia (Semeniuk, 1981), and the Gulf of California (Thompson, 1968) all fit into this mode of deposition.

The mid-flat environment occurs in the central portions of intertidal flats, is inundated for approximately 50% of a tidal cycle, and, therefore, experiences a near equal period of time for both suspension and bed-load sedimentation. This nearly equal alternation of bed-load and suspension deposition generates a mixed lithology (DeRaaf and Boersma, 1971) of interbedded sand and mud. Bedforms that develop include lower flow regime plane beds and current and interference ripples. Symmetrical ripples also occur, all within sand. Exposure features also are common. Internally, the alternating lithologies are organized into wavy bedding depending on the ratio of sand to mud. Thus if the sand to mud ratio is large, flaser bedding is favored, whereas lenticular bedding is favored if the sand to mud ratio is less than unity. Tidal bedding and wavy bedding are favored where the sand to mud ratio approaches unity. These general observations pertain in particular to the North Sea, but they have also been reported from northwestern Australia (Semeniuk, 1981), the Gulf of California (Thompson, 1968), and Boundary Bay (Kellerhals and Murray, 1969).

The low-tidal flat zone consists almost completely of sand of various size ranges that is fashioned into bedforms of variable size, including ripples, dunes, and sand waves. Current ripples often are superimposed on the surfaces of larger bedforms and oriented parallel, oblique, or opposite to the larger bedform. Internal cross-stratification, reactivation surfaces, micro-cross-laminae, and parallel laminae are the most common sedimentary structures observed in this setting. Bioturbation features are rare, primarily because of the high degree of instability of the sand substrate. The dominant mode of sediment transport in the low-tidal flat setting is bed-load transport by tidal currents that are characterized commonly by time-velocity asymmetry. This transport takes place during the period of submergence of almost two-thirds of a tidal cycle or more. Emergence runoff processes also are characteristic of this setting.

The seaward-coarsening textural distribution and associated bedforms are diagnostic of the intertidal-flat environment. It owes its origin to the combination of differential time of inundation and submergence of intertidal flats during a tidal cycle, associated changes in bottom-current velocities of tidal currents during a tidal cycle (being therefore concentrated over the low-tidal flats), and the dominance of suspension processes near the time of high tide and slack water, which favors preservation of mud in high-tidal flats.

The idealized presentation of a seaward-coarsening textural distribution on intertidal flats is not always identical to the North Sea (Van Straaten, 1961; Reineck, 1963; Evans, 1965), the Bay of Fundy (Klein, 1963, 1970a), or the northwest coast of Australia (Semeniuk, 1981). Instead, muddy intertidal flats from high to low water are common. Wells and Coleman (1981a,b) and Wells et al. (1980) demonstrated that the intertidal zone off the coast of Surinam is mud-dominated, with almost no sand occurring at all. High sediment yield comes mostly from the Amazon River, Brazil, and the continental shelf off Surinam, both of which provide sediment consisting of silt and clay. The intertidal flats of New Hampshire also comprise a mud-dominated system with almost no sand present (Anderson et al., 1981). Most of the sand there appears to have been confined to estuaries or transported offshore. Broad, sandy intertidal flats are also known, and I observed such a broad intertidal sand flat in Swansea Bay, Wales, just east of Swansea. In Turnagain Arm, Alaska, Ovenshine et al. (1975, 1976) reported muddy intertidal flats separated from large intertidal sand bodies by a channel, although some seaward coarsening of sediment sizes in the intertidal zones there was observed.

DEVELOPMENT OF INTERTIDAL FLAT SEQUENCES

Progradation of intertidal flats in a seaward direction depends largely on a moderate to high sediment yield. Many large tidal flats are known where they adjoin point sources (most riverine) or line sources (continental shelves) where a large volume of sediment is available. Thus, along the coast of western Korea, extensive intertidal flats occur with widths ranging from 5 km to 25 km (Wells and Huh, 1979) because of sediment yield from the Hwang Ho and Yangtze Rivers of China and smaller rivers in Korea (Chung and Park, 1977; Wells et al., 1981). The intertidal flats of northwestern Europe appear to owe their sediment availability to the high yields of the Rhine River and associated coastal current transport systems.

Rapid rates of intertidal flat progradation are documented directly from only two locations. Along the coast of northern France, LeFournier and Friedman (1974) determined an intertidal flat progradation rate of 1 km per century based on survey records of several centuries. Along the coast of Turnagain Arm, coastal subsidence of 1.5–2.0 m occurred following the 1964 Alaska earthquake. By 1974, an additional 2.0-m thick sequence of intertidal flat sediments prograded over a depositional zone that is 1.8 km (Table 1). The high sediment yield comes from the melting of the Portage Glacier (Ovenshine et al., 1975, 1976).

Along the coast of The Netherlands, the Holocene stratigraphy is well known (DeJong, 1965; Hageman, 1972), and within it, four intertidal flat successions are preserved, each separated by a ravinement surface associated with a transgression. The horizontal distance and the time interval of successive progra-

TABLE 1.—HOLOCENE RATES OF INTERNAL FLAT PROGRADATION.

Location	Rate	Reference
The Netherlands	4.9 m/yr	Hageman (1972), DeJong (1965)
Northwest France	1.0 m/yr	LeFournier and Friedman (1974)
Turnagain Arm, Alaska	12.0 m/yr	Ovenshine et al. (1975, 1976)

dational events are also known and indicate a progradational rate that averages 4.9 m/yr. In comparison to northwestern France and Alaska, these rates fall midway between low (1m/yr) for France, to extreme (12 m/yr) for Alaska (Table 1).

Sediment accumulation rates along tidal flats of the Yellow Sea of Korea are known to be high. Lead-Uranium radiometric dating established annual accumulation rates of 5–9 mm/yr on mid tidal flats there, whereas low tidal flats yielded annual accumulation rates of 1–2 mm/yr. (Alexander et al., 1991).

The consequence of progradation of tidal flats in the North Sea of western Europe is not only to shift shorelines in a seaward direction and thus displace laterally the components of intertidal flats, but also to generate a progradational vertical sequence of all components of the intertidal zone distributed vertically in a consistent and predicable way (Klein, 1971, 1972). When progradation occurs, each of the intertidal flat subenvironments oversteps the adjoining seaward-most subenvironment. Thus, high-tidal flat muds are observed to prograde over mid-flat interbedded, mixed lithologies of sand and mud, which in turn prograde over low-tidal flat sands. The supratidal zone will, in turn, prograde over the high-tidal flat. Continued progradation will generate a vertical sequence that fines upward (Klein, 1977a, p. 85, his Fig. 76; Klein, 1985a, p. 191, his Fig. 3-2). From the base upward, this vertical sequence (Klein, 1971, 1972) consists of lower intertidal flat sands, overlain by mid-flat interbedded sands and muds, and high-tidal flat muds. Such sequences may be overlain by supratidal marshes, mangrove roots, or evaporites (cf. Semeniuk, 1981). The thickness of these sequences (Klein 1985a, p. 191, his Fig. 3-2), coincides with Holocene tidal range (Klein, 1971, 1972) and may be used to approximate paleotidal range (see Klein, 1977a). Such a sequence fits the North Sea very closely. The nature of sequences in muddy intertidal flats, sand intertidal flats, or silty intertidal flats is less well known. In Turnagain Arm, Ovenshine et al. (1975, 1976) demonstrated that although the lithologies differ, a fining-upward grain-size trend was observed, fitting the general zonation of depositional processes.

PHASE II: DYNAMICS OF SEDIMENT TRANSPORT

The intertidal sand body system tends to occur along mesotidal and macrotidal coasts and also within some estuarine complexes. Some of these intertidal sand bodies are isolated from associated intertidal flats such as in parts of the Bay of Fundy (Klein, 1970a; Knight, 1980) and in Turnagain Arm (Ovenshine et al., 1975, 1976), whereas in other locations they are welded onto intertidal flats and comprise part of the low-tidal flat zone. Examples of such cases include part of the Bay of Fundy (Klein, 1970a; Knight and Dalrymple, 1975), the coast of western Korea (Chung and Park, 1977; Wells and Huh, 1979), and several locations in the North Sea.

Intertidal sand body systems have accumulated in areas of relatively strong bottom tidal current velocities, subjecting the seabed to a large rate of sand transport by bedload processes of deposition. Many of these current systems show evidence of the property of time-velocity asymmetry during which bottom current velocities are characterized by larger velocity magnitudes during a dominant phase of the tidal cycle and lower magnitudes during the subordinate phase of the tidal cycle (Klein, 1970a; Boersma and Terwindt, 1981; Knight, 1980; Dalrymple et al., 1978). Bedform migration and orientation of bedforms, topography of sand bodies, distribution of sediment facies and dispersal patterns of sand are all controlled largely by time-velocity asymmetry on these sand bodies (Klein, 1970a; Boersma and Terwindt, 1981; Balazs and Klein, 1972). The distance of migration, sand transport rates, and volume of preserved sediment also tend to be favored during times of spring tides in comparison to times of neap tides (Visser, 1980; Boersma and Terwindt, 1981), as discussed earlier (see Klein, 1985a, p. 196–197, his Figs. 3-6 and 3-7).

Intertidal sand bodies are linear in plan and asymmetrical in cross-section. Their orientation is parallel to the main flow of tidal currents. Sand on these sedimentary bodies ranges in size from very fine to very coarse, depending on availability and sorting. Bedforms developed on the sand bodies include current ripples, dunes and sand waves (see Klein, 1985a, p. 214, his Fig. 3-22). Migration of sand waves and dunes occurs only during times of dominant tidal flow (Klein and Whaley, 1972; Dalrymple et al., 1978), and much of it is accomplished during very short periods of time, perhaps as little as an hour. The orientation of internal cross-stratification of dunes and sand waves is in agreement with sand body alignment and dominant flow direction of tidal currents. Thus, unimodal orientations of cross-strata are in agreement with dominant tidal current flow and sand body alignment. Such cross-stratification is truncated by reactivation surfaces (see above discussion on bedload processes). Bundles of unimodally cross-stratified zones are grouped laterally into thicker units and thinner units, representing spring and neap tide sedimentation, respectively (Boersma and Terwindt, 1981; see Klein, 1985a, p. 197, his Fig. 3-7).

The distribution of different zones of time-velocity asymmetry over intertidal sand bodies appears to be dependent in part on sand body topography. Klein (1970a) reported that in the Minas Basin, Bay of Fundy, relatively narrow and steep zones of intertidal sand bodies are dominated by tidal currents characterized by a flood-tide, time-velocity asymmetry, whereas gentler sloping surfaces, showing a wider area of exposure, are shaped and reworked by ebb-dominated tidal currents. Consequently, sediment is dispersed around the sand body through alternating flood-dominated and ebb-dominated time-velocity asymmetry zones (Klein, 1970a, 1985a, p. 217, his Fig. 3-25). This elliptical dispersal pattern produces well-rounded sands on the sand bodies, as demonstrated by Balazs and Klein (1972).

Most of the Holocene intertidal sand bodies that have been described appear to be in areas of strong reworking (Klein, 1970a) rather than in areas of active progradation. Some such sand bodies welded to the intertidal zone comprise the lower parts (low-tidal flat) of fining-upward intertidal flat sequences (Klein, 1971). However, the nature of such sequences in macrotidal settings remains unknown simply because no areas of macrotidal settings that also are characterized by progradational history have been documented. This situation also exists for intertidal sand bodies that are free-standing from intertidal flats.

Nevertheless, one hypothetical vertical sequence was suggested assuming coastline progradation. It is based on Bay of Fundy observations by Knight and Dalrymple (1975; see Klein, 1985a, p. 218, his Fig. 3-26). In this hypothetical sequence, a vertical succession overlies subtidal sand, gravel, or bedrock and is in turn overlain by an intertidal sand body. It in turn is overlain by a braided bar tidal channel deposit, which grades upward into intertidal mudflats capped by supratidal salt marshes. Overall, this macrotidal flat sequence contains a thicker section of sand and includes braid-bar deposits. This intertidal sand body-macrotidal coastline sequence should be readily distinguishable from the normal intertidal flat sequence described above.

In subtidal areas, similar sand bodies, referred to as tidal current sand ridges, are common. In the North Sea, parallelism of orientation of sands bodies to tidal current flow was reported by Stride (1963), Houbolt (1968), and Terwindt (1971). Stride (1963, 1982) and others (Caston, 1972, Jordan, 1962, Off, 1963, Belderson et al., 1972, McCave, 1971) showed that the orientation of dunes superimposed on sand bodies also was aligned parallel to tidal current flow directions. Houbolt (1968), using both sediment coring and seismic profiling, provided insight into the internal anatomy of these tidal current sand ridges, and observed that on the gentle slopes of these sand ridges, bedforms migrated towards the crest, whereas on the steeper slopes, they migrated at right angles to the crest. He proposed a dispersal pattern suggesting that sand on tidal current sand ridges are transported alternatively through more gently sloping and more steeply sloping surfaces (See Klein, 1985b, p. 127 and p. 130, his figs. 6.8 and 6.11B, respectively). The resulting pattern is identical to what Klein (1970a) described from intertidal sand bodies in the Bay of Fundy. Internally, seismic profiles suggest a sequence of thick, cross-stratified intervals that become thinner towards the bar crest. These thick cross-beds dip in the same direction as the slip face of the linear sand ridge, suggesting that overall, ridge migration is in the direction of the steepest face, a finding confirmed later by Caston (1972) and Ludwick (1974). Ultimately, such migration leads to the development of a sand sheet as such tide-dominated linear sand ridges merge.

It should be observed that Houbolt's (1968) work lacked data on current velocities, and his sediment dispersal pattern could, perhaps, by comparison, be explained as being controlled by time-velocity asymmetry. Later work addressed this issue in the Yellow Sea of Korea. The Yellow Sea is the widest continental shelf in the world and, as a consequence, is tide-dominated, (Off, 1963, Niino and Emery, 1961; Chough, 1983), and its coast, as discussed earlier, is macrotidal (Wells and Huh, 1979). Tidal current sand ridges are extremely widespread (Off, 1963). Klein et al. (1982) mapped one of these sand bodies in detail with side-scan sonar. There, dunes and sand waves on the flanks of the sand ridges are oriented parallel to regional ebb and flood tidal flow directions. On the gentler surfaces of the ridge, bedforms are aligned parallel to the ridge crest, whereas on the steeper slope of the ridges, plane beds were observed. Measurements of bottom current velocities showed a strong time-velocity asymmetry, and regional north-northeast flood tidal flow, and south-southwest ebb tidal flow controlled bedform migration on the flanks. A rotary tidal flow at both high and low tide caused migration of bedforms towards the ridge crest. The resulting dispersal pattern was not elliptical, but more of a trapezoid (See Klein, 1985b, p. 130, his Fig 6.11D). Comparison of bathymetry between a 1964 Korean Navy chart and bathymetry reported by Klein et al. (1982) showed that the tide-dominated sand ridges migrated 600 m in the direction of the slip face during a 16-year period.

PHASE III. TIDAL BUNDLING AND NEAP-SPRING TIDE INFLUENCES

Boersma and Terwindt (1981) demonstrated that within estuarine tidal sand bodies, tidal current velocities, characterized by time-velocity asymmetry, show larger mean and maximum bottom-current velocities and sediment transport rates during the spring tide phase of the lunar tidal cycle than occurs during the neap phase. Within dunes and sand wave complexes in a flood-dominated area, where bottom-current velocity measurements were obtained, accretionary bundles of cross-strata with distinct bounding surfaces were observed (see Klein, 1985a, p. 196–199, his figs. 3-6, 3-7, 3-8). These could be correlated to both neap and spring tide sediment transport. Reactivation surfaces (which they termed pause planes because of lack of migration of bedforms) and associated structures graded laterally into a bundle of thick avalanche cross-strata organized into distinct laminae. These cross-strata are termed vortex structures by Boersma and Terwindt. Laterally, these structures grade into a terminal interval of cross-strata that contain less well-sorted sand with the angle of repose decreasing downcurrent. These are termed slackening structures. The entire sequence is overlain by an ebb-cap of sediments showing ebb-oriented cross-strata. This lateral change in structures records deposition during a single flood-dominated tidal phase with the reactivation structure representing nondeposition during a subordinate phase and the avalanche cross-strata representing the active phase of dune migration during the dominant flood phase. The slackening phase represents the diminishing of flow velocities toward the end of a tidal cycle. The internal organization of cross-strata and reactivation surfaces differ between the neap and spring phase because the neap phase shows thinner bundles and thinner cross-strata sets, reflecting smaller bottom current velocities, whereas the spring tidal phase shows thicker sets of cross-strata and longer bundles, reflecting greater velocities and greater sand transport rates. Thus the lateral dimensions and sediment volume permit recognition of spring and neap tidal phases in present-day intertidal and sand bodies. This interpretation also can be applied to ancient counterparts where the neap pause planes may be represented instead by mud flaser beds (Allen, 1982; Visser, 1980; Boersma and Terwindt, 1981).

PHASE IV: TIDAL BEDDING AND ASTRONOMICALLY-DERIVED TIDAL PERIODS

Tidal bedded couplets (tidal bedding of Reineck and Wunderlich, 1968a,b) of alternating parallel-bedded sand and clay have been observed in a variety of modern and ancient tidalites. They are stacked in a rhythmic mode, although the thickness of the individual couplets tends to differ. In Carboniferous rocks, these have been described by various workers including Archer (1991), Kvale et al. (1989), Kvale and Archer (1990), amongst others, and calibrated with ocean circulation models (Archer, 1996).

Measurement of the thickness variation of these Carboniferous, tidally bedded couplets in nearly complete sections shows a systematic variation (Kvale et al, 1989). Closer analysis of this variation shows a direct correlation to tidal variation, and is useful as an indicator of tidal harmonics, tidal dynamics, and tide-related astronomical forcing factors. Such observations also show that tidalites are rapidly preserved over a short period of time and therefore in such sequences, time-gaps represented by unconformities may be longer than previously supposed. Accompanying papers by Archer and Kvale (this volume) provide additional details, and the reader is referred to them for more information.

SUMMARY

This paper reviewed four phases of evolution of tidalite research during the past 60 years. Studies of tidalites evolved from descriptive and areal studies (Phase I) in different parts of the world, but mostly the North Sea, to development of a vertical sequence facies model. This model integrated considerable work during the first phase of research. Phase II focused on the dynamics of sediment transport involving derived transport models from bedform orientation and internal seismic character of tidal current sand ridges; and direct measurement of bottom current velocities; and comparison of their orientations and magnitudes with orientations of bedforms, facies, and other attributes on intertidal sand bars (Klein, 1970a; Knight, 1980; Knight and Dalrymple, 1975; Dalrymple et al., 1978). Extension of such an approach to tidal current sand ridges in the continental shelf zone of the Yellow Sea showed comparable observations and results (Klein et al., 1982). Concurrently, a period (Phase III) of detailed observations of the internal anatomy of bedforms resulted in the recognition of migration patterns and bundling of cross-strata controlled by neap-spring alternations of sediment transport. Currently, the most active new frontier of tidalite research, representing Phase IV, is the study of tidal bedding and its correlation to astronomically influenced tidal periods.

Space restrictions in this review did not permit an extensive review of ancient tidalites. Nevertheless, numerous examples exist (see Klein, 1977a, b; Klein and Ryer, 1978). Other papers in this volume address the current status of each of these four phases of tidalite research as further refinement, new insight, and more detailed analysis are presently underway.

ACKNOWLEDGMENTS

I wish to express my appreciation to R. A. Davis Jr., and V. J. Henry, co-conveners of the Tidalites '96 conference, for inviting me to prepare this paper as a keynote presentation at the conference. My deepest gratitude is extended also to R. N. Ginsburg, and the editors, C. R. Alexander, R. A. Davis Jr., and V. J. Henry, for their helpful and cogent comments on an earlier version of this manuscript. Lacking professional drafting and photographic services at my disposal, I chose to refer to existing published illustrations, where appropriate, rather than redraw existing illustrations that have appeared if not once, but several times, in the sedimentary geological literature during the last 40 years. I regret any inconvenience to the reader.

REFERENCES

Alexander, C. R., Nittrouer, C. A., DeMaster, D. J., Park, Y. A., and Park, C. S., 1991, Macrotidal mudflats of the southwestern Korean Coast: A model for interpretation of intertidal deposits: Journal of Sedimentary Petrology, v. 61, p. 805–824.

Allen, J. R. L., 1982, Mud drapes in sand-wave deposits: A physical model with application to the Folkestone Beds (Early Cretaceous, Southeast England): Royal Society of London, Philosophical Transactions, Series aaa, v. 306, p. 291–345.

Allen, J. R. L., and Friend, P. F., 1976, Changes in intertidal dunes during two spring-neap cycles, Lifeboat Station Bank, Wells-next-the-Sea, Norfolk (England): Sedimentology, v. 23, p. 329–247.

Anderson, F. E., Black, L., Watling, L. E., Mook, W., and Mayer, L. M., 1981, A temporal and spatial study of mudflat erosion and deposition: Journal of Sedimentary Petrology, v. 51, p. 729–736.

Archer, A. W., 1991, Modeling of tidal rhythmites using modern tidal periodicities and implications for short term sedimentation rates, *in* Franseen, E. K., Whatney, W. L., Kendall, C. G. st.C, and Ross, W., eds., Sedimentary modeling: Computer simulations and methods for improved parameter definition: Kansas Geological Survey Bulletin 223, p 185–194.

Archer, A. W., 1996, Panthalassa: Paleotidal resonance and a global paleocean-seiche: Paleoceanography, v. 11, p. 625–632

Bajard, J., 1966, Figures et structures sedimentaries dans la zone intertidale de la partie orientale de la Baie du Mont-Saint-Michel: Review. de Geographe. Physique et de Geologie Dynamique, v. 9, p. 39–111

Balazs, R. J., and Klein, G. deV., 1972, Roundness-mineralogical relations of some intertidal sands: Journal of Sedimentary Petrology, v. 42, p. 425–433.

Belderson, R. H., 1964, Holocene sedimentation in the western half of the Irish Sea: Marine Geology, v. 2, p. 147–169.

Belderson, R. H., Kenyon, N. H., Stride, A. H., and Stubbs, A. R., 1972, Sonographs of the Sea Floor: Amsterdam, Elsevier, 195 p.

Boersma, J. R., and Terwindit, J. H. J., 1981, Neap-spring tide sequences of intertidal shoal deposits in a mesotidal estuary: Sedimentology, v. 28, p. 151–170.

Boggs, Jr., S., 1974, Sand-wave fields in Taiwan Strait: Geology, v. 2, p. 251–253.

Boothroyd, J. C., and Hubbard, D. K., 1975, Genesis of bedforms in mesotidal estuaries, *in* Cronin, L.E., ed.: Estuarine Research, v. 2, p 217–234.

Caston, V. N. D., 1972, Linear sand banks in the southern North Sea; Sedimentology, v. 18, p. 63–78.

Chough, S. K., 1983, Marine geology of Korean seas: Boston, IHRDC, 160 p.

Chung, G. S., and Park, Y. A., 1977, Sedimentological properties of the Recent intertidal flat environment, southern Nam Yang Bay, west coast of Korea: Journal of the Oceanographic Society of Korea, v. 13, p. 9–18.

Coleman, J. M., 1980, Deltas: Processes of Deposition and Models for Exploration, second edition: Minneapolis, Burgess Publishing Company, 124 p.

Coleman, J. M., Gagliano, S. M. and Smith, W. G., 1970, Sedimentation in a Malaysian high tide tropical delta, *in* Morgan, J. P., and Shaver, R. H. eds., Deltaic Sedimentation: Tulsa, Society of Economic Paleontologists and Mineralogists Special Publication 15, p. 185–197.

Collinson, J. D., 1969, Bedforms of the Tana river, Norway: Geografiskar Annaler, v. 52, p. 31–56.

Cram, J., 1979, The influence of continental shelf width on tidal range: Paleoceanographic implications: Journal of Geology, v. 87, p. 441–447.

Dalrymple, R. W., Knight, R. J., and Lambiase, J. J., 1978, Bedforms and their hydraulic stability relationships in a tidal environment, Bay of Fundy, Canada: Nature, v. 275, p. 100–104.

Davies, J. L., 1964, A morphogenic approach to world shorelines: Zeitschrift der Geomorphologie, v. 8, p. 127–142.

DeJong, J. D., 1965, Quaternary sedimentation in the Netherlands, *in* Wright, H. E., and Frey, D. G. (eds.), International Studies on the Quaternary: Boulder, Geological Society of America Special Paper 84, p. 95–124.

DeRaaf, J. F. M., and Boersma, J. R., 1971, Tidal deposits and their sedimentary structures: Geologie en Mijnbouw, v. 50, p. 479–504.

Emery, K. O., 1945, Entrapment of air in beach sand: Journal of Sedimentary Petrology, v. 15, p. 39–49.

Evans, G., 1965, Intertidal flat sediments and their environments of deposition in The Wash: Geological Society of London Quarterly Journal, v. 121, p. 209–241.

Evans, G., 1975, Intertidal flat deposits of The Wash, western margin of the North Sea, *in* Ginsburg, R. N. ed., Tidal Deposits: New York, Springer-Verlag, p. 13–20.

Gellatly, D. C., 1970, Cross-bedded tidal megaripple from King Sound: Sedimentary Geology, v. 4, p. 185–192.

Groen, P., 1967, On the residual transport of suspended matter by an alternating tidal current: Netherlands Journal of Undersea Research, v. 3, p. 564–574.

Hageman, B. P., 1972, Sedimentation in the lowest part of river systems in relation to the post-glacial sea level rise in the Netherlands: XXIV International Geological Congress, v. 12, p. 37–47.

HANTZSCHEL, W., 1939, Tidal flat deposits (Wattenschlick), *in* Trask, P. D. (ed.), Recent Marine Sediments: Tulsa, Society of Economic Paleontologists and Mineralogists, Tulsa, OK, p. 195–196.

HANTZSCHEL, W., AND REINECK, H. E., 1968, Faziesunterschungen in Hettangium von Helstedt (Niedersachsen): Geol. Staatsinst. Hamburg Mitt., v. 7, p. 5–39.

HAYES, M. O., ed., 1969, Coastal environments of northeastern Massachusetts and New Hampshire, Eastern Section Guidebook: Society of Economic Paleontologists and Mineralogists., 462 p.

HAYES, M. O., 1975, Morphology of sand accumulation in estuaries, *in* Cronin, L. E., ed., Estuarine Research, v. 2, p. 3–22.

HAYES, M. O., 1979, Barrier island morphology as a function of tidal and wave regime, *in* Letherman, S. P., ed., Barrier Islands: New York, Academic Press, p. 1–27.

HOUBOLT, J. J. C., 1968, Recent sediments in the southern bight of the North Sea: Geologie en Mijnbouw, v. 47, p. 245–273.

JORDAN, G. F., 1962, Large submarine sand waves: Science, v. 136, p. 839–848.

KELLER, G. H., LAMBERT, D., ROWE, G., AND STARESINIC, N., 1973, Bottom currents in the Hudson Canyon: Science, v. 180, p. 181–183.

KELLERHALS, P., AND MURRAY, J. W., 1969, Tidal flats at Boundary Bay, Fraser River Delta, British Columbia: Canadian Petroleum Geology Bulletin, v. 17, p. 67–91.

KENYON, N. H., BELDERSON, R. H., STRIDE, A. H. AND JOHNSON, M. A., 1981, Offshore tidal sand banks as indicators of net sand transport and as potential deposits, *in* Nio, S. D., Schuttenheim, T. E. and van Weering, T. C. E., eds., 1981, Holocene Marine Sedimentation in the North Sea Basin: Oxford, International Association of Sedimentologists Special Publication 5, p. 257–268.

KINDLE, E. M., 1917, Recent and fossil ripple marks: Geological Survey of Canada Museum Bulletin 25, 56 p.

KLEIN, G. DEV., 1963, Bay of Fundy intertidal zone sediments: Journal of Sedimentary Petrology, v. 33, p. 844–854.

KLEIN, G. DEV., 1970a, Depositional and dispersal dynamics of intertidal sand bars: Journal of Sedimentary Petrology, v. 40, p. 1095–1127.

KLEIN, G. DEV., 1970b, Tidal origin of a Precambrian quartzite—the Lower Fine-grained Quartzite (Dalradian) of Islay, Scotland: Journal of Sedimentary Petrology, v. 40, p. 973–985.

KLEIN, G. DEV., 1971, A sedimentary model for determining paleotidal range: Geological Society of America Bulletin, v. 82, p. 2585–2592.

KLEIN, G. DEV., 1972, Determination of paleotidal range in clastic sedimentary rocks: XXIV XXIV International Geological Congress, v. 6, p. 397–405.

KLEIN, G. DEV., (ed.,), 1976, Holocene Tidal Sedimentation: Stroudsburg., Pennsylvania, Dowden, Hutchinson and Ross, Inc., 423 p.

KLEIN, G. DEV., 1977a, Clastic Tidal Facies: Continuing Education Publication Company, Champaign, Illinois., 149 p.

KLEIN, G. DEV., 1977b, Tidal circulation model for deposition of clastic sediments in epeiric and mioclinal shelf seas: Sedimentary Geology, v. 7, p. 1–12.

KLEIN, G. DEV., 1985a, Intertidal flats and intertidal sand bodies, *in* Davis, R. A. Jr., ed., Coastal sedimentary environments, 2nd Edition: New York, Springer-Verlag, p. 187–224.

KLEIN, G. DEV., 1985b, Sandstone depositional models for exploration for fossil fuels, 3rd Ed.: Boston, IHRDC Publishing Company, 209 p.

KLEIN, G. DEV., AND RYER, T. A., 1978, The circulation patterns in Precambrian, Paleozoic and Cretaceous epeiric and mioclinal shelf seas: Geological Society of America Bulletin, v. 89, p. 1050–1058.

KLEIN, G. DEV., PARK, Y. A., CHANG, J. H., AND KIM, C. S., 1982, Sedimentology of a sub-tidal, tide-dominated sand body in the Yellow Sea, southwest Korea: Marine Geology, v. 50, p. 221–240.

KLEIN, G. DEV., AND SANDERS, J. E., 1964, Comparison of sediments in tidal flats in the Bay of Fundy and the Dutch Wadden Zee: Journal of Sedimentary Petrology, v. 34, p. 18–24.

KLEIN, G. DEV., AND WHALEY, M. L., 1972, Hydraulic parameters controlling bedform migration on an intertidal sand body: Geological Society of America Bulletin, v. 83, p. 3465–3470.

KNIGHT, R. L., 1980, Linear sand bar development and tidal current flow in Cobequid Bay, Bay of Fundy, Nova Scotia, *in* McCann, S. B., ed., The Coastline of Canada: Ottawa, Geological Survey of Canada Paper 80-10, p. 123–152.

KNIGHT, R. L., AND DALRYMPLE, R. W., 1975, Intertidal sediments from the south shore of Cobequid Bay, Bay of Fundy, Nova Scotia, Canada, *in* Ginsburg, R. N., ed., Tidal Depositions: New York, Springer-Verlag, p. 47–56.

KVALE, E. P., AND ARCHER, A. W., 1990, Tidal deposits associated with low-sulfur coals, Brazil Formation (Lower Pennsylvanian), Indiana: Journal of Sedimentary Petrology, v.60, p. 563–574.

KVALE, E. P., ARCHER, A. W., AND JOHNSON, H. R., 1989, Daily, monthly, and yearly tidal cycle within laminated siltstones of the Mansfield Formation (Pennsylvanian) of Indiana: Geology. v. 17, p. 365–368.

LAMBIASE, J. J., 1980, Sediment dynamics in the macrotidal Avon River estuary, Bay of Fundy, Nova Scotia: Canadian Journal of Earth Sciences, v. 17, p. 1628–1641.

LARSONNIEUR, C., 1975, Tidal deposits, Mont-Saint-Michel Bay, France, *in* Ginsburg, R. N., ed., Tidal Deposits: New York, Springer-Verlag, p. 21–30.

LEFOURNIER, J., AND FRIEDMAN, G. M., 1974, Rate of lateral migration of adjoining sea-margin sedimentary environments shown by historical records, Authie Bay, France: Geology, v. 2, p. 497–498.

LONDSALE, P., AND MALFAIT, B., 1974, Abyssal dunes of foraminiferal sand on the Carnegie Ridge: Geological Society of America Bulletin, v. 85, p. 1697–1712.

LONSDALE, P., NORMARK, W. R., AND NEWMAN, W. A., 1972, Sedimentation and erosion on Horizon Guyot: Geological Society of America Bulletin, v. 83, p. 289–316.

LUDWICK, J. C., 1974, Tidal currents and zig-zag shoals in a wide estuary entrance: Geological Society of America Bulletin, v. 85, p. 717–726.

MACAR, P., AND EK, C., 1965, Un curieux phenomeme d'erosion fammeinnienne: Le "Pains de gres" de Chaambralles (Ardenne, Beldge): Sedimentology, v. 4, p. 53–64.

MCCAVE, I. N., 1970, Deposition of fine-grained suspended sediment from tidal currents: Journal of Geophysical Research, v. 75, p. 4151–4159.

MCCAVE, I. N., 1971, Wave-effectiveness at the sea bed and its relationship to bedforms and deposition of mud: Journal of Sedimentary Petrology, v. 41, p. 89–96.

MCCAVE, I. N., and GEISER, A. C., 1979, Megaripples, ridges and runnels on intertidal flats of The Wash, England: Sedimentology, v. 26, p. 353–369.

MCCAVE, I. N., and LANGHORNE, D. N., 1982, Sand waves and sediment transport around the end of a tidal sand bank: Sedimentology, v. 29, p. 95–110.

MCMULLEN, R. M., 1964, Modern sedimentation in the Mawddach Estuary, Barmouth, North Wales: Unpublished Ph.D. Dissertation, University of Reading (UK), 399 p.

NIINO, H., and EMERY, K. O., 1961, Sediments of shallow portions of East China Sea and South China Sea: Geological Society of America Bulletin, v. 72, p. 731–762.

OFF, T., 1963, Rhythmic linear sand bodies caused by tidal currents: American Association of Petroleum Geologists Bulletin, v. 47, p. 325–341.

OOMKENS, E., 1974, Lithofacies relations in the Late Quaternary Niger Delta Complex: Sedimentology, v. 21, p. 195–221.

OVENSHINE, A. T., BARTSCH-WINKLER, S. R., O'BRIEN, N. R., and LAWSON, D. E., 1975, Sediments of the high tidal range environment of Upper Turnagain Arm, Alaska, *in* Recent and Ancient Sedimentary Environments in Alaska: Alaska Geological Society, p. 1–40.

OVENSHINE, A. T., LAWSON, D. E., and BARTSCH-WINKLER, S. R., 1976, The Placer River Silt—an intertidal deposit caused by the 1964 Alaska Earthquake: U.S. Geological Survey Journal of Research, v. 4, p. 151–162.

PESTRONG, R., 1972, Tidal flat sedimentation at Cooley Landing, southwest San Francisco Bay: Sedimentary Geology, v. 8, p. 251–288.

POSTMA, H., 1954, Hydrography of the Dutch Wadden Sea: Unpublished Ph.D. Dissertation, University of Groningen, 106 p.

POSTMA, H., 1961, Transport and accumulation of suspended matter in the Dutch Wadden Sea: Netherlands Journal of Undersea Research, v. 1, p. 148–190.

REINECK, H. E., 1963, Sedimentgefuge im Bereich der Sudliche Nordsee: Abhandlen. Senckenbergiana Naturlisches Gesellschaft, v. 505, p. 1–138.

REINECK, H. E., 1967, Layered sediments of tidal flats, beaches and shelf bottoms of the North Sea, *in* Lauff, G. H., ed., Estuaries: American Association for the Advancement of Science Special Publication 83, p. 191–206.

REINECK, H. E., 1972, Tidal flats, *in* Rigby, J. K., and Hamblin, W. K., eds., Recognition of Ancient Sedimentary Environments: Society of Economic Paleontologists and Mineralogists. Special Publication 16, p. 146–159.

REINECK, H. E., and SINGH, I. B., 1973, Depositional Sedimentary Environments, 1st edition: New York, Springer-Verlag, 439 p.

REINECK, H. E., and SINGH, I. B., 1980, Depositional Sedimentary Environments, 2nd edition: New York, Springer-Verlag, 549 p.

REINECK, H. E., and WUNDERLICH, F., 1968a, Classification and origin of flaser and lenticular bedding: Sedimentology, v. 11, p. 99–104.

REINECK, H. E., and WUNDERLICH, F., 1968b, Zeitmessungen und Bezeitenschichten: Natur und Museum, v. 97, p. 193–197.

RHOADS, D. C., 1967, Biogenic reworking of intertidal and subtidal sediments in Barnstable Harbor and Buzzards Bay, Massachusetts: Journal of Geology, v. 75, p. 461–476.

SEMENIUK, V., 1981. Sedimentology and the stratigraphic sequence of a tropical tidal flat, northwestern Australia: Sedimentary Geology, v. 29, p. 195–221.

SHEPARD, F. P., DILL, R. F., and VON RAD, U., 1969, Physiography and sedimentary processes of LaJolla submarine fan and fan valley, California: American Association of Petroleum Geologists Bulletin, v. 53, p. 420.

SHEPARD, F. P., and MARSHALL, N. F., 1973, Currents along floors of submarine canyons: American Association of Petroleum Geologists Bulletin, v. 57, p. 244–264.

SHINN, E. A., 1968, Practical significance of birds-eye structures in carbonate rocks: Journal of Sedimentary Petrology, v. 38, p. 215–223.

STEWART, JR., H. B., 1956, Contorted sediments in modern coastal lagoon explained by laboratory experiments: American Association of Petroleum Geologists Bulletin, v. 40, p. 153–161.

STRIDE, A. H., 1963, Current-swept sea floors near the southern half of Great Britain Geological Society of London Quarterly Journal, v. 119, p. 75–197.

STRIDE, A. H., 1982, Offshore Tidal Sands: London, Chapman and Hall, 222 p.

TERWINDT, J. H. J, 1971, Sand waves in the southern bight of the North Sea: Marine Geology, v. 10, p. 51–68.

THOMPSON, R. W., 1968, Tidal flat sedimentation on the Colorado River Delta northwest Gulf of California: Geological Society of American Memoir 107, 133 p.

VAN STRAATEN, L. M. J. U., 1952, Biogene textures and the formation of shell beds in the Dutch Wadden Sea: Koninklijke Nederlandse Akadamie van Wetenschapen Proceedings, Series B, v. 55, p. 500–516.

VAN STRAATEN, L. M. J. U., 1953, Megaripples in the Dutch Wadden Sea and in the Basin of Archachon (France): Geologie en Mijnbouw, v. 15, p. 1–11.

VAN STRAATEN, L. M. J. U., 1954, Sedimentology of recent tidal flat deposits and the Psammites du Condroz: Geologie en Mijnbouw, v. 16, p. 25–47.

VAN STRAATEN, L. M. J. U., 1959, Minor structures of some Recent littoral and neritic sediments: Geologie en Mijnbouw, v. 21, p. 197–216.

VAN STRAATEN, L. M. J. U., 1961, Sedimentation of flat areas: Journal of the Alberta Society of Petroleum Geologists, v. 9, p. 203–226.

VAN STRAATEN, L. M. J. U., and KUENEN, P. D., 1957, Accumulation of fine-grained sediments in the Dutch Wadden Sea: Geologie en Mijnbouw, v. 19, p. 329–354.

VISSER, M. J., 1980, Neap-spring cycles reflected in Holocene subtidal large-scale bedform deposits; a preliminary note: Geology, v. 8, p. 543–546.

WELLS, J. T., and COLEMAN, J. M., 1981a, Periodic mudflat progradation, northeastern coast of South America: a hypothesis: Journal of Sedimentary Petrology, v. 51, p. 1069–1075.

WELLS, J. T., and COLEMAN, J. M., 1981b, Physical processes and fine-grained sediment dynamics, coast of Surinam, South America: Journal of Sedimentary Petrology, v. 51, p. 1053–1068.

WELLS, J. T., and HUH, O. L., 1979, Tidal flat muds in the Republic of Korea: Chinhae to Inchon: Office of Naval Research Science Bulletin, v. 4, p. 21–30.

SEDIMENT DISTRIBUTION PATTERNS IN THE BACK-BARRIER AREAS OF THE WADDEN SEA, SPIEKEROOG ISLAND, GERMANY

N. NYANDWI

Institute of Marine Sciences, P. O. Box 668, Zanzibar, Tanzania

ABSTRACT: Sedimentological studies have recently been carried out in the Spiekeroog back-barrier area (southern North Sea coast of Germany) to explore the interrelationship between the surficial sediment distribution patterns, energy levels, and transport processes. The sediment distribution patterns show that the sediment generally becomes finer landwards (north-south), irrespective of the tidal channel orientations (east-west). A closer examination, however, shows two distinct patterns. The first pattern is a general landward-fining of the sediments within the inlet from about 1.0 phi (0.50 mm) in the inlet throat to about 3.5 phi (0.088 mm) on the landward reaches of the inlet. This pattern is a result of the decrease in current velocity from the inlet throat landwards. The second pattern, which is the most conspicuous on the mean grain-size map, shows a distinct shore-normal (north-south), landward sediment fining across the tidal flats from about 2.0 phi (0.25 mm) on the islands to 2.5–3.0 phi (0.25–0.125 mm) on the tidal flats to as fine as 3.5 phi (0.088 mm) along the dike (mainland coast). This shore-normal sediment fining has been found to be a result of the shore-normal energy gradient (flow velocity) associated with overbank flow from one channel to the next.

The analysis of skewness evolution has been shown to be a powerful tool for the interpretation of transport pathways in a tidal environment. Skewness distribution patterns indicate that the inlet areas act as sources of sediment from which sediment is transported landwards during the flood tide and seawards during the ebb phase. In the whole area, however, the fine sediment fraction (population) whose energy niche is the landward margin of the backbarrier areas, and the coarse fraction characterizing the inlet throat areas, undergo a range of population mixing as well as progressive sorting of the individual populations. Progressive sorting appears to be dominant normal to the shore and across the tidal flats whereas mixing processes are more pronounced along the main channel of the tidal inlets.

INTRODUCTION

The island of Spiekeroog and its surrounding waters is perhaps the most studied area in the whole of the German Bight. The studies performed in the area form part of the worldwide effort to understand back-barrier as well as tidal environments. The study of modern tidal environments is important for understanding the deposition environments and processes of ancient tidalites. It is already known in the Wadden Sea for instance, that the sediment distribution pattern is characterized by a landward fining of sediment often culminating in thick mud deposits along the mainland coast or dike (e.g., Reineck and Singh, 1973; Figge, 1981; Ragutzki, 1982; Reineck et al., 1986; Grotjahn, 1990). On the basis of their mud content, sediments of the German part of the Wadden Sea have been divided into three types (Sindowski, 1973; Reineck and Siefert, 1980) that form coast-parallel belts. These belts are arranged seawards starting with mudflats (>50% mud), mixed flats (10–50% mud) and sand flats (5–10% mud). The efforts of many workers, however, have been dedicated to describing the textural trends and not the process.

The landward fining of sediment pertaining to the landward transport and deposition of suspended matter was explained long ago by several workers (Van Straaten and Kuenen, 1957, 1958; Postma, 1961; Groen, 1967). The general landward fining in the coarse sediments, however, has only recently been explained hydraulically (Nyandwi and Flemming, 1995). The importance of grain-size sorting and mixing processes has been demonstrated using high-resolution grain settling velocity analyses (Bartholomae and Flemming, 1994). The analysis of grain-size fractions has pointed to a depletion in the mass content of fine fractions of sediments close to the coast, which was shown to be due to human intervention in the system through land reclamation and dike construction (Flemming and Nyandwi, 1994). More studies and analysis of all granulometric parameters are needed in order to fully explain the sedimentation processes.

In this study, granulometric parameters of close-grid samples covering the whole Spiekeroog back-barrier area have been analyzed to provide information about the interrelationship between transport processes, energy levels, and sediment distribution patterns. Understanding the mechanisms that are responsible for the sediment distribution patterns observed in modern tidal environments could be useful for interpreting the mode of deposition in ancient tidalites.

STUDY AREA

The island of Spiekeroog is one of the barrier islands bordering the southern coast of the North Sea (Fig. 1). Actually, the complete chain of the barrier islands straddles the North Sea coasts of three countries, The Netherlands, Germany, and Denmark. Between the islands and the mainland coast lies an extensive tidal basin which, with the exception of the tidal channels, remains virtually dry at low water springs (Ehlers, 1988). The body of water separating the islands from the mainland coast is referred to as the Wadden Sea.

The hydrodynamics of the island of Spiekeroog and the Wadden Sea as a whole is linked to the circulation regime of the North Sea, which is mainly a result of tidal currents and atmospheric pressure fields. The tidal circulation is driven by a system of three amphidromic points (Huntley, 1980). The tidal wave propagates eastwards along the German North Sea coast (Fitzgerald and Penland, 1987), and the more eastward the island is situated within the barrier islands chain, the greater the delay in the time of high water. The variation in the tidal range within the back-barrier basins shows a northwest-southeast increase between the islands and the mainland as observed in the open North Sea (Fig. 2). This direction coincides with the main flow direction of the tidal currents and orientation of the inlets.

The current regime in the back-barrier areas is a combination of tidal currents and wind-driven currents. Current measurements during the period of calm weather give maximum velocities of 0.7–1.3 m s^{-1} in major channels (Davis and Flemming, 1989) and only up to 0.3 m s^{-1} on the tidal flats (Bartholomae, 1993). In general, the tidal currents show a pronounced time-velocity asymmetry characterized by a shorter and faster flood phase (Kock and Luck, 1975; Fitzgerald and Penland, 1987). The currents decrease in magnitude away from the inlet throat areas. During stormy weather conditions, wind-driven currents

Tidalites: Processes and Products, SEPM Special Publication No. 61

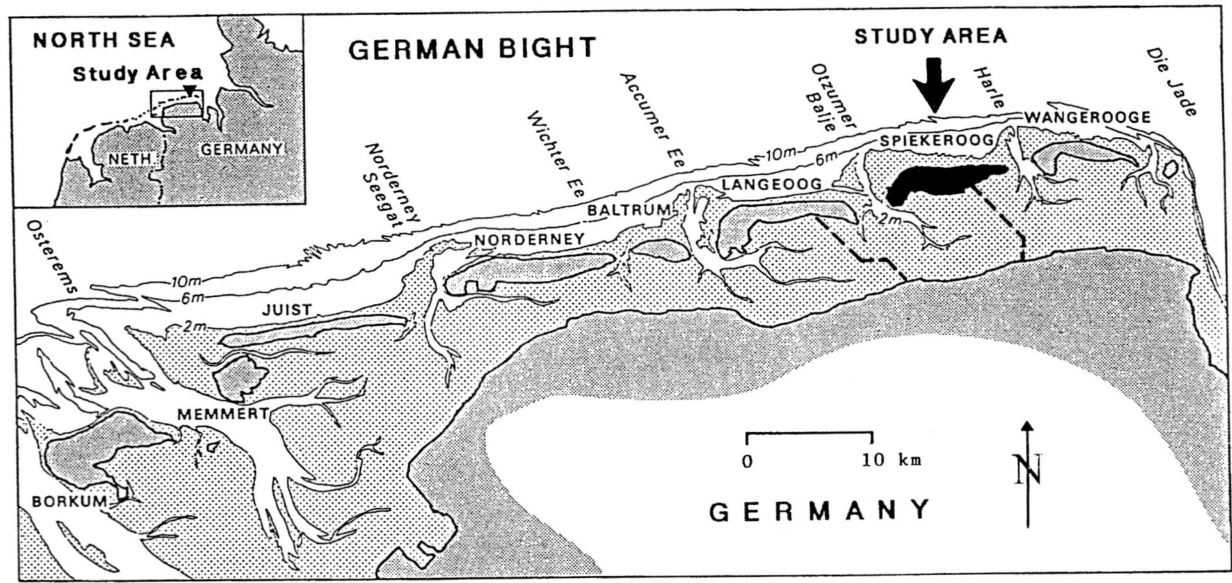

FIG. 1.—Part of the German North Sea Coast showing the study area, near Spiekeroog Island.

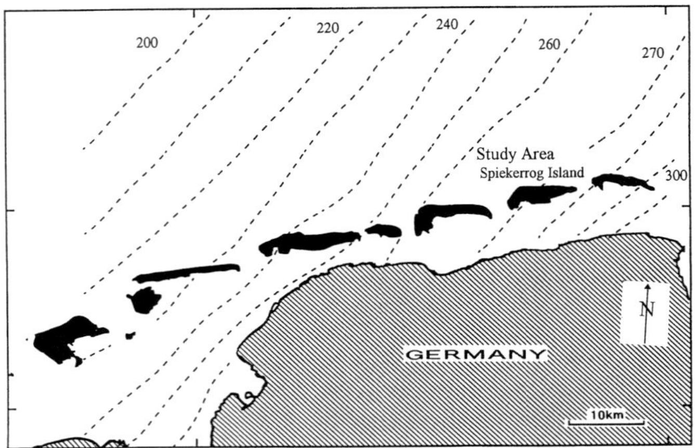

FIG. 2.—Variation of the tidal range (cm) along the German North Sea Coast (modified after Lassen, 1989).

amplify or reduce the velocity of the tidal current depending on whether the wind blows in the same or opposite direction as the tidal current (Koch and Niemeyer, 1978). An analysis of meteorological data from between 1965 and 1985 shows that winds stronger than 10 Beaufort (24.5–28.4 m/s at 10 m height) do occur largely (85%) during winter months (Antia, 1993). These winds blow from the west with the northwest contributing 47%, followed by west (30%), southwest (19%) and north (4%).

METHODS

The present study is part of a major project aimed at understanding the hydrodynamics, morphodynamics, and sediment dynamics of back-barrier areas (Nyandwi, 1995). As such, a close-grid sampling (275 × 275 m) of surficial sediment was carried out covering the whole back-barrier area of Spiekeroog Island (Fig. 3). The field and laboratory procedures followed Flemming and Ziegler (1995). A portable (DECCA) navigator was used to fix the sampling positions. In the laboratory, the samples were desalinated and the mud (<4.0 phi or 0.063 mm) was determined by wet sieving. Sample splits of the remaining sand fraction were run through a high-resolution, automatically recording, macrogranometer (Brezina, 1979). The recorded settling velocities were converted into equivalent settling diameters using a glass sphere standard. The mean, standard deviation, skewness and kurtosis were all calculated following Folk and Ward (1957).

RESULTS

In addition to mud content, four granulometric parameters were calculated: the mean, sorting, skewness and kurtosis, and the percent contribution of various size fractions in a sample at 0.25 phi, 0.5 phi and 1.0 phi intervals. These data were obtained from settling-tube analyses. Presentation of the spatial distribution of the parameters has been reported in various forms elsewhere (Flemming and Davis, 1994; Flemming and Nyandwi, 1994; Flemming and Ziegler, 1995; Nyandwi, 1995; Nyandwi and Flemming, 1995). The present analysis focuses on the processes responsible for the sediment distribution patterns.

Distribution of the Mean Grain Size

The spatial distribution of the sand-fractions is presented in fig. 4. The mean grain sizes of the samples were plotted on a map and contoured at an interval of 0.25 phi. In the mean value range of 1.0–2.5 phi, however, the resolution was poor and the contour interval was changed from 0.25 to 0.5 phi and 1.0 phi for clarity. The results show that the back-barrier sediment is largely made up of medium to very fine sand with the mean grain size ranging from about 3.25–3.5 phi (0. 105–0.088 mm) at the landward margin of the study area to 1.0–2.0 phi (0.25–0.50 mm) in the tidal inlets. The distribution shows two

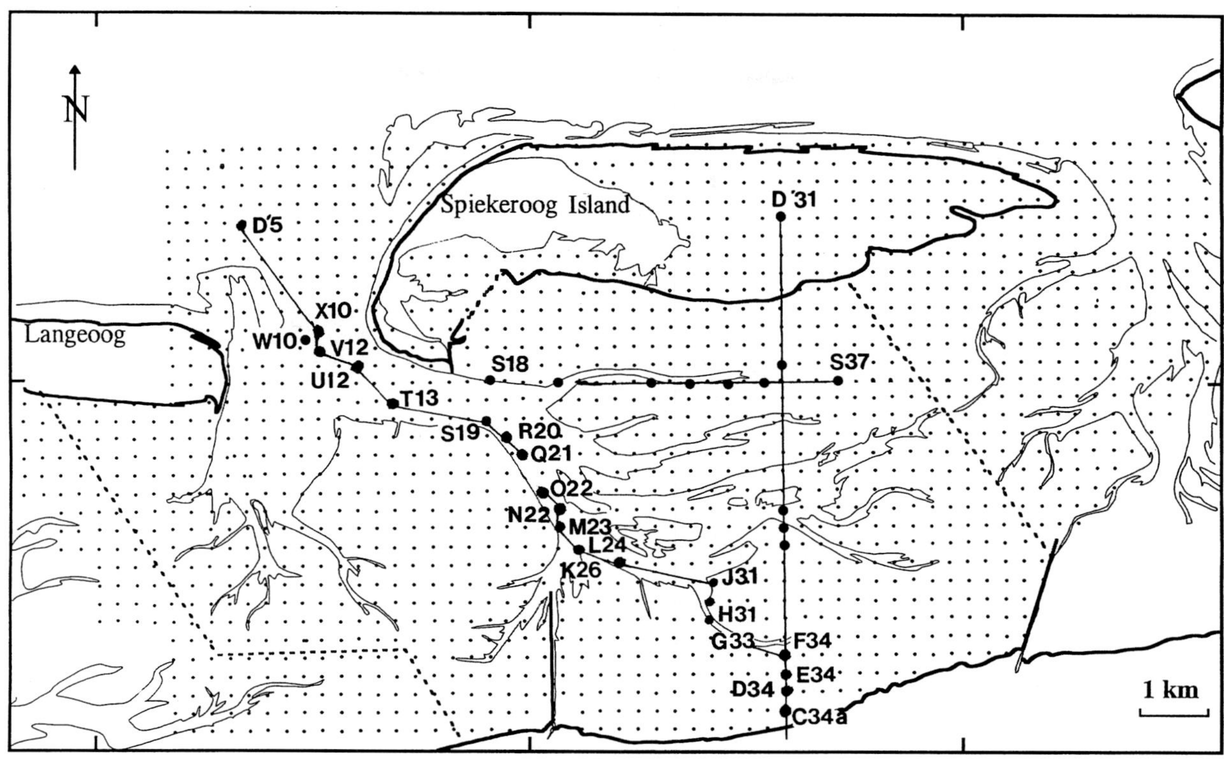

FIG. 3.—Physiographic map of the study area showing sampling grid and selected transects along the channel (C34a—D′5), normal to the shore (C34a—D′31) and an east-west (S18–S37) transect.

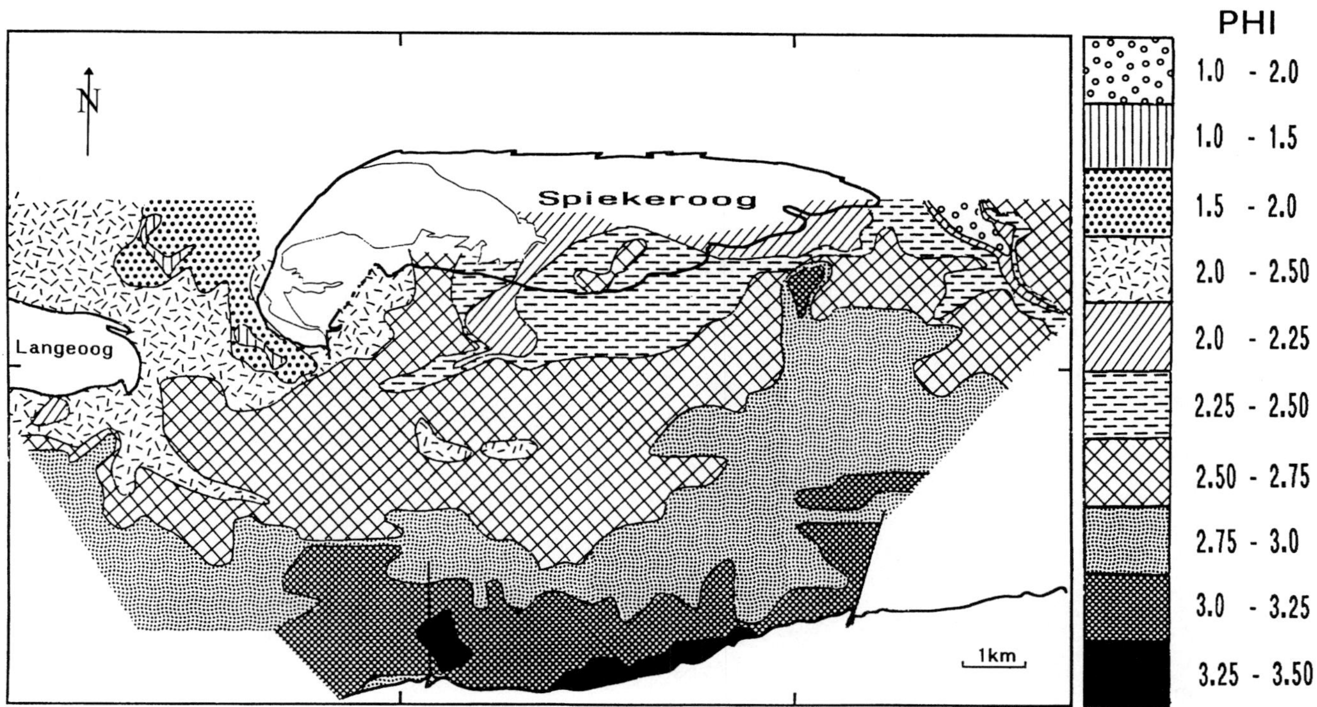

FIG. 4.—Spatial distribution of the mean sediment grain-size in the back-barrier area of Spiekeroog Island. Note the landward-fining trend.

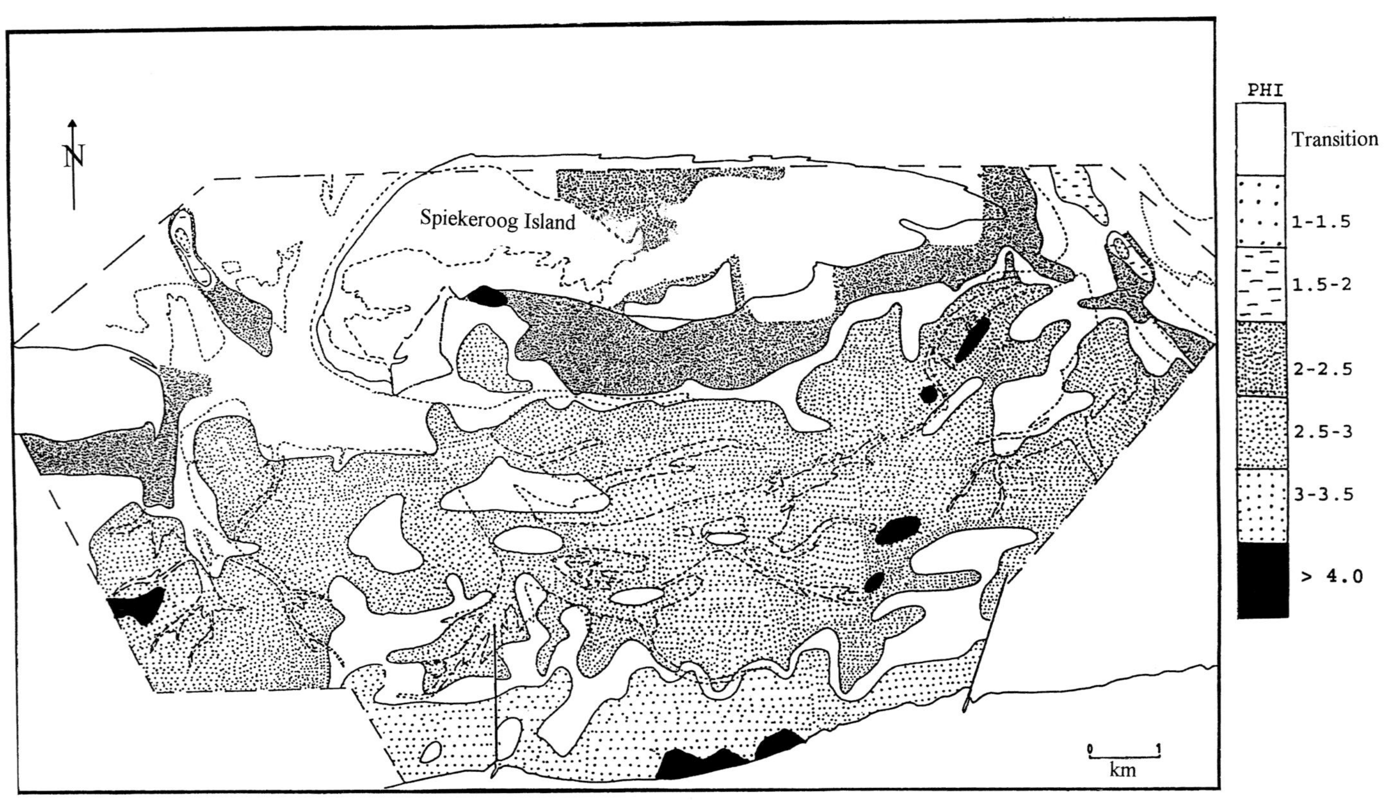

FIG. 5.—Spatial segregation of the dominant sand fractions. The occurrence of transition zones indicates that the energy gradient responsible for their segregation is continuous.

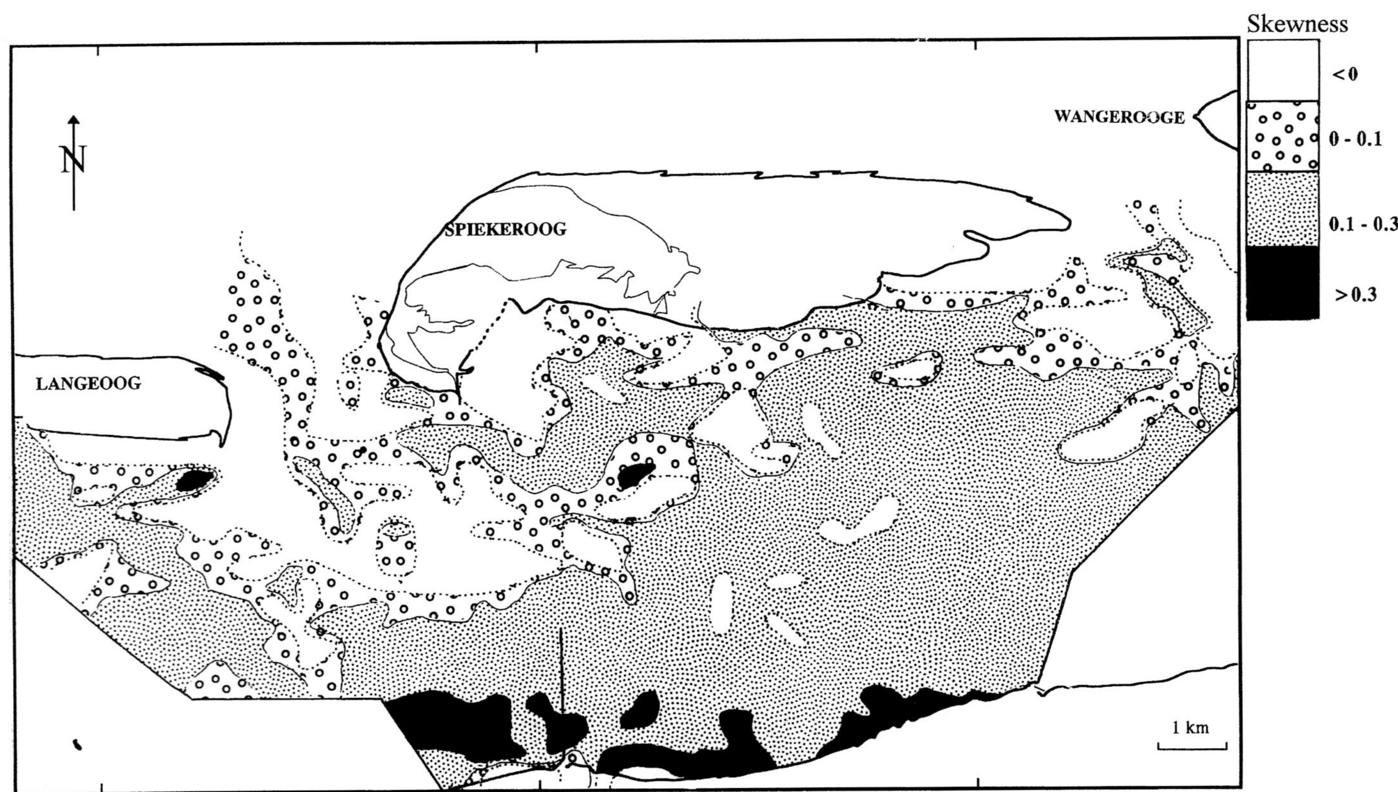

FIG. 6.—Distribution of the skewness polarity for the sediments adjacent to Spiekeroog Island indicating the sediment source-deposit relationship.

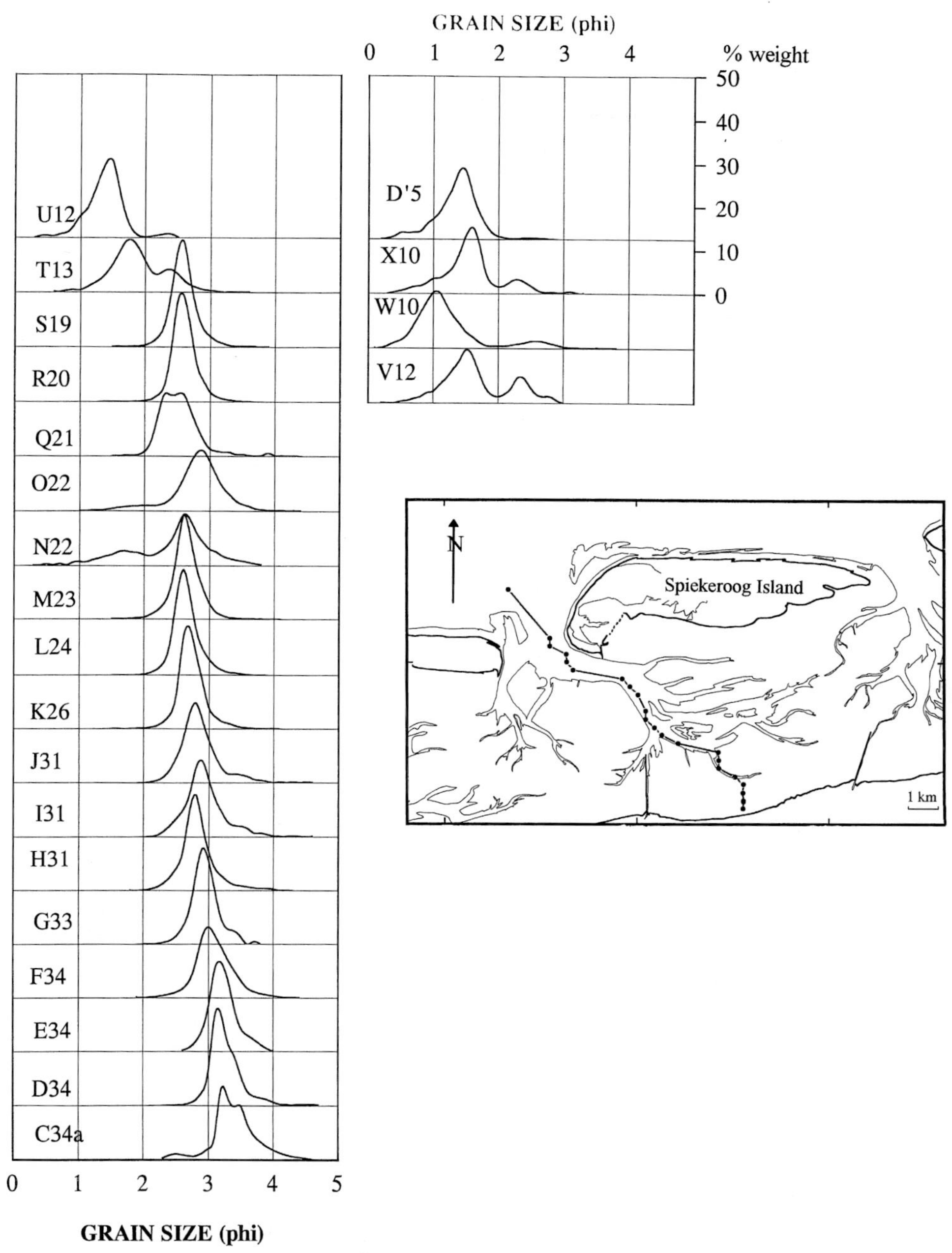

FIG. 7.—Selected grain-size frequency curves (samples D′5—C34A) showing various distribution shapes along the inlet (sample positions are shown on Fig. 3).

patterns—a general landward-fining of the sediment along the inlet from as coarse as 1.0–2.0 phi in the inlet throat area to 3.25–3.5 phi at the landward reaches of the inlet, and a shore-normal (north-south) landward sediment fining across the tidal channels from about 2.0 phi (0.25 mm) on the island to 2.5–3.0 phi (0.177–0.125 mm) on the tidal flats to as fine as 3.25–3.5 phi along the mainland coast. This latter trend is the most conspicuous on the grain size distribution map and points to a shore-normal energy gradient. The former trend, sediment fining along inlets, is expected from the decreasing flow velocities landwards from the inlet mouth.

Distribution of the Dominant Fractions

The concept of dominant fractions can be used by determining the percentage composition of different size fractions in a sample at one-quarter, one-half, and whole phi intervals. A dominant fraction is defined as the class interval that contributes 50% or more by weight to the sample. Inspection of the data shows that this interval was 0.5 phi. This means that if, for instance, the dominant fraction of a sample is 2.0–2.5 phi, then smaller classes within the interval contribute less than 50% each. Fig. 5 shows the spatial distribution of the dominant fractions. The dominant fractions map, just like the mean grain size,

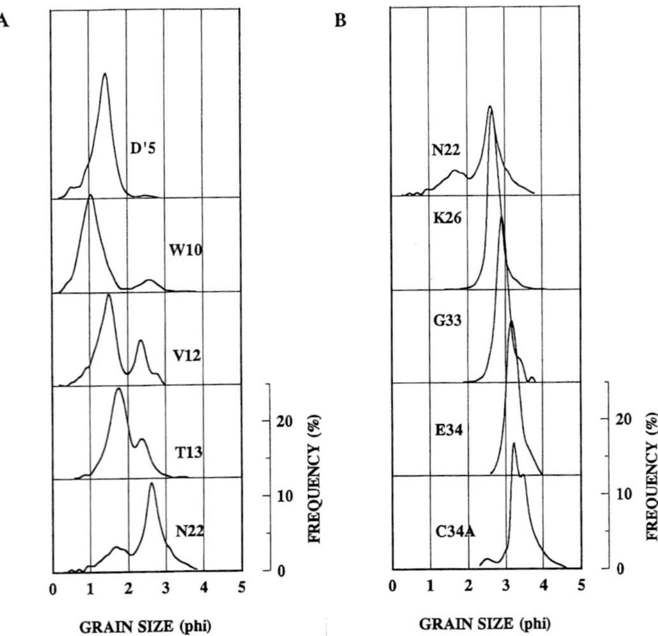

FIG. 8.—Selected grain-size frequency curves from Fig. 7 showing both mixing of subpopulations (A) and progressive size-sorting (B). Progressive size sorting of the primary mode is observed both landwards and seawards of the throat area (W10), whereas mixing with a secondary mode is observed landwards between W10 and N22 before progressive size sorting of the secondary mode takes place further landwards.

shows shore-parallel belts of grain fractions coarsening from the mainland seawards. At the landward margin of the tidal flats, the coarser fraction of the very fine sand (3.0–3.5 phi, that is 0.125–0.088 mm) predominates. Seawards of the 3.0–3.5 phi belt lies the next belt, predominated by 2.5–3.0 phi (0.177–0.125 mm) size fraction. At the seaward margin of the study area and on the island itself, the 2.0–2.5 phi (0.250–0.177 mm) size fraction predominates. The coarser fractions (the fine and coarse fractions of the medium sand, that is 1.5–2.0 and 1.0–1.5 phi) were found to be typical in the high-energy, inlet mouth areas. One important feature that cannot be observed when using the mean grain size only is the presence of narrow zones separating the belts dominated by different grain sizes. Such zones are composed of sediment from both belts. The zones are interpreted to represent transitional energy levels between low and high energy levels, thus suggesting a continuous energy gradient across the tidal flats.

Sorting and Mixing Processes

Having established the distinct segregation of size fractions according to the energy level, it is interesting to know how the various spatially segregated fractions relate to one another. The map of skewness distribution (Fig. 6) was found to shed some light on the transport-deposition relationships. The landward change in skewness values from more negative to more positive values suggests the transport direction, the more negative areas acting as sources for the more positive areas. The winnowing of fines from a deposit renders the lag coarser and more negatively skewed than the source, whereas the new deposit will be finer and more positively skewed.

In the course of transport from source to deposit in a tidal environment, both progressive sorting and mixing processes occur. Whereas transport in one direction of the tidal flow would result in progressive sorting of the grain subpopulations, deposition by the reverse flow may result in mixing. A selection of grain size frequency distributions along the main inlet (Fig. 7) demonstrates not only the landward decreasing energy gradient as mentioned above, but also the sorting and mixing processes. Whereas the bimodality of some of the frequency distributions demonstrates that mixing processes of different hydraulic populations occur, the progressive shift in the modal size points to a progressive size sorting. Fig. 8 shows progressive size-sorting of the primary coarse mode from the inlet

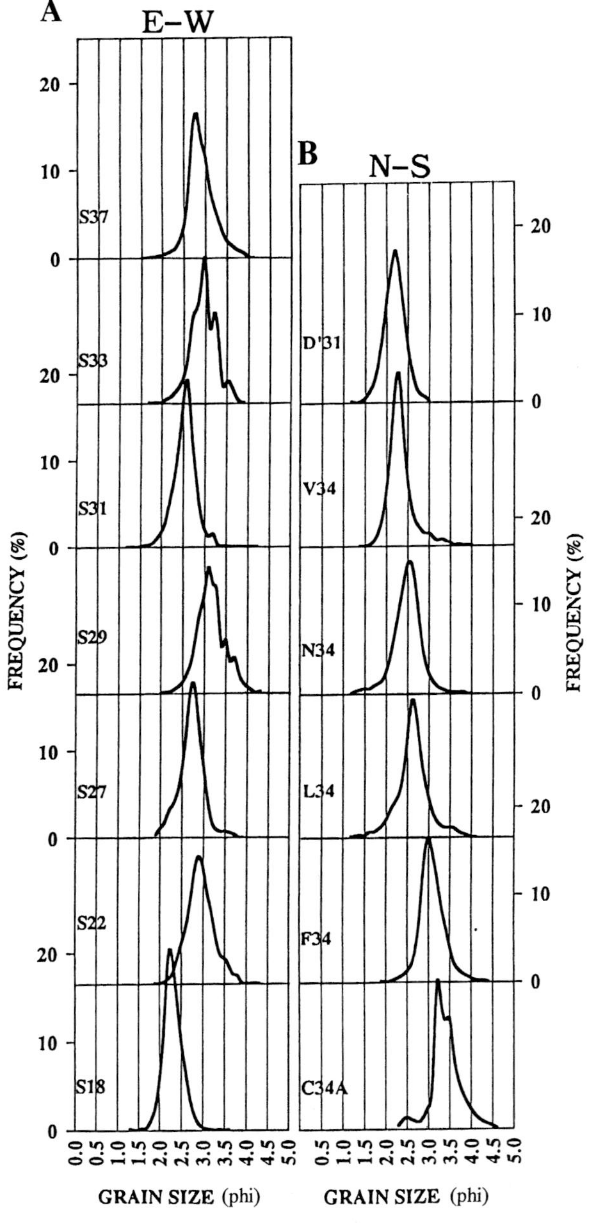

FIG. 9.—Comparison of the grain-size frequency curves along east-west (A) and north-south (B) transects (see Fig. 3). There is a clear modal shift in the north-south transect indicating a well-defined energy gradient.

throat area towards the tidal flats and mixing with the secondary finer mode (Fig. 8). Beyond a certain point on the tidal flat the secondary mode predominates and is progressively size-sorted landwards (Fig. 8).

DISCUSSION

The mixing process is usually guided by fluctuations in energy level (Flemming, 1977, 1988). The transitional zones depicted by the distribution of dominant size fractions noted above may represent the zones of mixing of different hydraulic populations, whereas the directional grain-size (mean) decrease observed elsewhere is attributed to progressive size sorting (Nyandwi, 1995). Between two mixing end-members (one coarser than the other) the mixing proportion of each increases towards its end-member. Therefore, along the transport path from the coarse to the finer end-member, the coarser mode is size-sorted downstream as the current velocity decreases whereas the finer mode is increasing in quantity until it predominates before it, too, starts to undergo size sorting (Folk and Ward, 1957; Folk and Robles, 1964). Away from the inlet mouth area, therefore, the decreasing energy level leads to a progressive decrease in the grain size of the sediment deposited. Sediment is transported seawards during the ebb tide and landwards during flood. Within the central region of the back-barrier area, coarse sediments from the inlet and finer sediment of the tidal flats are mixed whereas progressive size-sorting of the tidal-flat sediments resumes further landwards. Thus, in effect, the back-barrier sediment comprises two or more hydraulic populations, each undergoing progressive size-sorting in the course of its transport, and each overlapping with the next finer population before dropping out further down the transport path (Fig. 8).

The use of the evolution of the skewness polarity to infer transport pathways (source/deposit relationship) has been successfully applied in several other works (McLaren, 1981; McLaren and Bowles, 1985; Antia, 1993; Nyandwi, 1995). The fact that the sediments in the inlet areas are more negatively skewed than those in the tidal flats shows that the bulk of the sediment that is generated in the inlets and offshore is transported landwards during flood tides and deposited on the tidal flats. The exception to the landward transport occurs seawards of the inlet mouths owing to strong ebb current jets (Antia, 1993) at isolated, elevated locations on the tidal flats, where biological sedimentation mainly of mud occurs (Flemming and Delafontaine, 1994), and along the mainland coast, where the finest fine-sand fraction is resuspended and transported offshore (Flemming and Nyandwi, 1994).

The shore-normal energy gradient, (Fig. 9), on the other hand does not represent a transport direction; rather it is a result of the decrease in the tidal prisms of intertidal channels (and hence, the peak velocities) the more landwards a channel is situated (Nyandwi and Flemming, 1995). This is further supported by the lack of a clear energy gradient (lateral shift in the mode) along the tidal channels (Fig. 9).

CONCLUSION

There are two main distribution patterns of the sand-sized sediment in the Spiekeroog back-barrier area. The first pattern of landward sediment fining follows the inlet orientation and is limited to the areas in and around the inlets. This pattern is governed by the landward change in flow velocity along the inlet. The second pattern however, is defined by a shore-normal distribution of landward-fining sediment bands across the tidal flats. This pattern is a result of the landward decrease in overbank flow velocities from one tidal channel to the next in the landward direction and could only be detected by high density sampling.

Although generally the decreasing grain size in a given direction would suggest a transport direction, the observed shore-normal grain-size distribution across the tidal flats cannot be thus interpreted. This calls for more care in interpreting transport directions (pathways) based on analysis of granulometric parameters.

The skewness distribution shows a better agreement with the general transport direction and could be used to analyze grain-size trends and transport pathways. This follows the fact that, generally, the sediment becomes more positively skewed if it is transported from a location and deposited downcurrent. This could be useful in studying ancient tidalites.

Fine sediment from the landward margins of the back-barrier areas are mixed in varying proportions with coarse sediments from the sea and the inlet. At the same time, however, each sub-population experiences progressive size-sorting in the direction of decreasing energy.

ACKNOWLEDGMENTS

My stay at the Forschungsinstitut Senckenberg in Wilhelmshaven, Germany, where the work was done, was financed by the German Academic Exchange Service (DAAD) to whom I am greatly indebted. Thanks are extended also to the Forschungsinstitut Senckenberg for its hospitality and to B. W. Flemming in particular for giving me access to the facilities.

REFERENCES

ANTIA, E. E., 1993, Sedimentology, Morphodynamics and facies association of a mesotidal barrier island shoreface (Spiekeroog, southern North Sea): Berichte, Fachbereich Geowissenschaften, Universität Bremen, Nr. 32, 370 p.

BARTHOLOMAE, A., 1993, Zeitliche Variabilitaet und Raeumliche Inhomogenitaet in den Substrateigenschaften und der Zoobenthosbesiedlung im Umfeld von Miesmuschelbaenken. Teil D., Hydrodynamik: Bericht Senckenberg am Meer, Nr. 93/1, p. 117–123.

BARTHOLOMAE, A., AND FLEMMING, B. W., 1994, Massenbilanz und Transportsortierung der Sedimente auf der Groeninger Plate, *in* Umweltbundesamt (Deutschland), ed., Berichte aus der Oekosystemforschung, No. 4(2), p. 218–220.

BREZINA, J., 1979, Particle size and settling rate distributions of sand-sized materials: Proceedings of the 2nd European symposium on particle characterization, Nürnberg, 21 p.

DAVIS, R. A., JR., AND FLEMMING, B. W., 1989, Time series study of mesoscale tidal bedforms, Martens Plate, Wadden Sea, Germany: Memoirs of the Canadian Society of Petroleum Geologists, v. 16, p. 275–282.

EHLERS, J., 1988, The morphodynamics of the Wadden Sea: Rotterdam, A. A. Balkema, 397 p.

FIGGE, K., 1981, Sedimentverteilung in der Deutschen Bucht: Deutsches Hydrographisches Institut, Hamburg Blatt Nr. 2900.

FITZGERALD, D. M., AND PENLAND, S., 1987, Backbarrier dynamics of the East Frisian islands: Journal of Sedimentary Petrology, v. 54, p. 746–754.

FLEMMING, B. W., 1977, Depositional processes in Saldanha Bay and Langebaan Lagoon: Professional Research Series No. 2, National Research Institute of Oceanology, South Africa, 215 p.

FLEMMING, B. W., 1988, Process and pattern of sediment mixing in a microtidal coastal lagoon along the west coast of South Africa, *in* de Boer, P. L., van

Gelder, A., and Ńio, S-D., eds., Tide-influenced Sedimentary Environments and Facies: Dordrecht, D. Reidel, p. 275–288.

Flemming, B. W., and Davis, R. A. Jr., 1994, Holocene evolution, morphodynamics, and sedimentology of the Spiekeroog barrier island system (southern North Sea): Senckenbergiana Maritima, v. 24(1/6), p. 117–155.

Flemming, B. W., and Delafontaine, M. T., 1994, Biodeposition in a juvenile mussel bed of the East Frisian Wadden Sea (southern North Sea): Netherlands Journal of Aquatic Ecology, v. 28, p. 289–297.

Flemming, B. W., and Nyandwi, N., 1994, Land reclamation as a cause of fine-grained sediment depletion in backbarrier tidal flats (southern North Sea): Netherlands Journal of Aquatic Ecology, v. 28, p. 299–307.

Flemming, B. W., and Ziegler, K., 1995, High-resolution grain size distribution patterns and textural trends in the backbarrier environment of Spiekeroog Island (southern North Sea): Senckenbergiana Maritima, v. 26, p. 1–24.

Folk, R. L., and Robles, R., 1964, Carbonate sediments of Isla Perez, Alacran reef complex, Yucatan: Journal of Geology, v. 72, p. 255–292.

Folk, R. L., and Ward, W. C., 1957, Brazos River bar. A study in the significance of grain size parameters: Journal of Sedimentary Petrology, v. 27, p. 3–26.

Groen, P., 1967, On the residual transport of suspended matter by an alternating tidal current: Netherlands Journal of Sea Research, v. 3, p. 564–574.

Grotjahn, M., 1990, Sedimente und Mikrofauna der Watten bei der Insel Spiekeroog. Untersuchungen im Rahmen des "Sensitivitätesrasters Deutsche Nordsee Küste": Forschungsstelle für Insel- und Küstenschutz, Norderney, Jahresbericht, No. 39, p. 97–119.

Huntley, D. A., 1980, Tides on the northwest European continental shelf, *in* Banner, F. T., Collins, M. B., and Massie, K. S., eds., The Northwest European Shelf Seas: The Sea Bed and the Sea in Motion II. Physical and Chemical Oceanography, and Physical Resources: Amsterdam, Elsevier, p. 301–351.

Koch, M., and Luck, G., 1975, Untersuchung zu den Stromungsverhaltnissen auf dem westlichen Juister Watt: Forschungsstelle für Insel-und Küstenschutz, Norderney, Jahresbericht, No. 13, p. 29–33.

Koch, M., and Niemeyer, H. D., 1978, Sturmtiden-Strommessungen im Berreich des Norderneyer Seegats: Forschungsstelle für Insel- und Küstenschutz, Norderney, Jahresbericht, No. 29, p. 1–108.

Lassen, H., 1989, Oetliche und zeitliche Variationen des Meeresspiegels in der suedostlichen, Nordsee: Die Kueste, v. 50, p. 65–95.

McLaren, P., 1981, An interpretation of trends in grain-size measures: Journal of Sedimentary Petrology, v. 51, p. 611–624.

McLaren, P., and Bowles, D., 1985, The effects of sediment transport on grain-size distributions: Journal of Sedimentary Petrology, v. 55, p. 457–470.

Nyandwi, N., 1995, The Nature of the Sediment Distribution Patterns in the Spiekeroog Backbarrier Area, the East Frisian Islands: Berichte, Fachbareich Geowissenschaften, Universitaet Bremen, No. 66, 162 p.

Nyandwi, N., and Flemming, B. W., 1995, A hydraulic model for the shore-normal energy gradient in the East Frisian Wadden Sea (Southern, North Sea): Senckenbrgiana Maritima, v. 25, p. 163–171.

Postma, H., 1961, Transport and accumulation of suspended matter in the Dutch Wadden Sea: Netherlands Journal of Sea Research, v. 1, p. 148–190.

Ragutzki, G., 1982, Verteilung der Oberfläschensedimente auf den Niedersächsischen Watten: Forschungsstelle für Insel- und Küstenschutz, Norderney, Jahresbericht, No. 32, p. 55–67.

Reineck, H.-E., and Siefert, W., 1980, Factoren der Schlickbildung im Sahlenburger und Neuwerker Watt: Die Küste, v. 35, p. 26–51.

Reineck, H.-E., and Singh, I. B., 1973, Depositional Sedimentary Environments (2nd edition): Berlin, Heidelberg, Springer-Verlag, 439 p.

Reineck, H.-E., Chen, C., and Wang, S., 1986, Die Rueckseitenwatten zwischen Wangerooge und Festland, Nordsee: Senckenbergiana Maritima, v. 17, p. 241–252.

Sindowski, K.-H., 1973, Das ostfriesische Kuestengebiet. Sammlung Geologischer Fuehrer, v. 57: Berlin, Borntraeger, 162 p.

Van Straaten, L. M. J. U., and Kuenen, Ph. H., 1957, Accumulation of fine grained sediment in the Dutch Wadden Sea: Geologische en Minjnbouw, v. 19, p. 329–354.

Van Straaten, L. M. J. U., and Kuenen, Ph. H., 1958, Tidal action as a cause of clay accumulation. Journal of Sedimentary Petrology, v. 28, p. 406–413.

FACIES CHARACTERISTICS OF BACK-BARRIER TIDAL FLATS OF THE EAST FRISIAN ISLAND OF SPIEKEROOG, SOUTHERN NORTH SEA

GÜNTHER HERTWECK

Forschungsinstitut Senckenberg, Schleusenstrasse 39A, D-26382 Wilhelmshaven, Germany

ABSTRACT: A survey of tidal flat structure in the back-barrier area of Spiekeroog Island, Germany, southern North Sea, has provided a variety of morphologically distinct sediment bodies which, however, reveal only a limited number of definite structural units. These exhibit specific facies profiles visible in sediment cores, particularly in relief casts.

The most conspicuous structural and facies units, as seen from the margins at low water line towards the central parts of the tidal flats, are: (1) fine-sandy edges and spits where strong physical reworking by tidal currents results mainly in laminated stratification and complete absence of bioturbation; (2) edge-gully zones with channel fill structures; (3) trough channels with intense erosion of *Mya arenaria* shells originating from ancient colonies; (4) sandy tidal flats with ripple stratification showing sporadic and transient bioturbation by *Arenicola marina*; (5a) sandy tidal flats with ripple stratification, sporadic *Arenicola* bioturbation and sparsely spaced tubes of *Lanice conchilega*; (5b) sandy platforms with densely spaced tubes of *Lanice conchilega* and indistinct remnants of physical stratification; (6) *Mytilus* beds underlain by shell layers representing remnants of former *Mytilus* colonies; (7) muddy channel zones crossing the elevated watershed area of the back-barrier system totally bioturbated by *Heteromastus filiformis*, additionally populated by *Mya arenaria*; and (8) old semi-consolidated mud banks totally bioturbated and with shells of *Mya arenaria* in life position, representing ancient deposits in former watershed channel zones.

The time covered by the facies profiles increases from the marginal zones towards the central platforms of the tidal flats, that is, from several tides to several decades or even centuries. In addition to the succession of facies types in the horizontal direction, vertical successions are visible, as well, according to Walther's facies rule.

INTRODUCTION

The starting-point of this study is a survey of tidal flat structure carried out between 1988 and 1992 in the back-barrier area of Spiekeroog Island, Germany, in the southern North Sea (Hertweck, 1995; cf. Grotjahn, 1990; Flemming and Davis, 1994). The results of this survey (Fig. 1) present a variety of morphologically distinct sediment bodies with only a limited number of defined structural units.

The most conspicuous structural units, as seen from the margins at low water line towards the central parts of the tidal flats, are: (1) sandy edges and spits showing strong physical reworking; (2) edge-gully zones usually developed at stoss sides of main channels, effecting drainage of the adjacent tidal flats; (3) trough channels, where shells from former endobenthic colonization periods eroded and accumulated; (4) sandy tidal flats mainly populated by *Arenicola marina*; (5a) sandy tidal flats with sparse colonization by *Lanice conchilega* in addition to *A. marina*; (5b) sandy platforms with densely spaced colonies of *L. conchilega*; and (6) *Mytilus* beds (cf. Grotjahn, 1990; Knecht et al., 1990; Knecht, 1991). In the elevated watershed area, separating the catchments of the channel systems between the barrier islands, muddy channel zones (7) occur that are mainly populated by *Heteromastus filiformis* and *Mya arenaria*. Having the lowest hydrodynamic energy level, these zones function as preferential sites for the deposition of fine-grained sediment. In places, older semi-consolidated mud banks (8) are eroded in stoss sides of the main tidal channels. These banks represent deposits of former watershed channel zones.

The distinctness of these structural units and the fact that they occupy well delimited subareas of the back-barrier area investigated gave rise to the problem of facies typification in vertical sediment profiles. A particular element emerged; some benthic species occur or are absent for longer time periods (fluctuation and patchiness), whereas others are more or less permanent elements of the intertidal fauna. For example, the polychaete *L. conchilega* and the blue mussel *Mytilus edulis* developed into representative colonies during the time of survey, 1988–1992. However, in prior years, *L. conchilega* was not present in the Spiekeroog back-barrier area and *M. edulis* had disappeared almost completely during the following period. For the facies typification, the question arose whether and how such fluctuations would be represented in the stratigraphic profile. The present study is a first approach to these problems focusing especially on those structural units and facies, respectively, that occupy the major part of the study area.

METHODS

Area of Investigation

The area of investigation, Spiekeroog Island, Germany, is located in the Wadden Sea, which borders the coastal region of the southern North Sea (Fig. 1). Here, a chain of barrier islands extends from Den Helder in The Netherlands to the Weser mouth in northern Germany. These are separated from the mainland by an extended tidal flat zone that represents a mesotidal environment *sensu* Hayes (1979). The tidal range increases from 1.5 m at the West Frisian island of Texel, The Netherlands, to 3 m at the East Frisian island of Wangerooge, Germany. In the internal part of the German Bight low-macrotidal conditions (Hayes, 1979; Table 4) are realized. The highest tidal-range values are known from the tidal estuary of Jade Bay (3.84 m at Dangast near Wilhelmahaven, Germany) and from the river estuaries of the Weser (4.1 m at Bremen, Germany) and the Elbe (3.5 m at Hamburg, Germany).

In the Spiekeroog area (Fig. 1) the tidal range is 2.7 m near the western part of the island and 2.9 m at the mainland. The back-barrier areas of these islands are characterized by extended tidal flats that are dissected by systems of subtidal channels. The morphology is mainly controlled by tidal currents. In the main channel (Otzumer Balje), extending from southeast to northwest, current velocities of 0.7–1.0 m/s were measured (Bartholomä and Flemming, 1993), whereas for the tidal flats only 0.3 m/s are known (Bartholomä, 1993; Bartholomä and Flemming, 1993). However, wave action may also play an important role in tidal flats adjacent to the main channel, especially those exposed to storm winds from the western sector.

The sediment distribution of the backbarrier tidal flat area of Spiekeroog displays a coarsening succession from south to

Tidalites: Processes and Products, SEPM Special Publication No. 61

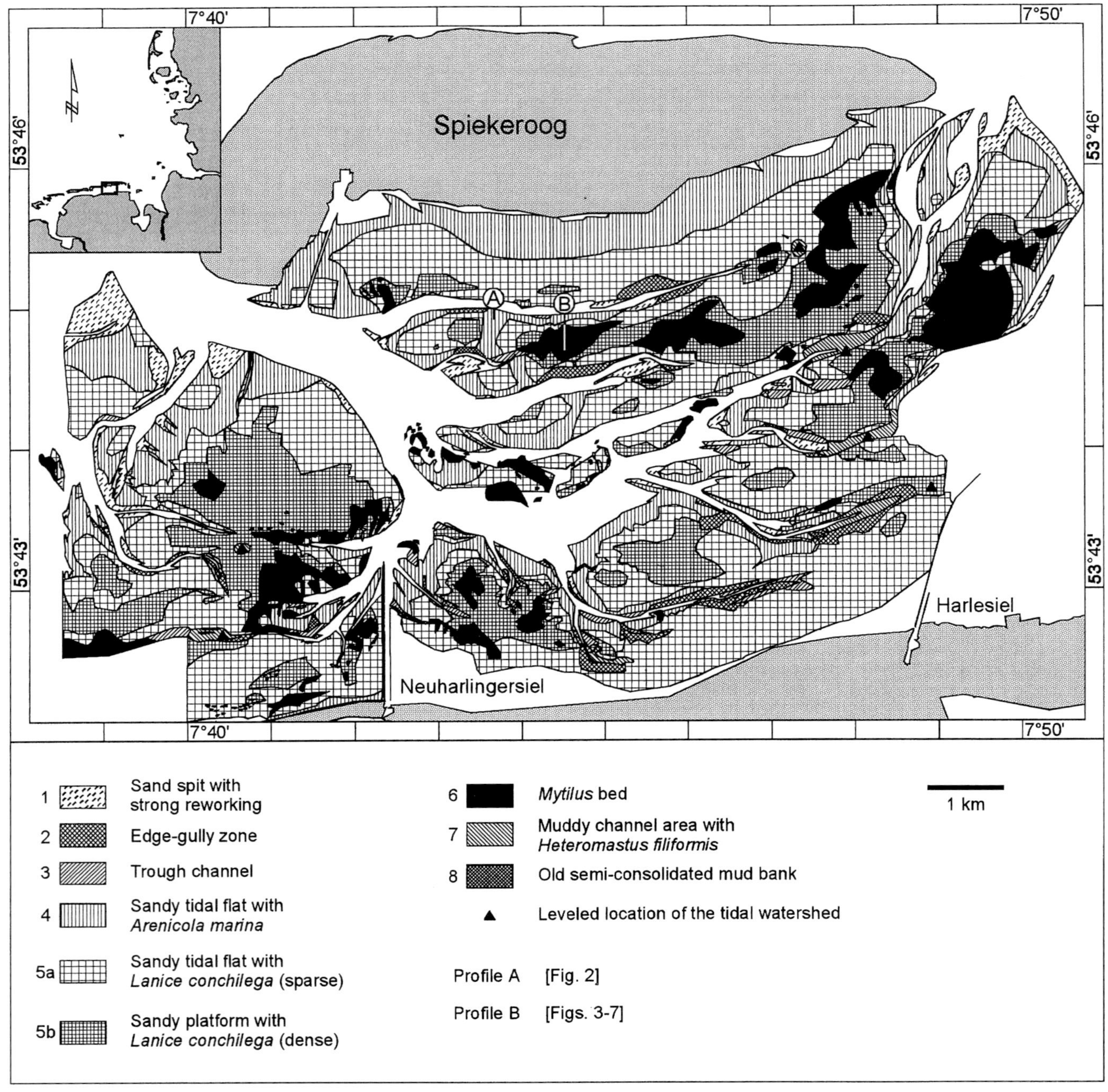

FIG. 1.—Tidal flat units in the back-barrier environment of the East Frisian Island of Spiekeroog, Germany, southern North Sea, showing the distribution of characteristic sediment bodies and benthic organisms. Modified from Hertweck, 1995.

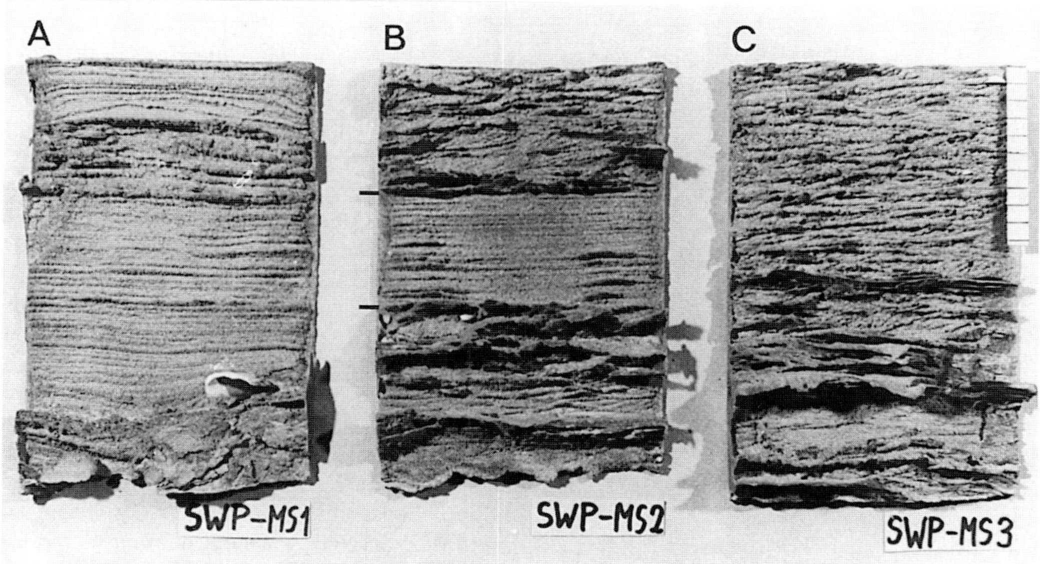

FIG. 2.—Depositional sedimentary structures in the strongly reworked sandy zone at the northern margin of the Swinnplate back-barrier area of Spiekeroog Island. Relief casts from box cores taken at distances of 10 m and within a height range of 1.6 m. (A) laminated stratification at the Low Water Line. (B) alternations of laminated sand and ripple cross-stratification with intercalated muddy lenses, at the slope of tidal flat edge. (C) ripple cross-stratification originating from alternating current and wave reworking on top of the tidal flat edge. Scale units = 1 cm.

FIG. 3.—Relief cast from the Swinnplate, back-barrier area of Spiekeroog Island, southern North Sea showing a vertical profile from the marginal zone of a sandy tidal flat populated by the lugworm *Arenicola marina,* the cockle *Cerastoderma edule* and the polychaete *Scoloplos armiger.* The upper half of the profile is well stratified by permanent physical reworking. Typical bends and angles in the stratification (center of photo) indicate profile sections close to *Arenicola* burrows. In the mid portion of the sediment column the stratification is slightly disturbed by *Scoloplos* burrows. The *Cerastoderma* shell layer originates from physical reworking by wave action during storm surges. Scale units = 1 cm.

FIG. 4.—Relief cast from the Swinnplate, back-barrier area of Spiekeroog Island, southern North Sea showing densely spaced tubes of a living *Lanice* colony in the upper part of the profile. In the middle, between two reworked *Cerastoderma* shell layers, tubes originating from an older colonization phase. Scale units = 1 cm.

FIG. 5.—Relief cast from the Swinnplate back-barrier area of Spiekeroog Island, southern North Sea showing the vertical profile of a mature *Mytilus* bed with two generations of *M. edulis:* adults (6–7 cm length); juveniles (3.2 cm length). Space between individual mussels is filled with muddy sediment, below the live horizon a 4-cm thick biosedimentary mud layer: fine sand with approximately 44% mud (<63 μm). The deeper parts of the profile exhibit two reworked shell layers originating from former *Mytilus* colonies. Scale units = 1 cm.

north (Ziegler, 1989). The size fractions contributing >50 weight-% to the total sand fraction increase from very fine sand (3.0–3.5 phi) in the mixed flat area in front of the mainland to coarser fine sand (2.0–2.5 phi) on the sand flat bordering the island at the south. The central belt of the back-barrier area, a variety of east-west oriented sand flats, is dominated by finer fine sand (2.5–3.0 phi) (Flemming and Davis, 1994, Fig. 31; Flemming and Ziegler, 1995).

Field Work and Laboratory Methods

From 1988 to 1993, the whole back-barrier area of Spiekeroog was mapped at scale 1:25,000 (Hertweck, 1992, 1993a, 1993b, 1995) in order to delimit sediment bodies and habitats of macrobenthic invertebrates producing distinct lebensspuren.

For the investigation of vertical sediment profiles, box cores of 25–35 cm length and 1 m tube cores were taken at low tide in the northern part of the Swinnplate. From the sediment column of each core, relief casts were produced in the laboratory by hardening them with artificial resin (Reineck, 1970). Sediment and shell samples were taken from every characteristic layer of the cores.

RESULTS AND DISCUSSION

Description of the Cores

The characteristic structural units of the tidal flat surface exhibit a variety of specific facies visible in the relief casts studied.

In places, marginal parts of the tidal flats show fine sand bodies that are either sandy spits at exposed ends of tidal flats or point bars of the slightly meandering main channels cutting through the flats. The northernmost part of the Swinnplate (Fig. 1) is formed by a 20-m wide, steep edge sloping towards the low water line within a vertical range of 1.6 m. Three box cores taken here at distances of 10 m reveal variations in sedimentary structures (Fig. 2) caused by the gradation of hydrodynamic energy in the marginal zone of tidal flats (Bartholomä, 1993; Bartholomä and Flemming, 1993). At the low water line, laminated stratification is developed, reflecting high current velocity in the channel. In the middle of the slope, laminated sand still prevails; only the top portion of the core shows ripple cross-stratification. The sediment column of the sloping edge zone exhibits a decimeter-thick oxidation layer. Also, this zone is almost devoid of endobenthic macrofauna and bioturbation, as well. Both features, together with the mainly laminated stratification, reflect a high-energy depositional and/or reworking regime. On top of the tidal flat edge, ripple cross-stratification originating from alternating tidal currents and wave reworking is most conspicuous. Thin mud flasers are sometimes visible.

Opposite the point-bar sand-bodies, the stoss sides of the channels are usually bordered by edge-gully zones much like those recently described by Reineck (1995). These gullies are transient transversal drainages of the neighboring tidal flats. Thus, small-scale channel fill structures are frequently produced here. Other small-dimensional structures are trough channels occurring particularly as drainages of adjacent *Mytilus* beds. In these shallow troughs, shells originating from ancient colonies, especially *M. arenaria*, are eroded and condensed in place. Later, they may be reburied as characteristic shell layers.

The normal marginal zone of the tidal flats is, however, represented by sand flats primarily populated by the lugworm *Arenicola marina*, and in some periods also by the cockle *Cerastoderma edule*. Also, the polychaete *Scoloplos armiger* frequently occurs. The sediments of this zone undergo permanent bioturbation by lugworms. However, physical reworking is the dominant process. This is due to the effect of tidal currents displaying velocities of 40–70 cm/s in the marginal zones of tidal flats (Brandt et al., 1995a, 1995b). In the cross-sectional profile (Fig. 3), a brownish oxidation layer of 3-cm thickness is visible on top, followed by a stratified grayish reduction layer. Usually, bioturbate structures of *A. marina* are visible only in places and are represented by slight bends and angles in the stratification. Occasionally the core and the relief cast may hit a longer portion of a burrow in its vertical extension. In many cores taken in the whole area, layers of *Cerastoderma* shells, single valves and sub-articulated pairs of valves as well, are visible. These represent accumulations or shell lag deposits due to winnowing by wave action during storm surges.

The internal areas of the tidal flats are populated by the tube-building polychaete *Lanice conchilega*. The outer zone of its habitat shows sparse colonies of up to 100 specimens per m, and is additionally populated by *A. marina* and *C. edule*, as in the marginal zone. The inner parts of the tidal flats are repre-

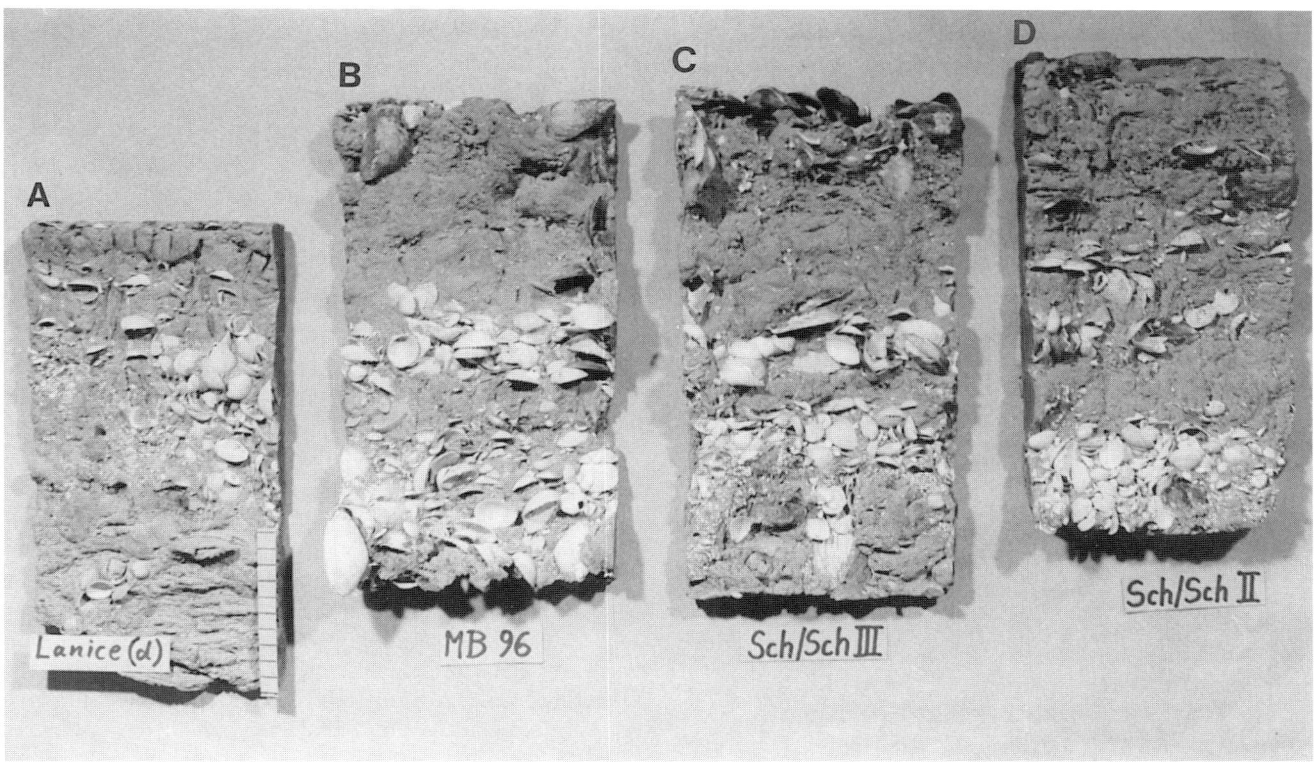

FIG. 6.—Relief casts from the Swinnplate back-barrier area of Spiekeroog Island, southern North Sea, showing correlation of shell layers and unconformities at different sites of the tidal flat studied. A) densely spaced living *Lanice conchilega.* B) remnants of *Mytilus* bed with dead but sub-articulated mussels and the biosedimentary mud layer still in place. C) *Mytilus* shell pavement enriched on tidal-flat surface after complete destruction of a mussel bed. D) biosedimentary mud layer still in place after erosion of the *Mytilus* shells (the coring station is 1 m away from sample C). Scale units = 1 cm.

sented by sandy platforms with densely spaced colonies of *L. conchilega*, viz. 10,000 per m. The tubes of *L. conchilega*, densely spaced as well, are visible throughout the cores taken in this zone (Figs. 4–7). Repeatedly, shell layers with dominating *Cerastoderma* shells occur, as seen in the *Arenicola* zone.

The central parts of the tidal flat platforms are, besides the *Lanice* colonization, covered by mussel beds in many places, that is, *Mytilus edulis*. These appear in various densities (cf. Flemming and Delafontaine, 1993), ranging from extended, densely spaced populations to patches showing extensions of a few decimeters to meters. Spaces between such *Mytilus* patches are usually inhabited by *L. conchilega*. In some places mussel beds also occur in marginal position of the tidal flats. At the time of the survey (Hertweck, 1992, 1993a, 1993b, 1995) *Mytilus* beds were covering approximately 10% of the back-barrier area of Spiekeroog Island. Cores taken in the mussel beds show the living *M. edulis* population on top (Fig. 5). Spaces between the individual shells are filled with biosedimentary muddy material, and the colony is underlain by a biosedimentary layer of 4-cm thickness (see Hertweck and Liebezeit, 1996, Fig. 3). This material originates from feces and pseudofeces, both provided by the mussels, and from fine-grained material trapped in the rugged bottom relief of the *Mytilus* colony. Its mud content (<63 μm) amounts up to 49 weight-percent. The shell layers in the 6–9 cm and 14–17 cm levels of the profile (Fig. 5) are rich in *Cerastoderma* and *Mytilus* shells. Among the latter, pairs of sub-articulated shells frequently occur. This suggests that they were embedded soon after death of the organisms, thus indicating the former existence of *Mytilus* beds at this site. In the deepest section of the core *Lanice* tubes of an older colonization phase are visible.

A particular low-energy environment occurs in the topographically highest subarea, that is, the watershed that separates the western and eastern catchments of the back-barrier channel systems (Fig. 1). The head regions of the channels draining the watershed area in oposite directions initially form shallow, elongated troughs that emerge at low tide. Here, the gradual flow-off of the ebb-currents in both directions can be observed until the troughs fall dry. At high tide, the flooding waters of both catchments meet here, thus losing their current energy. In this way, these sites function as sediment traps and the most muddy material is deposited here. The endobenthic fauna of these zones consists mainly of dense colonies of both the polychaetes *Heteromastus filiformis* and *Nereis diversicolor.* Additionally, a dense population of juvenile *M. arenaria* was present in 1988.

Processes of Facies Development

The sediment complexes that reflect the highest reworking rates, that is, tidal flat edges and exposed spits, are related to highly dynamic current regimes comprising only one or a few tides. They are characterized by well-aerated sediments and the sparseness of macro-endobenthos (Fig. 2).

Oxidation and reduction zones can be distinguished in the marginal zone of the sandy tidal flats populated by *A. marina*. The less-aerated sediments of the reduction zone are also

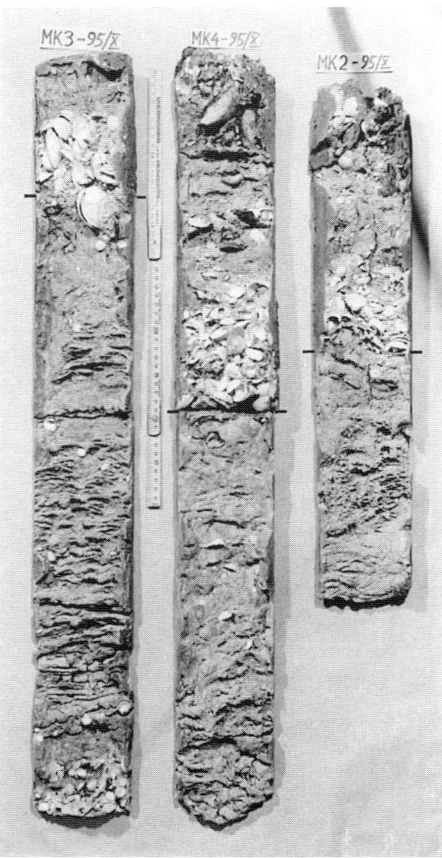

FIG. 7.—Relief casts of long cores from the Swinnplate back-barrier area of Spiekeroog Island, southern North Sea, showing correlation of facies down to 90 cm below the tidal flat surface. Upper portions: shell layers reflecting *Cerastoderma* and *Mytilus* colonization. Lower portions: previous stratification, partly destroyed by *Arenicola* and *Scoloplos* bioturbation. Scale = 50 cm.

stratified and are bioturbated only in places where living lugworms occur (Fig. 3). The prevalence of physical sedimentary structures is due to occasional storm events affecting deeper parts of the sediment (cf. Davis and Flemming, 1991, 1992, 1995). Bioturbation by *S. armiger* producing tiny irregular burrows 1 mm in diameter does not completely obliterate the preexisting stratification.

A long-term stratigraphic record is realized in *Mytilus* bed profiles in the central parts of tidal flats. For example, on the Swinnplate a time span of 21 to 35 (possibly up to 50) years is documented in the upper 23 cm of the sediment column, which shows a seven-year-old living mussel colony on top and exhibits two deeper shell layers containing high amounts of subarticulated *Mytilus* shells that have reached the same age at least (Hertweck and Liebezeit, 1996). Estimation of the time span indicated by the *Mytilus* bed facies is based on three assumptions: (1) the maximum age of the adult mussels in each layer; (2) recolonization by younger generations at the half life-cycle of the respective adult specimens (cf. Delafontaine, 1993); and (3) interruptions in *Mytilus* colonization of several years between phases with fully developed mussel beds at the respective site.

The processes leading to the embedding of former *Mytilus* beds in the sediment column are connected to the destructional phase of this epibenthic system after death of the mussels. Living *Mytilus* colonies may be interred only exceptionally by a mud avalanche and thus killed and preserved *in situ*. Normally, the late phase of the system results in restricted patches of mussel beds covered with macro-algae and extended shell pavements containing mainly *Mytilus* and *Cerastoderma* valves. Biosedimentary mud may persist below the shell pavement for some time. In places, the receding *Mytilus* patches are surrounded by mud halos of about 1-m width. Therein, the thickness of the mud layer decreases from 5 cm near the mussel patch to 0 cm at the outer rim. All these are transient situations finally resulting in erosion and export of the fine-grained biosedimentary material as soon as the retaining capacity of the mussel beds is terminated (see Hertweck and Liebezeit, 1996). Alternations of layers containing different material and reflecting different colonization phases (see particularly Fig. 5), as well as local unconformities truncating lebensspuren of macrobenthos (Fig. 6), constitute the facies profiles characteristic for the central areas of tidal flat environments (Figs. 6–7).

After establishing the facies types with 25-35 cm box-cores, a deeper stratigraphic approach was made with 1-m coring tubes. Evidence shows that beds do not extend deeper than 23 cm below the sediment surface on the Swinnplate shell layers reflecting former *Mytilus* (Fig. 7). Below this level the *Mytilus* facies is underlain by the typical *Arenicola* facies, characterized by stratified sand layers with inconspicuous bends and, in places, bioturbation and tiny burrows of *S. armiger.* This succession suggests that the establishment of mussel beds occurred after a critical elevation of the tidal flat had been reached. Such vertical accretion can result from local sediment deposition or from a lateral shifting of the intertidal sediment body (Fig. 8) (cf. Brosinsky et al., 1990; Brosinsky, 1991).

The most stable depositional environment of a back-barrier tidal flat area is represented in the muddy sediments of the watershed channel zones that originate from slow deposition of fine-grained material under low energy conditions. The old mud banks exposed at some channel margins were deposited under similar conditions in time periods when the back-barrier watershed was located at the respective site. The shifting of the watershed from west to east since 1650 was deduced from old nautical charts (Homeier and Luck, 1969). Today, these muds are semiconsolidated.

The horizontal facies succession from the marginal zone of the tidal flat towards the central platform reflects an increase in the time span covered by the profiles from several tides to several decades and even centuries (Fig. 8).

CONCLUSIONS

The structural units characterizing the sediment surface of the East Frisian back-barrier tidal flats exhibit a variety of specific facies visible in vertical relief cast profiles.

The formation of the individual facies is controlled, to a different extent, by hydrodynamic energy and biogenic activity of benthic organisms. In the marginal zones of tidal flats, physical sedimentary structures are predominant. Bioturbate structures and shells, in *post mortem* life position and in reworked shell layers as well, prevail in the inner and central parts.

The increase of bioturbation from margin to center of the intertidal sediment body is associated with a decrease in physi-

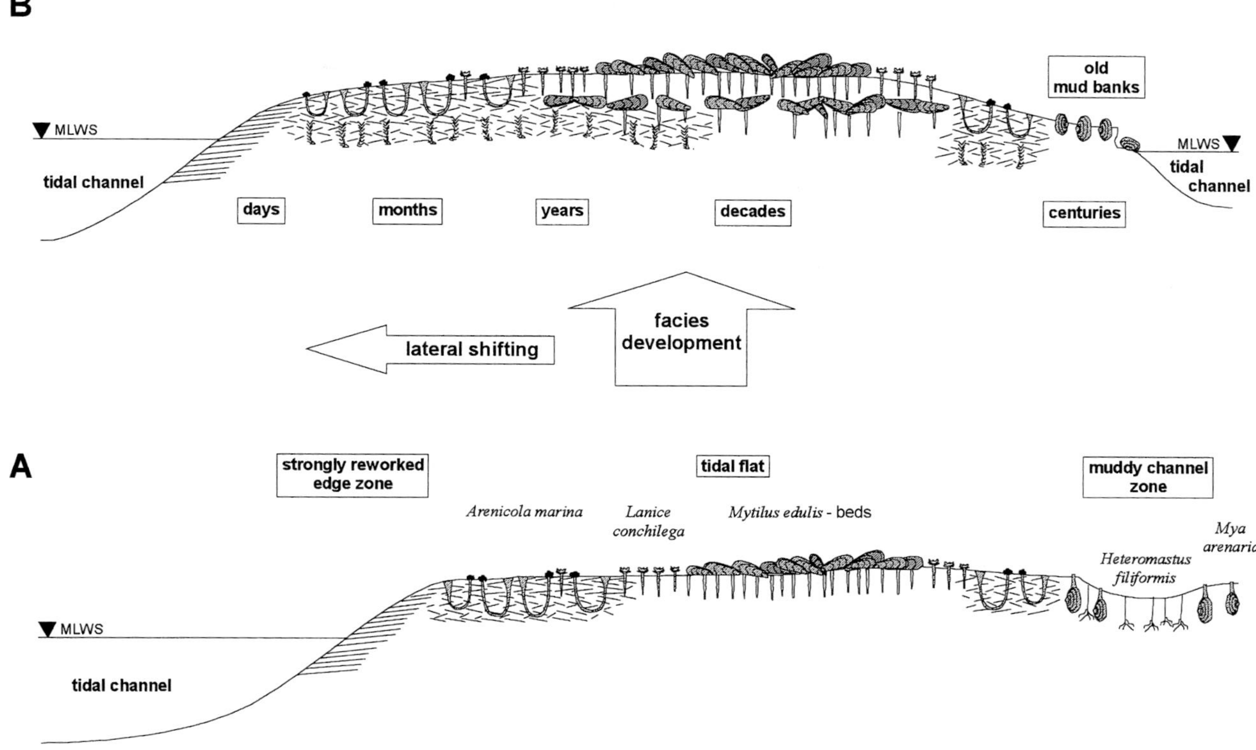

FIG. 8.—Development of facies by lateral shifting of tidal flats in the back-barrier area of the East Frisian Islands, southern North Sea. A) Horizontal succession of hydrodynamic zones and benthic habitats populated by various organisms. Depositional and reworking processes, as well as bioturbation activity and skeletal remains of organisms, lead to the formation of distinct facies across the tidal flat. B) Lateral shifting of the tidal flats during long-term changes of the back-barrier geographical configuration leads to horizontal displacement and vertical superposition of the different facies, in the sense of Walther's facies rule. The results are different stratigraphic sequences at characteristic sites of the tidal flat area reflecting different time spans of development. MLWS = mean low water spring (chart datum).

cal sediment reworking due to progressively lower hydrodynamic energy.

As a result of the gradation in intensity and depth of physical reworking, the time span covered by the facies profiles increases from the marginal zones towards the central platforms of the tidal flats, that is, from several tides to several decades or even centuries (Fig. 8).

The vertical facies succession and the corresponding horizontal facies zonation is a good example of Walther's facies rule.

ACKNOWLEDGMENTS

I am grateful to my colleagues Gerd Liebezeit, Achim Wehrmann and Ingrid Kröncke, who critically read the text, and to Burg Flemming, H.-E. Reineck and F. Wunderlich for valuable discussions and advice. Thanks are also given to two anonymous reviewers who contributed to the improvement of the manuscript. The map, Fig. 1, has been provided by the ÖSF Niedersächsisches Wattenmeer—GIS 1997. I am indebted to A. Wehrmann for the final design of Figs. 1 and 8, and to A. Hartkopf and M. Stamm for the photo prints.

This work was financially supported by the Bundesministerium für Forschung und Technologie, Bonn, under grant 03F0112A. The responsibility for the contents of this publication rests with the author. This is contribution No. 332 of the German Ecosystem Research Program Wadden Sea.

REFERENCES

BARTHOLOMÄ, A., 1993, Zeitliche Variabilität und räumliche Inhomogenität in den Substrateigenschaften und der Zoobenthosbesiedlung im Umfeld von Miesmuschelbänken. D. Hydrodynamik: Bericht Senckenberg am Meer, Wilhelmshaven, v. 93/1, p. 117–123.

BARTHOLOMÄ, A., AND FLEMMING, B. W., 1993, Zeitliche Variabilität in den Sedimentparametern und der Morphologie auf der Gröninger Plate (Spiekerooger Watt): Bericht Senckenberg am Meer, Wilhelmshaven, v. 93/1, p. 54–48.

BRANDT, G., FLESSNER, J., GLASER, D., KAISER, R., KAROW, H., MÜNKEWARF, G., AND NIEMEYER, H. D., 1995a, Dokumentation zur hydrographischen Frühjahrs-Meßkampagne 1994 der Ökosystemforschung Niedersächsisches Wattenmeer im Einzugsgebiet der Otzumer Balje, *in* Berichte zur Ökosystemforschung—Hydrographie, v. 8: Norderney, Niedersächsisches Landesamt für Ökologie, Forschungsstelle Küste, p. 1–94.

BRANDT, G., FLESSNER, J., GLASER, D., KAISER, R., KAROW, H., MÜNKEWARF, G., AND NIEMEYER, H. D., 1995b, Dokumentation zur hydrographischen Frühjahrs-Meßkampagne 1994 der Ökosystemforschung Neidersächsisches Wattenmeer im Bereich des Janssand, *in* Berichte zur Ökosystemforschung—Hydrographie, v. 9: Norderney. Niedersächsisches Landesamt für Ökologie, Forschungsstelle Küste, p. 1–23.

BROSINSKY, D., 1991, Morphodynamische Prozesse und Sedimentverteilung im Bereich der Muschelbalje (östliches Spiekerooger Watt): Unpublished Diplomarbeit Universität Gießen, p. 1–77.

BROSINSKY, D., FLEMMING, B. W., AND STEIN, R., 1990, Morphologische Prozesse und Sedimentverteilung im Bereich der Muschelbalje (östliches Spiekerooger Watt): Sediment 90, 5. Sedimentologen-Treffen Bonn 1990, Poster Volume, 3 p.

DAVIS, R. A., AND FLEMMING, B. W., 1991, Time-series study of mesoscale tidal bedforms, Martens Plate, Wadden Sea, Germany: Memoirs Canadian Society of Petroleum Geologists, v. 16, p. 275–282.

Davis, R. A., and Flemming, B. W., 1992, Stratigraphy of a wave-dominated tidal flat: Martensplate, German Wadden Sea, *in* Flemming, B. W., ed., Tidal Clastics 92, 3rd International Research Symposium on Modern and Ancient Clastic Tidal Deposits, Courier Forschungsinst Senckenberg, v. 151, p. 22–23.

Davis, R. A., and Flemming, B. W., 1995, Stratigraphy of a combined wave- and tide-dominated tidal flat system: Martens Plate, German Wadden Sea, *in* Flemming, B. W., and Bartholomä, A., eds., Tidal Signatures in Modern and Ancient Sediments: Oxford, of the International Association of Sedimentologists Special Publication 24, p. 121–132.

Delafontaine, M. T., 1993, Zeitliche Variabilität und räumliche Inhomogenität in den Substrateigenschaften und der Zoobenthosbesiedlung im Umfeld von Miesmuschelbänken. B. Aspects of a shell budget: production, burial and interspecific competition: Bericht Senckenberg am Meer, Wilhelmshaven, v. 93/1, p. 64–63.

Flemming, B. W., and Davis, R. A., 1994, Holocene evolution, morphodynamics and sedimentology of the Spiekeroog Barrier Island system (southern North Sea): Senckenbergiana Maritima, v. 24, p. 117–155.

Flemming, B. W., and Delafontaine, M. T., 1993, Zeitliche Variabilität und räumliche Inhomogenität in den Substrateigenschaften und der Zoobenthosbesiedlung im Umfeld von Miesmuschelbänken. C. Morphodynamik und Biodeposition: Bericht Senckenberg am Meer, Wilhelmshaven, v. 93/1, p. 99–116.

Flemming, B. W., and Ziegler, K., 1995, High-resolution grain size distribution patterns and textural trends in the backbarrier environment of Spiekeroog Island (southern North Sea): Senckenbergiana Maritima, v. 26, p. 1–24.

Grotjahn, M., 1990, Sedimente und Makrofauna der Watten bei der Insel Spiekeroog—Untersuchungen im Rahmen des Sensitivitätsrasters Deutsche Nordseeküste: Jahresbericht 1987 Forschungsstelle Küste, Norderney, v. 39, p. 97–119.

Hayes, M. O., 1979, Barrier island morphology as a function of tidal and wave regime, *in* Leatherman, S. P., ed., Barrier Islands from the Gulf of St. Lawrence to the Gulf of Mexico: New York, Academic Press, p. 1–27.

Hertweck, G., 1992, Distribution pattern of benthos and lebensspuren in the Spiekeroog backbarrier tidal flat area (southern North Sea), *in* Flemming, B. W., ed., Tidal Clastics 92, Abstract Volume, Courier Forschungsinstitut Senckenberg, v. 151, p. 40–42.

Hertweck, G., 1993a, Zeitliche Variabilität und räumliche Inhomogenität in den Substrateigenschaften und der Zoobenthosbesiedlung im Umfeld von Miesmuschelbänken. A. Biofazies und Aktuopaläontologie: Bericht Senckenberg am Meer, Wilhelmshaven, v. 93/1, p. 49–63.

Hertweck, G., 1993b, Faziesentwicklung im biosedimentären System *Mytilus*-Bank: Sediment 93, 8. Sedimentologen-Treffen Marburg 1993, Abstract Volume, p. 44–45.

Hertweck, G., 1995, Verteilung charakteristischer Sedimentkörper und Benthossiedlungen im Rückseitenwatt der Insel Spiekeroog, südliche Nordsee. I. Ergebnisse der Wattkartierung 1988–92: Senckenbergiana Maritima, v. 26, p. 81–94.

Hertweck, G., and Liebezeit, G., 1996, Biogenic and geochemical properties of intertidal biosedimentary deposits related to *Mytilus* beds, *in* Dworschak, P. C., Stachowitsch, M., and Ott, J. A., eds., Proceedings of the 29th European Marine Biology Symposium 1994, Vienna: Pubblicazioni della Stazione Zoologica di Napoli I: Marine Ecology, v. 17, p. 131–144.

Homeier, H. and Luck, G., 1969, Das Historische Kartenwerk 1:50 000 der Niedersächsischen Wasserwirtschaftsverwaltung als Ergebnis historisch-topographischer Untersuchungen und Grundlage zur kausalen Deutung hydrologisch-morphologischer Gestaltungsvorgänge im Küstengebiet: Veröffentlichungen des Niedersächsischen Instituts für Landeskunde, Landesentwicklung an der Universität Göttingen, Series A, Forschungen zur Landes- und Volkskunde, I. Natur, Wirtschaft, Siedlung und Planung, v. 93: Göttingen-Hannover, Verlag Wurm, 36 p., Appendix 28 p.

Knecht, S., 1991, Morphologische Entwicklung, Sedimentverteilung und Faziesdifferenzierung der Martensplate (östliches Spiekerooger Watt): Unpublished Diplomarbeit Universität Gießen, 67 p.

Knecht, S., Flemming, B. W., and Stein, R., 1990, Morphologische Entwicklung, Sedimentverteilung und Faziesdifferenzierung der Martensplate (östliches Spiekerooger Watt): Sediment 90, 5. Sedimentologen-Treffen Bonn 1990, Poster Volume, 3 p.

Reineck, H.-E., 1970, Reliefguß und projizierbarer Dickschliff: Senckenbergiana Maritima, v. 2, p. 61–66.

Reineck, H-E., 1995, Randpriele: Senckenbergiana Maritima, v. 26, p. 37–43.

Ziegler, K., 1989, Holozäne Entwicklung, Aktuogeologie und Sedimentdynamik des Spiekerooger Rückseitenwattes: Unpublished Diplomarbeit Technische Universität Clausthal, 129 p.

EVIDENCE FOR TEMPERATURE-ADJUSTED SEDIMENT DISTRIBUTIONS IN THE BACK-BARRIER TIDAL FLATS OF THE EAST FRISIAN WADDEN SEA (SOUTHERN NORTH SEA)

F. KRÖGEL
26316 Varel, Rahlinger Strasse 2, Germany
AND
B. W. FLEMMING
Senckenberg Institute, Department of Marine Research, 26382 Wilhelmshaven, Schleusenstrasse 39A, Germany

ABSTRACT: Theoretical considerations suggest that, while changes in hydrodynamic conditions are an important factor in mobilizing and resuspending sediments, seasonal shifts in spatial sediment distributions in shallow water environments of intermediate to high latitudes should be influenced by variations in settling velocities and critical shear strengths induced by changes in water temperature and fluctuations in salinity. In the Wadden Sea of the southern North Sea, water temperatures range from <4°C in January to almost 24°C in July, and salinities from 2.6% to 3.3%. In accordance, the kinematic viscosity of sea water varies from 1.5728–0.9096 cSt (centi-Stokes). Under these circumstances sediment particles respond to the changes in viscosity with corresponding changes in settling velocity and particles of a given size thus have as many settling velocities as there are viscosity changes. In hydraulic terms a particle can therefore have many effective grain sizes. To facilitate the calculation of critical shear velocity on the basis of settling velocity, the following equation was developed: $U_{*cr} = (0.482\,[((\delta_s - \delta_f)/\delta_f)\,\nu\, g]^{0.282}) * (0.15\, w_s^{0.5}) + 0.61$.

The back-barrier tidal basin investigated in this study is characterized by a shoreward-decreasing energy gradient, as documented by progressively decreasing grain sizes and the resulting landward-fining sequence of shore-parallel facies belts. Applying the viscosity principle, one should expect the facies belts to show a seasonal temperature adjustment. Time-series studies, however, have demonstrated that the sediment zonation pattern remains stable throughout the year; that is, a seasonal temperature-induced shift of the belts could not be confirmed.

This apparent contradiction can be explained if one assumes that the sediment distribution pattern is permanently adjusted to winter conditions at which the settling velocities are lowest (the sediment is hydraulically finest) and the energy input is highest. Subsequent summer conditions are unable to reverse the depositional pattern produced in winter because settling velocities are higher, making the sediment hydraulically coarser. This interpretation is supported by the observed seasonal dynamics of suspended matter. During winter, muddy sediments occupy relatively small areas adjacent to the mainland dike. In summer, on the other hand, mud contents of local sediments were found to increase and muddy sediments occupied much larger areas. This pattern is explained by the higher settling velocities of suspended matter in summer as required by the viscosity principle. Field data therefore support the theoretical considerations and demonstrate that the kinematic viscosity of the fluid plays an important role in depositional processes.

INTRODUCTION

Sediment distributions in the Wadden Sea have been interpreted in the past mainly as a product of varying hydrodynamic conditions (e.g., van Straaten and Kuenen, 1957; Postma, 1961). Published data, hence, infer a wide range of hydrodynamic settings (e.g., Groen, 1967). This should not appear unusual because hydrodynamic forces are commonly considered to be the dominating factor controlling erosion, transportation, and deposition. In this context it has also been suggested that first-order variations in erosion and deposition were controlled by temporal changes in tidal current strength and wave action (e.g., Frostick and Mc Cave, 1979).

In the Wadden Sea the level of energy flux is strongly seasonal, being higher in winter due to storm-amplified tidal currents and increased wave action. The seasonal variation in energy flux is revealed in the intensity of intertidal sediment reworking which, in turn, drives the morphodynamics of the system (e.g., Bartholomä and Flemming, 1993). Nevertheless, in the course of this study the variations in the spatial distribution of surficial sediments are difficult to explain by variations in energy flux alone. It is hence suggested that certain aspects of sediment distribution are instead the result of seasonal changes in the physical properties of sea water associated with changes in water temperature. This concept implies that sediment distribution, especially in shallow-water environments of intermediate to high latitudes, is at least partly controlled by the different settling rates of sediment particles induced by seasonal changes in water temperature. Furthermore, because shear velocities are known to respond sensitively to variations in the physical properties of a fluid (e.g., Komar and Clemens, 1986), one should expect corresponding changes in the transport behavior of sediment particles. The objective of this paper, therefore, is to demonstrate that the temperature effect is more substantial than hitherto recognized, being strong enough to be identified in local sediment distribution patterns on both seasonal and longer time scales.

PHYSICAL SETTING

The study area is situated along the mesotidal East Frisian barrier island coast of northwest Germany and occupies the tidal basin between the islands of Baltrum and Langeoog (Fig. 1). The sediment sample grid straddles the tidal basin draining through the Accumer Ee tidal inlet, which is situated between the catchment divide of Baltrum in the west and that of Langeoog in the east. It covers an area of about 106 km². The meteorological setting is characterized by relatively calm conditions in summer and more stormy conditions in winter, being dominated by northwesterly winds between January and April. The mean tidal range in the study area is approximately 2.6 m, with the tidal wave propagating from west to east along the coast (e.g., Flemming and Davis, 1994). The energy flux in the tidal catchment is controlled by three factors: (A) semi-diurnal tidal currents, (B) wave action generated within the tidal basin, and (C) swells penetrating the inlet from the open North Sea. Of these, the tides and local wind-generated waves are considered to be less important. Externally generated swells dissipate most of their energy by breaking along the seaward margin of the ebb delta and about 10% of the energy is penetrating the inlet to be dispersed across the back-barrier tidal basin. Thus, the mixture of local waves and external swells results in an energy level that progressively decreases in the direction of the mainland coast throughout the year (Krögel, 1997a).

Tidalites: Processes and Products, SEPM Special Publication No. 61

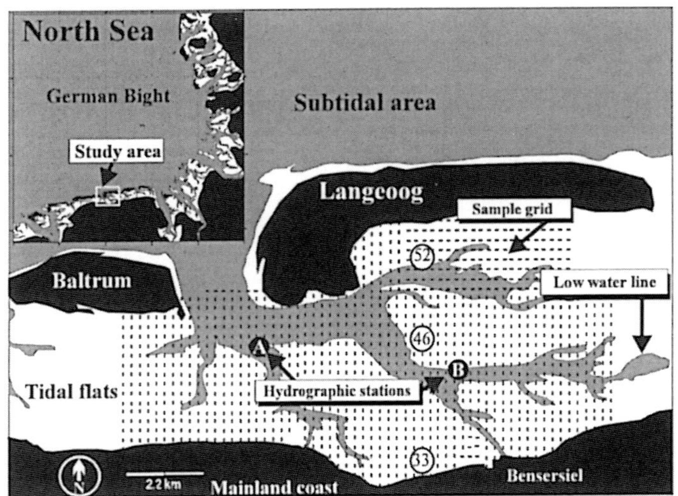

FIG. 1.—Location map of the study area in the East Frisian Wadden Sea of northern Germany. The grid of the surficial sediment samples (sample grid) has a density of 250 m. Wave measurements were carried out at station A and current measurements at station B.

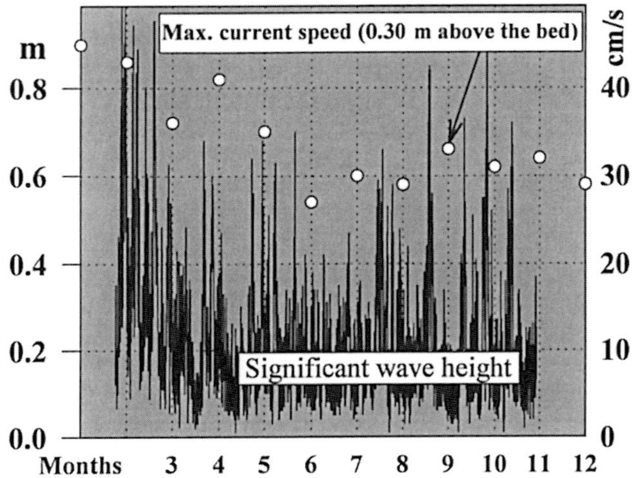

FIG. 2.—Significant wave heights and maximum current speed in 1994 (measuring location cf. 1).

METHODS

Surficial sediments were initially sampled in May 1993 in the course of a 17-day continuous field survey. Following the procedure described in Flemming and Ziegler (1995), sampling was arranged on a regular grid defined by geographic intervals of 0.15′ latitude and 0.25′ longitude. This corresponds to a spacing of roughly 275 × 275 m. The positions were fixed by means of a portable differential-Global Positioning System navigator. In all, 1500 sediment samples were collected and processed in this first campaign. Over the following two years, the sampling procedure was repeated monthly at 74 stations aligned along six transects that were selected on the basis of the textural data generated after the initial survey. Thin marker stakes were driven into the tidal flat at each station in order to allow subsequent sampling at identical positions. In each case a total sample of about 250 g was recovered from the upper 1.5–2 cm of the tidal flats, comprising a mixture of five subsamples taken at random from an area of about 4 m^2 around the marker stakes. Care was taken that samples included material from both ripple crests and troughs. In this way, the probable sampling error was minimized (e.g., Krumbein, 1934). In addition, the relative elevation of the intertidal flat was determined at each stake during the monthly sampling campaigns, thus providing some indication of physical reworking and sediment turnover.

The samples were dialyzed to remove the salt and then washed through a 0.063 mm sieve to separate sand and mud. Both fractions were subsequently dried and weighed to determine the relative contribution of sand and mud to the total sample. The mud fractions were stored. The sand fractions were repeatedly split by means of a mechanical sample divider until subsamples of suitable mass for settling tube analyses were obtained, generally 0.5–1.0 g. All grain size analyses of the sand fractions were carried out by measuring their settling velocity distributions on a computer-controlled settling tube system, and then converting the settling velocities into equivalent settling diameters using the glass sphere standard of Gibbs et al. (1971) as modified by Brezina (1979).

Following the international engineering convention, settling velocities are commonly computed on the basis of a water temperature of 24°C (e.g., Rouse, 1936, 1937; Rubey, 1933). In order to determine the particle size deposited under actual *in situ* conditions, settling velocities were in this case also computed for the local water temperatures measured during the various sampling campaigns. Differences in salinity were compensated for in a similar way (e.g., Gibbs, 1972). Textural parameters were calculated using percentile statistical procedures of the type described in Folk and Ward (1957) and Folk (1964). More detailed information and discussions concerning grain size analyses by settling tube can be found in Flemming and Thum (1978).

To register the seasonal variability in the physical properties of the local sea water, measurements of temperature, salinity and density were carried out in the study area at monthly intervals using a standard current temperature density-probe that was expanded to incorporate an acoustic doppler current

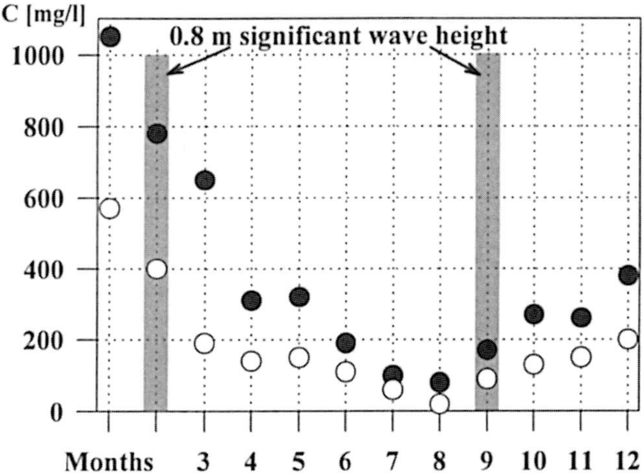

FIG. 3.—Minimum (open circles) and maximum (filled circles) turbidity in mg/l 0.30 m above the bed. Notice the difference between the winter and the summer concentration dominated by the same significant wave height.

meter and a light-attenuating transmissiometer (670 nm). The transmissiometer was carefully calibrated in the laboratory using concentrated suspended matter collected in the study area. Continuous vertical profiles through the water column were in each case measured at half-hourly intervals over a full tidal cycle at a hydrographic reference station located in one of the larger back-barrier channels (cf. Fig. 1). For the purpose of this paper, the monthly CTD-data were time- and depth-averaged over the full tidal cycle. Little variation was observed over depth at any particular time, indicating a well-mixed water body. Processed wave data were available from a back-barrier measuring station which, at the time, was jointly operated by the Forschungsstelle Küste (Norderney, Germany) and Delft Hydraulics (Delft, The Netherlands). It was located within the tidal basin just south of the tidal inlet (cf. Fig. 1).

RESULTS

Hydrodynamics

Depositional processes and the distribution of sediments ultimately depend on the impact of hydrodynamic forces. This is especially true in environments characterized by a relatively high temporal variability in tidal and wave energy levels, as is the case in the Wadden Sea. One would thus expect any temporal variability in sediment distribution to be closely related to variations in wave action and/or current strength. In the course of a tidal cycle, depth-averaged tidal current speeds in the channels close to the inlet fluctuate from 0–70 cm/s and on the tidal flats from 0–25 cm/s. At corresponding times of the tidal cycle mean current speeds at spring tide are on average about 10% higher than at neap tide. This range remains almost constant throughout the year, as demonstrated by two almost identical time histories recorded in the course of two measuring campaigns carried out in 1995 at corresponding tidal phases—one in July and the other in October. It can thus be assumed that, by and large, tidal currents on the intertidal flats follow a similar temporal pattern as those in the channels throughout the year, albeit with different velocity amplitudes. Flow velocities outside this observed range can be regarded as exceptional.

Whereas seasonal current variations were found to be small, the influence of seasonal fluctuations in wave action on sedimentary processes were expected to increase in relative importance. Indeed, the high-frequency wave data do reveal a seasonal pattern (Fig. 2). Thus, in January and February 1994, maximum significant wave heights at the measuring station within the tidal basin (cf. Fig. 1) exceeded 0.8 m. A similar pattern is observed in autumn, although peaks around 0.8 m were much less frequent. By contrast, the maximum significant wave heights in June and July of the same year were over 50% lower, fluctuating from 0.3–0.4 m. Furthermore, it is clearly evident that the period of least wave action (July) was very short.

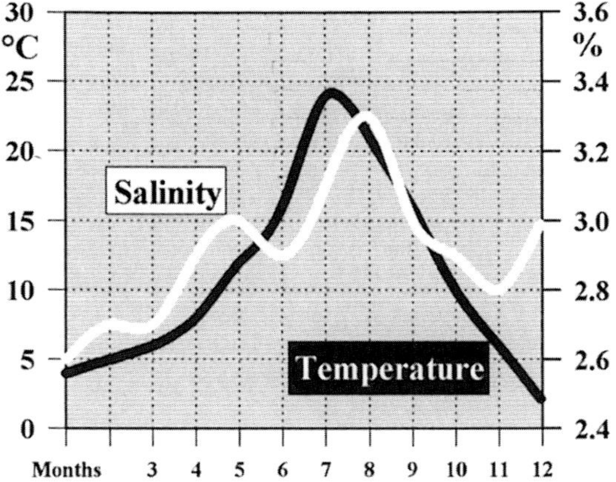

FIG. 4.—Variability of water temperature and salinity in 1994 (for location of measurements cf. 1).

In comparison, the monthly turbidity records show a much better defined seasonal signal than either the wave or the current data (Fig. 3). In winter, a maximum suspended sediment concentration of more than 1000 mg/l was measured, whereas the concentration in summer rarely exceeded 400 mg/l. The lowest turbidity levels reveal an identical picture, with values of 300 mg/l in winter and less than 10 mg/l in summer. Although the turbidity data show some degree of correlation with the corresponding wave data, the former have a much stronger seasonal signal than the waves. In particular, it can be shown that turbidity levels in winter were not only substantially higher than in summer under similar wave conditions, but that this discrepancy was still apparent at much lower winter wave conditions (Fig. 3). This would suggest that some other factor than currents and waves must have been responsible for the grossly different seasonal turbidity patterns observed in the study area. The turbidity data presented above were thus a first indication that the assumption of hydrodynamic forces being the overriding factor in local depositional processes may have been premature.

Physico-chemical Properties of Sea Water

The physico-chemical properties of the sea water were in each case determined at the same monthly intervals in which the sediment samples were collected and the current and turbidity measurements were made. They may thus be assumed to approximate the conditions under which the uppermost sediment layer, from which the samples were recovered, was deposited. Each temperature and salinity data set was collected over the same 12-hour tidal cycle during which the sediment samples were taken and the parameters therefore reflect the average conditions at the time of sampling. As expected, water temperature reveals a distinct seasonal trend. It begins with a seasonal low of 3°C in January and gradually increases over the course of the following months to reach 12°C in May.

During warm summers the surface of the dark brown tidal flats can heat to 30°C within the few hours around low tide. The water flowing across such warm surfaces is then heated both from above and of below and can reach temperatures up to 24°C in August. Following the annual maximum, the temperature decreases to <5°C in December. Vertical temperature profiles have confirmed the existence of a well-mixed water column.

The seasonal trend in salinity is similar to that of temperature, with the lowest values (2.6%) occurring in winter and the highest (3.2%) in summer. In contrast to the temperature, the salinity curve is less stable due to numerous influences such as evaporation effects in hot summers or highly variable fresh water

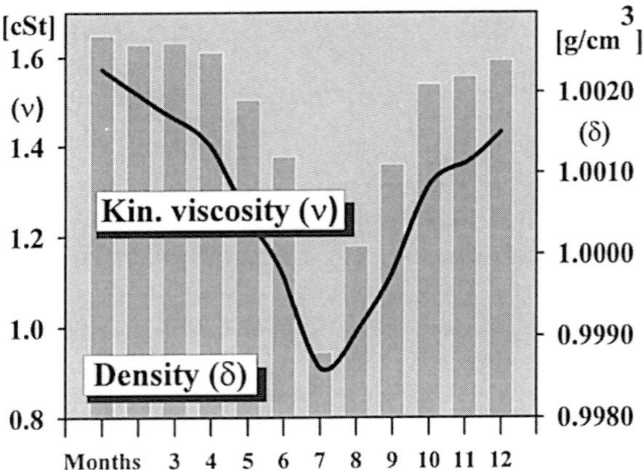

FIG. 5.—Variability in kinematic viscosity and density of sea water in the Accum tidal catchment in 1994, calculated after Krögel (1997a).

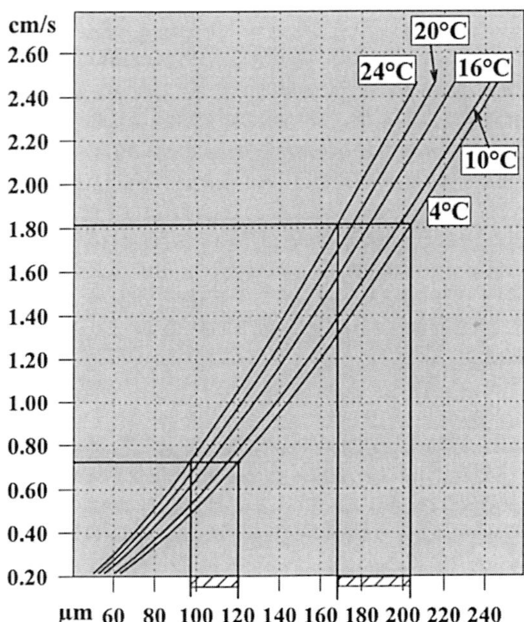

FIG. 6.—Settling velocity for quartz particles at various temperatures calculated after Brezina (1979).

discharges in the wake of strong rainfalls. Furthermore, salinities probably also increase in cold winters due to ice formation, as was the case in 1994 (Fig. 4).

In order to develop a simple equation for calculating the varying densities of sea water for each month, salinity- and temperature-dependent density values were obtained from the literature (e.g., Riley and Chester, 1971). In similar manner the kinematic viscosity of the sea water, which has a much greater influence on depositional processes than the fluid density alone, was calculated (Lax, 1967; Korson et. al., 1969; van Rijn, 1993). The seasonal trends illustrated in Fig. 5 show that both the density and the kinematic viscosity vary over a considerable range in the study area, that is from 0.9988–1.0027 g/cm^3 in the case of the density and from 1.5728–0.9096 cSt in the case of the kinematic viscosity (Fig. 5).

Settling Velocity and Critical Shear Velocity

The temperature- and salinity-induced variability in viscosity and density of sea water results in a wide range of settling velocities for sediment particles of any given diameter. This relationship (Fig. 6) has significant consequences for the depositional process. At a water temperature of 24°C, for example, a spherical quartz particle with a diameter of 0.093 mm has a settling velocity of about 0.70 cm/s. At a temperature of 4°C, on the other hand, the same settling velocity is reached by a particle with a diameter of 0.116 mm, the difference amounting to 0.023 mm. This temperature effect increases towards higher settling velocities. The difference between the corresponding particle sizes at a settling velocity of 1.80 cm/s at 24°C (0.168 mm) and that at 4°C (0.203 mm), for example, is 0.035 mm. The difference in diameter between particles with identical settling velocities deposited in winter and summer therefore increases with increasing settling velocity. This effect is due to the non-linear relationship between settling velocity and grain size as a function of water temperature.

Applying this principle to environments exposed to strong seasonal temperature fluctuations, the consequences of this temperature effect are immediately evident. Assuming constant energy conditions, then the position at which a given particle is in depositional equilibrium will be different in summer than in winter. Furthermore, the critical shear velocity of a given sediment particle will also change with a change in temperature (Bogardi, 1974; Inman, 1949; Komar and Miller, 1975). A recalculation of critical shear velocities as a function of temperature confirms that it increases with increasing temperature due to a corresponding decrease in viscosity. Following the approach of Komar and Clemens (1986), Krögel (1997a) developed a more accurate version of the equation for the calculation of critical shear velocities on the basis of settling velocities. Because it also incorporates the kinematic viscosity, this equa-

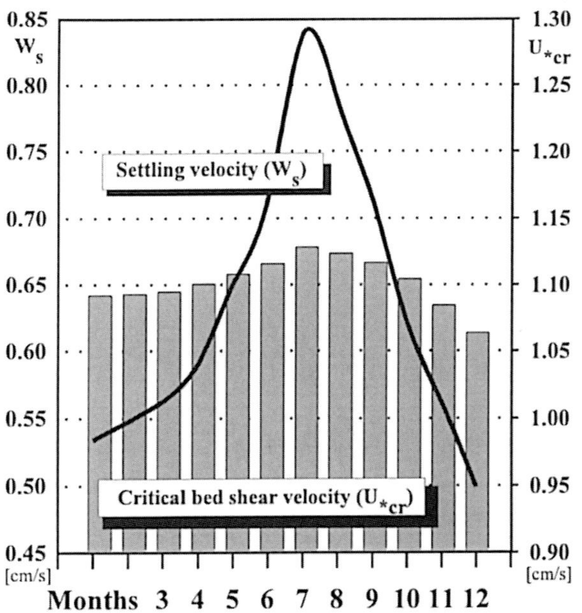

FIG. 7.—Variability of settling velocity for a 105-µm particle due to temperature and salinity variations observed in the field.

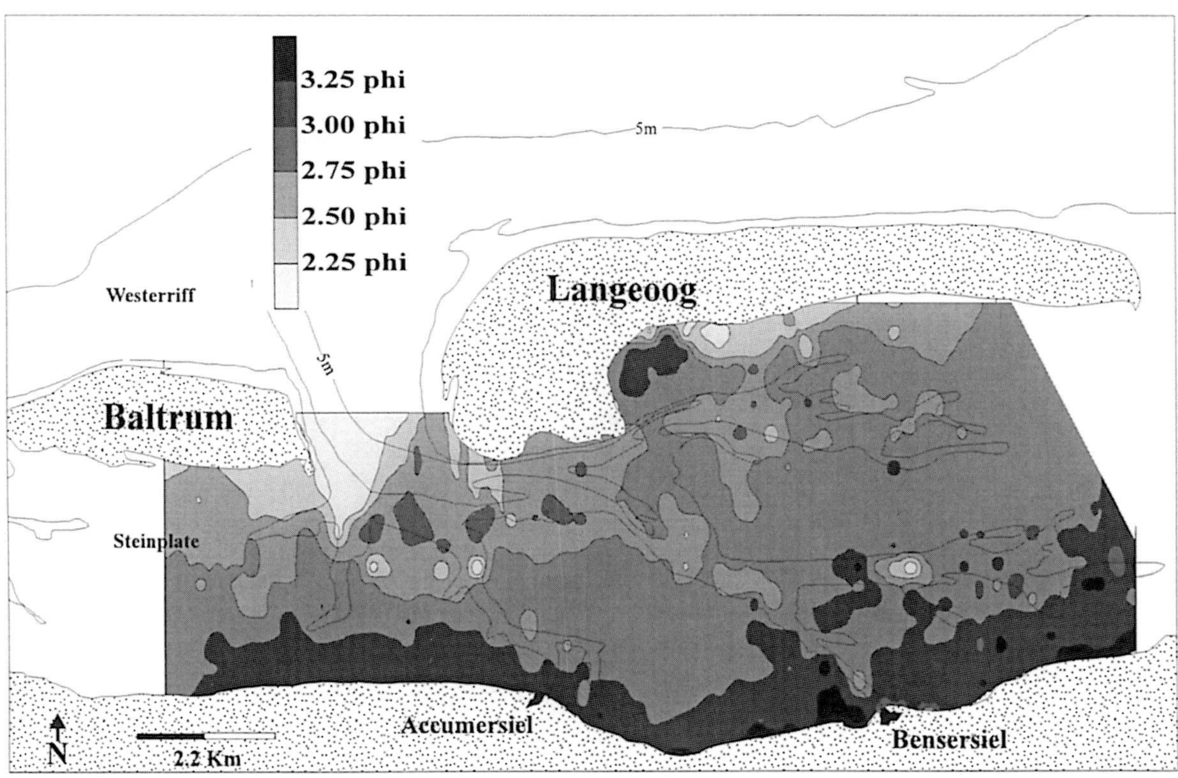

FIG. 8.—Spatial pattern of mean grain size (phi) of the sand fraction representing the conditions in July 1993. Conversions were made at a grain density of d_s = 2.65 g/cm³, a water temperature of T = 24°C (fresh water) and a hydraulic shape factor of SF′ = 1.18 (smooth glass beads).

tion allows the calculation of critical shear velocities for any water temperature. Thus:

$$U_{*cr} = (0.482\ [((\delta_s - \delta_f)/\delta_f)\ \nu\ g]^{0.282}) * (0.15\ w_s^{0.5}) + 0.61$$

where U_{*cr} is the critical shear velocity [cm/s], δ_s is the density of the particle [g/cm³], δ_f is the density of the fluid [g/cm³], ν is the kinematic viscosity [cSt], g is the acceleration due to gravity [cm/s], and w_s is the settling velocity of the particle [cm/s]. The equation is valid for settling velocities of up to 8.0 cm/s, which corresponds to a particle diameter of 0.4726 mm at 24°C in fresh water, a particle density of 2.65 g/cm³ and a shape factor of 1.18 (smooth glass spheres). Critical shear velocities calculated on this basis are in excellent agreement with the Shields Criterion for the initiation of sediment transport (Shields, 1936). Representing a more easily applied alternative that is valid for a substantial range of equivalent settling diameters, at the same time being more accurate than the equation of Komar and Clemens (1986), the settling-velocity-based equation formulated above can be used to replace the rather awkward Shields criterion.

In Fig. 7 the relative effects of water temperature on settling velocity and shear velocity of particles with a diameter of 0.105 mm are compared. It is immediately evident that the relative increase in shear velocity is much lower than the relative increase in settling velocity. The much higher settling velocities in summer thus override the effect that slightly higher viscosities may have on shear velocities, and hence affect erosional or depositional processes involving noncohesive sediments. As a result, sediment particles settle more rapidly and are entrained less readily in summer because of higher critical shear velocities, whereas the opposite applies in winter. The increasing discrepancy between settling velocity and shear velocity with increasing temperature clearly indicates that for the initiation of transport, a given particle will require higher relative shear stresses in summer than in winter. Furthermore, once resuspended, a particle will remain longer in suspension and hence be transported over longer distances at a greater height above the bed in winter than in summer, again assuming constant energy conditions.

Grain-size and Settling Velocity Distributions

The results of textural sediment analyses are usually presented as graphic representations of grain-size distributions (e.g., Martinez and Harbough, 1993). In the case of settling-tube analyses, the settling velocities are commonly converted into equivalent settling diameters at a standard water temperature of T = 24°C and a salinity of S = 0.0%. In the present study, a freshwater temperature of 24°C, a particle density of δ = 2.65 g/cm³, and a hydraulic shape factor of SF = 1.18 (smooth glass spheres) were used in the conversion (cf. Brezina, 1979). The resulting distribution of mean grain sizes of the sand fraction in the back-barrier tidal basin shows a distinct and coherent pattern (Fig. 8). A progressive fining of sediments in a more or less shore-normal direction is clearly evident. The coarser sediments occur along the island shore in the north, whereas the finer grain sizes are concentrated adjacent to the mainland dike. The pattern defines a number of shore-parallel grain-size belts with irregular boundaries. Pure mud flats com-

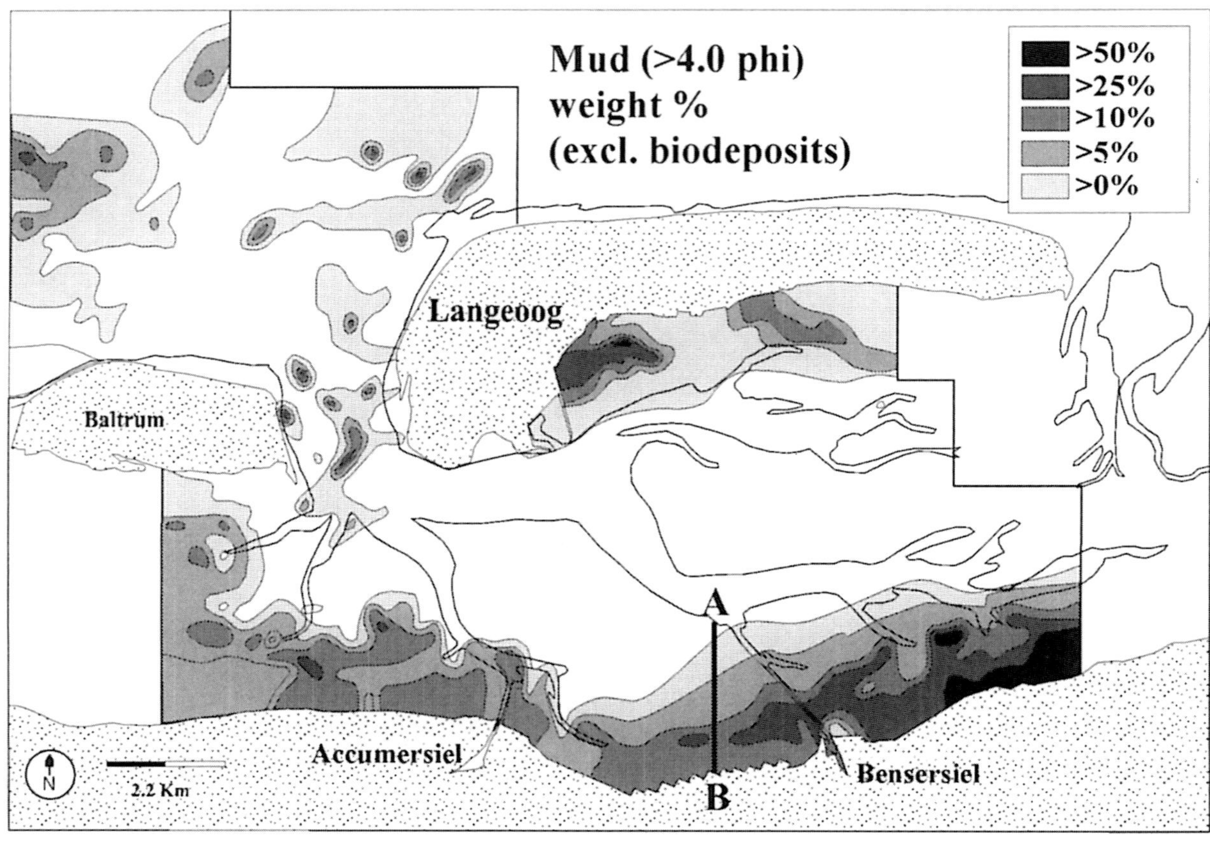

FIG. 9.—Spatial pattern of the mud fraction (grain size >0.063 mm or <4.0 phi) A-B) transect illustrated in Fig. 15.

prising mud contents >50% are rare. This general pattern is not unique to the study area but corresponds to that observed in other tidal basins along the East Frisian coast (e.g., Flemming and Ziegler, 1995; Krögel, 1997b).

The grain sizes of the surficial sands in the tidal basin range from 2.25–3.70 phi (0.210–0.088 mm), making them the finest sediments on the tidal flats adjacent to the mainland dike with settling velocities around 0.65 cm/s. Particles with lower settling velocites evidently lack an appropriate depositional zone. As a result, they are constantly being resuspended until they are exported during winter storms. Because the finer sediments are deposited under lower energy conditions than the coarser sediments, the shoreward-fining trend reflects a shore-normal energy gradient. Every energy level along this gradient corresponds to a particular settling velocity of the sediment which, in turn, depends on the kinematic viscosity and the density of the fluid, as well as the density and the diameter of the particle. The progressive grain-size sequence can be visualized more effectively by splitting it up into arbitrary ranges, each representing a discrete size fraction.

Particle aggregates, that is, mud flocs and fecal pellets, which are predominantly composed of particles <0.016 mm (>6.0 phi), occur in the form of low-density particle sizes of up to 0.3 mm, with settling velocities corresponding to quartz spheres up to 0.16 mm in diameter (e.g., Nichols and Biggs, 1985; Eisma, 1993). As shown by Flemming and Nyandwi (1995) and Flemming and Ziegler (1995) the deposition of these larger aggregates are confined to the tidal flats along the mainland coast, where they are associated with quartz grains that have the same settling velocities. The spatial distribution of mud, that is, grain sizes <0.063 mm (>4.0 phi), fits well into this pattern (Fig. 9). Following the energy gradient, the highest mud contents are found in narrow belts close to the high water line along the mainland dike and, to a lesser extent, also along the back-barrier island shore. Areas with mud contents >25% are very limited and even contents of 5–10% are confined to narrow belts in close proximity to the respective coastlines.

In contrast to the grain size distribution patterns calculated for standard parameters (Figs. 8 and 9), the sampling campaign took place in May 1993, when the water temperature was 12°C. The spatial distribution pattern shown in Fig. 10 hence shows mean grain-size trends calculated from settling velocities for a water temperature of T = 12°C and a salinity of S = 3.1%, as measured during the sampling campaign (cf. Fig. 4). Because the particle parameters used for the conversion procedure are the same as those for the conversion at 24°C it is not surprising that the general trends of the two patterns are very similar. Nevertheless, it is quite evident that corresponding grain-size belts are displaced in a landward direction. One consequence of this displacement is the elimination of the finest grain-size fractions at the foot of the dike, whereas the coarsest grain-size belts near the inlet have expanded considerably. The spectrum of mean grain sizes now ranges from approximately 2.0–3.25 phi (0.25–0.105 mm). Grain sizes of 3.25 phi and finer are absent. At a water temperature of 12°C, particles smaller than 3.25 phi have settling velocities <0.65 cm/s. Because these particles lack a

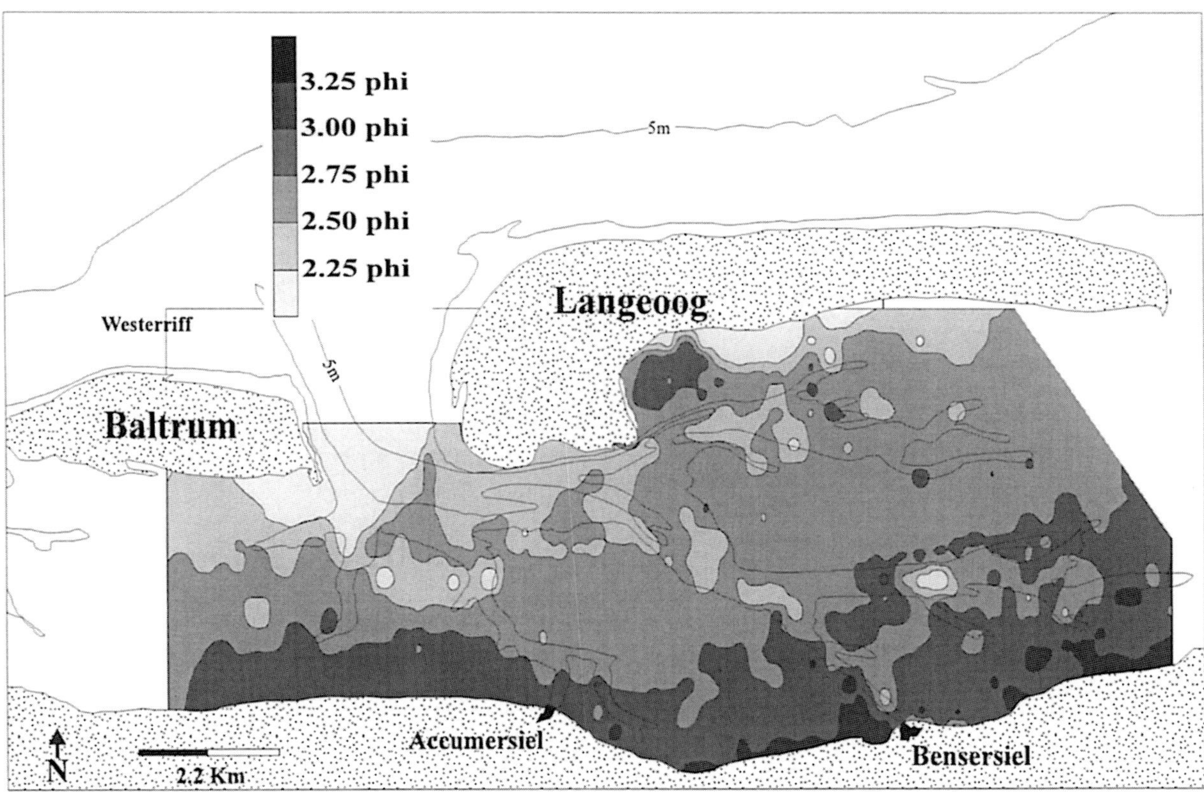

FIG. 10.—Spatial pattern of mean grain size (phi) of the sand fraction representing the conditions in May 1993. Conversions were made at a grain density of d_s = 2.65 g/cm³, a water temperature of T = 12°C, a salinity of S = 3.0% and a hydraulic shape factor of SF′ = 1.18 (smooth glass beads).

depositional zone that is in hydraulic equilibrium, they are constantly being resuspended until they are eliminated from the system as in the case of the finest sand fraction with grain sizes >3.50 phi (<0.088 mm) at a temperature of 24°C. This model is compatible with the settling lag/scour lag mechanism first introduced by van Straaten and Kuenen (1957) and later refined by Postma (1961). It simply adds an export loop at the rigid landward boundary produced by dike construction to indicate the elimination mechanism for those grain sizes that are too fine to be deposited.

DISCUSSION

Whereas settling velocities vary as water temperatures change, sediment particles transported in suspension would be deposited in energy zones corresponding to their seasonal settling velocities. Following the general energy gradient that characterizes the study area, this basic physical principle requires that the winter settling position of a particle must be located closer to the dike than its summer position. Depending on the location within the tidal basin and the particle size, the seasonal equilibrium positions could theoretically lie as much as 500 m apart (Fig. 11). In reality, however, such temperature-related adjustments would only be observed if the energy fluxes were seasonally reversed. Complete adjustment would require further that the summer energy input be large enough to compensate for the greater sediment mobility in winter. Because one would not expect the opposing energy fluxes to be so finely balanced, the seasonal position of a particle along the opposing gradients would depend on the net effect between sediment mobility and energy balance. Particles with higher settling velocities tend to be transported as bedload (Krögel, 1997a). As a result, the transport distance achieved during a single tidal cycle is very short. In winter the probability of transport increases because of lower critical shear velocities and lower settling velocities.

As pointed out above (cf. Fig. 2), the hydrodynamic energy flux in the Wadden Sea is distinctly higher in winter as compared to summer. Due to the regional weather pattern, the nature of tidal energy dissipation (Groen, 1967), and the general transport behavior of sediment particles in the Wadden Sea (van

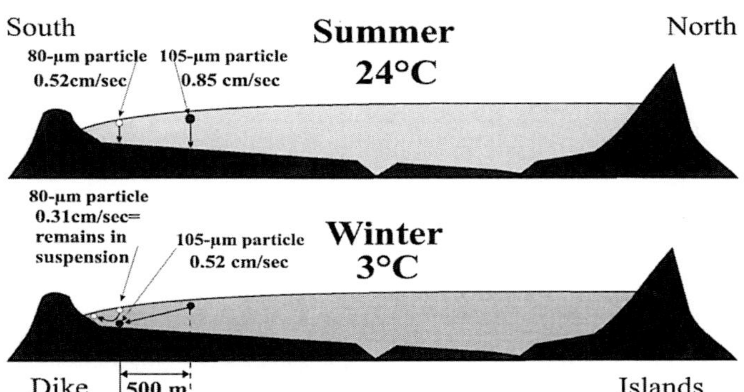

FIG. 11.—Schematic sketch illustrating the difference between the summer and winter settling processes in the Accum tidal catchment.

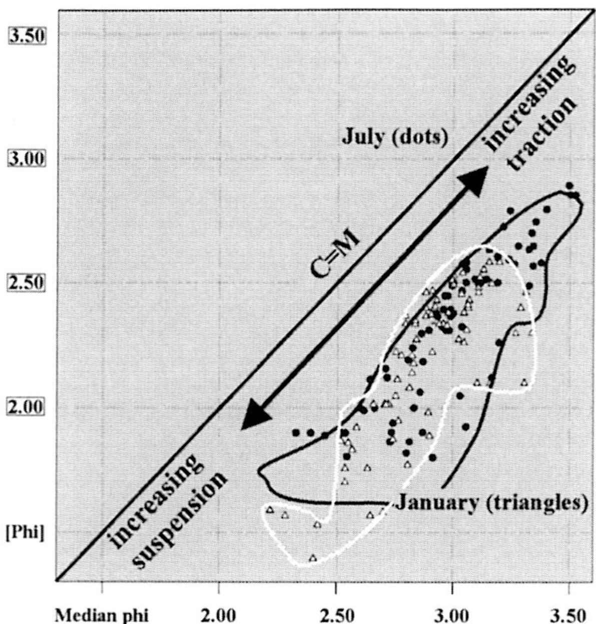

FIG. 12.—C/M diagram modified after Passega (1957) illustrating the transport mode of the sediments. In a fluid of lower viscosity due to a higher temperature, the particles tend to be transported as bedload.

Straaten and Kuenen, 1957; Postma, 1961), the net energy gradient remains directed towards the mainland coast (southwards), irrespective of the season. Under these circumstances a reversal of the southward winter adjustment to a northward summer adjustment of the grain-size belts is rather unlikely. For this reason, the theoretical model outlined above—in which a seasonal coarsening or fining of grain sizes in response to changing water temperatures is postulated—is not conspicuous. Because particle mobility is greatly reduced in summer, and considering the unidirectional energy gradient, the sediment distribution observed in the Wadden Sea evidently corresponds to a winter-adjusted depositional system.

Besides affecting the settling velocities and the critical shear velocities of sediment particles, the temperature effect must also affect the mode of sediment transport. Because both the kinematic viscosity and the density of the fluid are higher in winter, a sediment that is preferentially transported as bedload at higher water temperatures should, under the same hydrodynamic conditions, have a greater tendency for suspension transport at lower water temperatures. To highlight this differential transport behavior, a number of samples collected at identical locations in January and July are plotted in a modified Coarsest grain size/Median grain size-diagram (cf. Passega, 1957, 1964) (Fig. 12). The empirical C/M-model, which in a qualitative sense illustrates the mode of transport of a sediment on the basis of specific textural relationships, confirms that a substantial change in water temperature results in a significant shift along the scale between preferential bedload transport and dominant suspension transport (Krögel, 1997a). The July samples occupy positions closer to the bedload mode, whereas the January samples are displaced in the direction of preferential suspension transport because the sediments are entrained into higher levels of the near-bottom water column where they remain longer in suspension and are hence transported over greater distances. Because the transport mode is also dependent on the hydrodynamic regime, it can be assumed that a continuous transition should exist throughout the year between particles preferentially transported in suspension and others preferentially transported as bedload. This transition should be characterized by particles that pass through a range of transport modes in response to the seasonal changes in kinematic viscosity. In this context, an increase in temperature implies a strong decrease in particle mobility. This might explain why temperature adjustments in sediment distributions are inherently difficult to verify. Every sediment comprises a mixture of particles with different settling velocities. In the samples from the Accumer Ee tidal basin, the modal settling velocities (cf. Friedman, 1962) range from 0.84–2.0 cm/s, which corresponds to a grain size range of 0.10–0.17 mm at 24°C. In this range, the temperature effect might locally be recorded by a change in the weight percentages contributed by individual grain size fractions. The lower the settling velocity, the more pronounced the effect should become. In principle, any sediment whose settling velocity is too

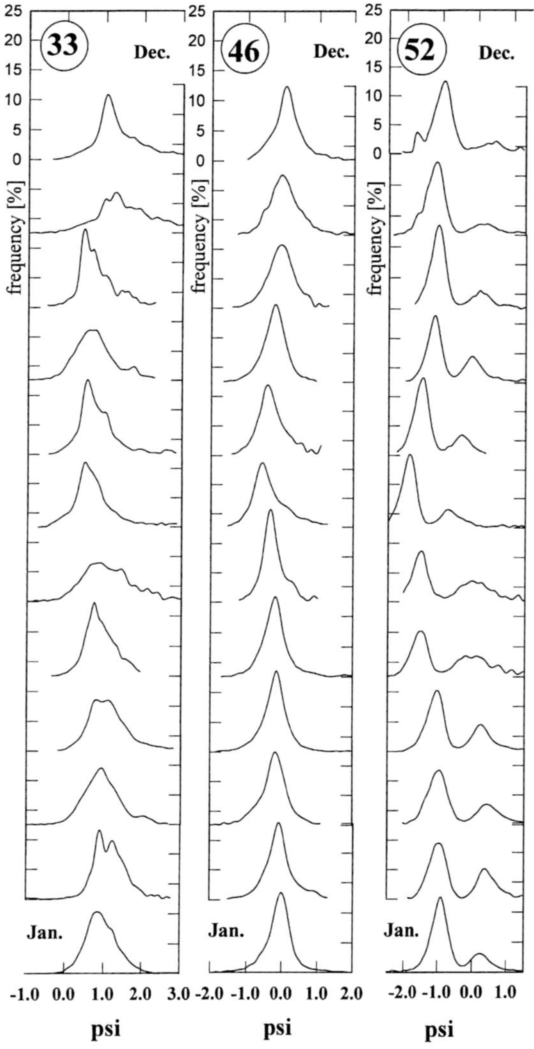

FIG. 13.—Distribution curves of settling velocity in psi (= − log2; v in cm/s) (for location of sample stations cf. Fig. 1).

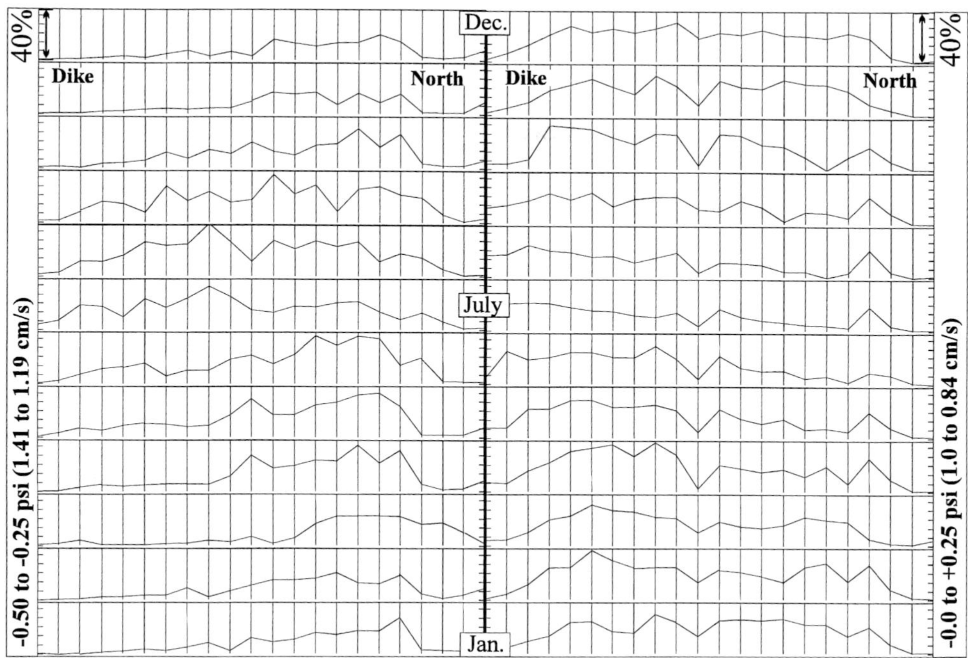

FIG. 14.—Temporal-spatial variability of weight percent of the settling velocity fraction −0.50 to −0.25 psi and 0.0 to +0.25 psi (for location of transect cf. Fig. 9).

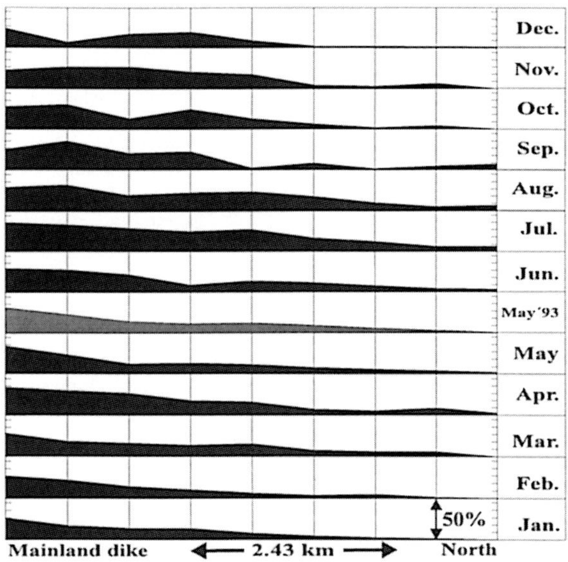

FIG. 15.—Temporal-spatial variability of weight percent of mud in the surface samples, January to December 1994, (for location of transect cf. Fig. 9) May 1993 for comparison.

low to be deposited must ultimately be eliminated from the system, a process recognized by their absence in surficial sediments. In coarser sediments, by contrast, the temperature effect would only be recognizable if the seasonal energy gradients were reversed. As outlined above, this is not the case in the Wadden Sea. If at all, the effect of temperature adjustment should be most conspicuous along the fine-grained boundary adjacent to the mainland dike, where the concentration of very fine sand and mud is highest.

The model presented in this study requires that if the settling velocity at a fixed point increases due to a temperature-induced reduction in viscosity, then the particle diameters able to settle at that location will decrease. This effect is illustrated by the three time-series shown in Fig. 13, which cover a period of 12 months at three typical sample locations. The highest variability is observed at position 52, which is located at the greatest distance from the mainland dike. It also has the highest settling velocities of the three stations. The modal settling velocity is about 2.0 cm/s (−1.0 psi) in January. In subsequent months, and corresponding to the gradual increase in water temperature, settling velocities progressively increase to reach maximum values around 4.0 cm/s in July. Thereafter, it again decreases to reach <2.0 cm/s in December. By comparison, the modal settling velocity of the sediment at station 33 is more stable, except for in July and October. This indicates that the variability increases with distance from the dike. Measurements of sediment turnover confirm that the sediments are more mobile in winter, especially on the distal flats.

In summer a similar sediment turnover rate was not evident. The monthly samples from the more distal flats are hence assumed to consist of material deposited during the previous winter. By computing the settling velocity of the January sample for a water temperature of 4°C and the July sample for a temperature of 24°C, the trend towards higher summer settling velocities can be demonstrated. On the mixed flats near the mainland dike the situation is different. In winter, when temperatures are low, particles with low settling velocities remain in suspension and are hence depleted in the sediment. As the water gradually becomes warmer towards summer, increasingly more suspended particles settle out, thereby reflecting the seasonal temperature effect without requiring a reversal of the energy gradient.

Whereas individual energy zones of the tidal basin were shown to be characterized by discrete grain-size spectra, the mean settling velocities at the various sample stations are almost constant. In this situation, the evidence for temperature effects would be reflected by changes in mean grain size. A closer look at individual settling velocity fractions shows that the 0.0–0.25 psi fraction (3.17–3.32 phi or 0.11–0.10 at 24°C) is hydraulically most stable, its contribution remaining almost constant throughout the year (Fig. 14). By contrast, the −0.50 to −0.25 psi fraction, which corresponds to the 2.88–3.03 phi fraction (0.13–0.12 mm at 24°C), shows strong spatial variability in the course of the year, its contribution increasing during summer and decreasing in winter. This trend is particularly prominent near the mainland dike. The same applies to the sand fraction <100 μm (at 24°C). As in the case of mud, this finest sand fraction also occupies a much larger area in summer as compared to winter.

The more widespread occurrence of mud during the summer months can only be explained by higher summer settling velocities of suspended matter in the water column. In winter, suspended mud particles can only settle out in the shallowest and most remote parts of the tidal basin, where energy levels are at a minimum. Beyond these areas, the winter energy levels are too high to allow sustained deposition (e.g., Nielsen, 1984). The higher summer settling velocities of suspended sediments have two effects. They allow muds to occupy much larger areas in summer and, at the same time, produce higher mud contents in surficial sediments (Fig. 15). Corresponding to their high mobility, the fine-grained suspended sediments show the greatest spatial variability of all back-barrier sediments.

CONCLUSIONS

The main results of this study may be condensed into the following conclusions:

1. Settling velocities of sediment particles are strongly controlled by the kinematic viscosity of a fluid, which, in turn, is a function of its temperature. Fluctuations in salinity, by contrast, have much less influence.

   In hydraulic terms, every particle has as many effective grain sizes as it has temperature-controlled settling velocities. At low water temperatures a particle has a lower settling velocity, that is, it responds as finer particles in higher water temperatures do. This suggests that grain size is not a useful concept in a hydraulic context.

   The same sediment is therefore more mobile at low water temperatures than at high temperatures. As a result, sediment distributions in the Wadden Sea probably reflect winter-adjusted zonation patterns.
2. In regions with high seasonal temperature fluctuations, sediment dynamic processes should respond sensitively to the associated fluctuations in the kinematic viscosity. Changes in water temperature also cause changes in critical shear velocities. Sediment particles thus not only settle more rapidly, but are also entrained less readily in summer than in winter.

   Accurate temperature-corrected critical shear velocities (U_{*cr}) can be calculated on the basis of the equation: $U_{*cr} = (0.482 [((\delta_s - \delta_f)/\sigma_f) \nu g]^{0.282}) * (0.15 w_s^{0.5}) + 0.61$
3. Under the same energy conditions a resuspended particle will remain longer in suspension and hence be transported over greater distances at a greater height above the bed in winter than in summer.

ACKNOWLEDGMENTS

The authors thank the captain and crew of the research vessel Senckenberg for their excellent navigation during fieldwork in the narrow back-barrier channels. We furthermore acknowledge the support of numerous students during sampling campaigns in the muddy tidal flats and their assistance with sample processing in the laboratory.

REFERENCES

BARTHOLOMÄ, A., AND FLEMMING, B. W., 1993, Zeitliche und räumliche Variabilität in den Sedimentparametern und der Morphologie auf der Gröninger Plate (Spiekerooger Watt): Senckenberg am Meer Bericht, v. 93/1, p. 5–48.

BOGARDI, J., 1974, Sediment Transport in Alluvial Streams: Budapest, Akademiai Kiado, 231 p.

BREZINA, J., 1979, Particle size and settling rate distributions of sand-sized material. 2nd European Symposium on Particle Characterisation (London), Proceedings, p. 1–44.

EISMA, D., 1993, Suspended Matter in the Aquatic Environment: Amsterdam, Springer, 315 p.

FLEMMING, B. W., AND DAVIS, R. A., JR., 1994, Holocene evolution, morphodynamics and sedimentology of the Spiekeroog barrier island system (southern North Sea): Senckenbergiana Maritima, v. 24, p. 117–155.

FLEMMING, B. W., AND NYANDWI, N., 1995, Land reclamation as a cause of fine-grained sediment depletion in backbarrier tidal flats (southern North Sea): Netherlands Journal of Aquatic Ecology, v. 28, p. 299–307.

FLEMMING, B. W., AND THUM, A. B., 1978, The settling tube—a hydraulic method for grain size analysis of sands: Kieler Meeresforschung, v. 4, p 82–95.

FLEMMING, B. W., AND ZIEGLER, K., 1995, High-resolution grain size distribution patterns and textural trends in the backbarrier environments of Spiekeroog Island (southern North Sea): Senckenbergiana Maritima, v. 26, p. 1–24.

FOLK, R. L., 1964, A review of grain size parameters: Sedimentology, v. 6, p. 73–93.

FOLK, R. L., AND WARD, W. C., 1957, Brazos River bar: A study in the significance of grain-size parameters: Journal of Sedimentary Petrology, v. 27, p. 3–26.

FRIEDMAN, G. M., 1962, On sorting, sorting coefficients, and the lognormality of the grain size distribution of sandstones: Journal of Geology, v. 70, p. 737–756.

FROSTICK, L. E., AND MCCAVE, I. N., 1979, Seasonal shifts of sediment within an estuary mediated by algal growth: Estuarine Coastal Marine Science, v. 9, p. 569–576.

GIBBS, R. J., 1972, The accuracy of particle-size analyses utilizing settling tubes: Journal of Sedimentary Petrology, v. 42, p. 141–145.

GIBBS, R. J., MATTHEWS, M. D., AND LINK, D. A., 1971, The relationship between sphere size and settling velocity: Journal of Sedimtary Petrology, v. 41, p. 7–18.

GROEN, P., 1967, On the residual transport of suspended matter by an alternating tidal current: Netherlands Journal of Sea Research, v. 3, p. 564–574.

INMAN, D. L., 1949, Sorting of sediments in the light of fluid mechanics: Journal of Sedimtary Petrology, v. 19 p. 51–70.

KOMAR, P. D., AND CLEMENS, K. E., 1986, The relationship between a grain's settling velocity and threshold of motion under unidirectional currents: Journal of Sedimtary Petrology, v. 56, p. 258–266.

KOMAR, P. D., AND MILLER, M. C., 1975, On the comparison between the threshold of sediment motion under waves and unidirectional currents with a discussion of the practical evaluation of the threshold: Journal of Sedimtary Petrology, v. 45, p. 362–367.

KORSON, L., DROST-HANSEN, W., AND MILLERO, F. J., 1969, Viscosity of water at various temperatures: Journal of Physical Chemistries, v. 73, p. 34–39.

KRÖGEL, F., 1997a, Einfluß von Viskosität und Dichte des Seewassers auf Transport und Ablagerung von Wattsedimenten (Langeooger Rückseitenwatt): Berichte aus dem Fachbereich 5 der Universität Bremen, v. 102, 168 p.

Krögel, F., 1997b, The Europipe Project: Effects of Dredging on Bedform Dynamics in the Accum Tidal Inlet: Berichte aus dem Forschungszentrum Terramare, v. 1: First International Symposium on Large Scale Coastal Environments, p. 43.

Krumbein, W. C., 1934, The probable error of sampling for mechanical analysis: American Journal of Science, v. 27, p. 204–214.

Lax, E., 1967, Taschenbuch für Chemiker und Physiker Bd. 1: Makroskopische physikalisch-chemische Eigenschaften (ed.): Heidelberg, Springer, 1522 p.

Martinez, P. A., and Harbaugh, J. W., 1993, Simulating Neareshore Environments: Oxford, Pergamon, 265 p.

Nichols, M. M., and Biggs, R. B., 1985, Estuaries, *in* Davis, R. A., Jr., ed., Coastal Sedimentary Environments: New York, Springer, p. 77–186.

Nielsen, P., 1984, On the motion of suspended sand particles: Journal of Geophysical Research, v. 89, p. 122–137.

Passega, R., 1957, Texture as characteristic of clastic deposits: American Association of Petroleum Geologists Bulletin., v. 41, p. 1952–1984.

Passega, R., 1964, Grain size representation by CM patterns as a geological tool: Journal of Sedimtary Petrology, v. 34, p. 830–847.

Postma, H., 1961, Transport and accumulation of suspended matter in the Dutch Wadden Sea: Netherlands Journal of Sea Research, v. 1, p. 148–190.

Riley, J. P., and Chester, R., 1971, Introduction to Marine Chemistry, eds.: London, Academic Press, 465 p.

Rouse, H., 1936, Nomogram for the settling velocity of spheres: Washington D. C., National Research Council, Committee on Sedimentation Publications, p. 57–64.

Rouse, H., 1937, Modern conceptions of the mechanics of fluid turbulence: American Society of Civil Engineers, v. 102, p. 463–505.

Rubey, W., 1933, Settling velocities of gravel, sand and silt particles: Journal of Sedimentary Petrology, v. 36, p. 403–413.

Shields, K.-H., 1936, Anwendungen der Ähnlichkeitsmechanik und der Turbulenz Forschung auf die Geschiebebewegung: Mittlungen der Preußischen Versuchsanstalt für Wasserbau und Schiffbau, v. 26.

Van Rijn, L., 1993, Principles of sediment transport in rivers, estuaries and coastal seas, ed.: Amsterdam, Aqua Publications, 1120 p.

van Straaten, L. M. N. U., and Kuenen, P., 1957, Accumulation of fine grained sediments in the Dutch Wadden Sea: Geologie end Mijnbouw, v. 19, p. 329–354.

TIDAL DYNAMICS AND SEASONAL DEPENDENT IMPORT AND EXPORT OF FINE-GRAINED SEDIMENT THROUGH A BACK-BARRIER TIDAL CHANNEL OF THE DANISH WADDEN SEA

J. BARTHOLDY AND D. ANTHONY*

Institute of Geography, University of Copenhagen, Øster Voldgade 10, DK-3050 Copenhagen, Denmark

ABSTRACT: In the main channel of the typical T-shaped tidal area of Grådyb (mean tidal range ≈ 1.5 m), the maximum tidal currents are highest in both ebb ($V_{1\%} \approx 1.5$ m/s; $V_{50\%} \approx 0.75$ m/s) and flood ($V_{1\%} \approx 1.2$ m/s; $V_{50\%} \approx 0.65$ m/s) in the central part where the inlet splits up into two main tributaries behind the barriers. The ebb dominates over the flood in the outer parts and visa versa. The transport of fine-grained sediment in the area is extremely dependent on the weather conditions. In general, during windy periods the concentration level is above 40 mg/l with mean values over the tidal period reaching a maximum of approximately 150 mg/l. In fair-weather periods the typical mean concentration in the main channel is between 15 mg/l and 30 mg/l and the *in situ* median grain size is surprisingly stable with a mean value of 26 μm. Recordings of the transport of suspended fine-grained sediment over 180 tidal periods, covering all seasons with typical weather conditions, showed that the important exchange of fine-grained material between the Wadden Sea and the open North Sea, is episodic. The investigated tidal area is exporting during stormy periods, concentrated in the winter term, where large amounts of fine-grained material are mobilized and apparently lost through the exchange of local turbid water with relatively clean water from the North Sea, The tidal area is importing during and after windy periods, following long periods of calm weather. This is speculated to be the outcome of fine-grained sediment settling on the shelf during long periods of calm weather, which increase the potential for high concentrations and large fall-velocities of suspended sediment. During subsequent windy periods those deposits are reworked and brought into the tidal area by the flood current. These conditions are most likely to appear in the summer.

INTRODUCTION

Tidal areas in the Wadden Sea and in similar estuarine environments are known to act as sinks for fine-grained sediments (e.g., McCave, 1973; van Es, 1977; Eisma, 1981; Postma, 1981; Bartholdy and Madsen, 1985; Ke et al., 1996). Although local sources such as rivers and coastal erosion may be extremely important, many tidal areas receive a major part of their fine-grained sediments through the tidal-induced exchange of estuarine and ocean water. The classical mechanisms controlling this shoreward transport were first described by van Straaten and Kuenen (1958) and Postma (1961, 1967). A recent update is given in several textbooks, e.g., Dyer (1994). Even with these well-documented mechanisms and the resulting deposition verified in numerous tidal areas, many questions about exactly how and when this import takes place still remain to be answered.

Along the Wadden Sea, a relatively weak northward residual current is superimposed on the tidal currents in the German Bight. This current carries water, with typical concentration levels of suspended sediment on the order of 5 mg/l-10 mg/l (Eisma, 1981; Eisma and Kalf, 1987). According to Sündermann (1994), the input of fine-grained material to the North Sea ($26 \cdot 10^6$ metric tons/yr.) is dominated by a net contribution from the Atlantic Ocean of $21 \cdot 10^6$ t/y of which $14 \cdot 10^6$ t/y enters through the English Channel. Together with the contributions from the Dutch-German inflows, it is transported northward along the coast. According to observations by Eisma (1981), Eisma and Kalf (1987), and Südermann (1994) the concentration decreases gradually in the transport direction.

Concentration levels of approximately 5 mg/l, are typical in the North Sea adjacent to the Danish Wadden Sea. The corresponding typical concentration level in the Grådyb tidal inlet is on the order of 15 mg/l-100 mg/l (Bartholdy, 1993a). A steady import from the North Sea, therefore, has to work against a relatively steep concentration gradient.

Based on an estimation of the size of local sediment sources and sedimentation rates (^{210}Pb age determinations) measured in the different accumulation areas of the tidal area, a sediment budget (Bartholdy and Madsen, 1985) shows a total fine-grained accumulation in the tidal area of approximately $0.14 \cdot 10^6$ t/yr, of which $0.12 \cdot 10^6$ t/yr, or 85%, is derived from the North Sea. These average budgets give good information about the overall sedimentation rates and the general transport directions, but they are unable to reveal anything about when and how the corresponding transport takes place. Is the import a result of the combined effect of numerous small daily inputs, or is it, rather, of an episodic nature? Questions like these can only be answered through long-term records of the sedimentation dynamics. The purpose of this paper is to contribute to this answer and examine the associated dynamics based on measurements in the main channel of Grådyb.

Related to a planned dredging of the navigation channel leading to Esbjerg (Fig. 1), a large data base with records of current velocity and turbidity was established as part of an environmental investigation financed by the harbor authorities in Esbjerg (Bartholdy, 1993a; Bartholdy, 1993b; Bartholdy and Anthony, 1994). The database was built on records taken every 5 minutes from 12 stations in the main channel. It comprises recordings of more than 500 tidal periods and provides a unique possibility to examine long term trends in the transport of fine-grained sediment. The results presented in this paper, primarily based on this database, show that the important exchange of material between this part of the Wadden Sea and the North Sea is episodic. Long periods with low transport rates and an insignificant small gain or loss of material are succeeded by short periods of rapid change. During stormy periods concentrated in the winter, large amounts of fine-grained material are mobilized and apparently lost through the exchange of local turbid water with relatively clean water from the North Sea. Import is found to take place during and after windy periods following long periods of calm weather conditions. These conditions are concentrated in summer.

AREA OF STUDY

Grådyb tidal area is located in the northern part of the Wadden Sea (Fig. 1). An overview of the physical setting and sedimentological conditions in the Danish Wadden Sea is given in Bartholdy and Pejrup (1994). The average tidal range in Grådyb is approximately 1.5 m and the tidal prism is in the order of $150 \cdot 10^6$ m^3. The total tidal area (including the salt marsh) is

*Present address: Geological Survey of Denmark and Greenland, Thoravej 8, DK-2400 Copenhagen NV.
Tidalites: Processes and Products, SEPM Special Publication No. 61

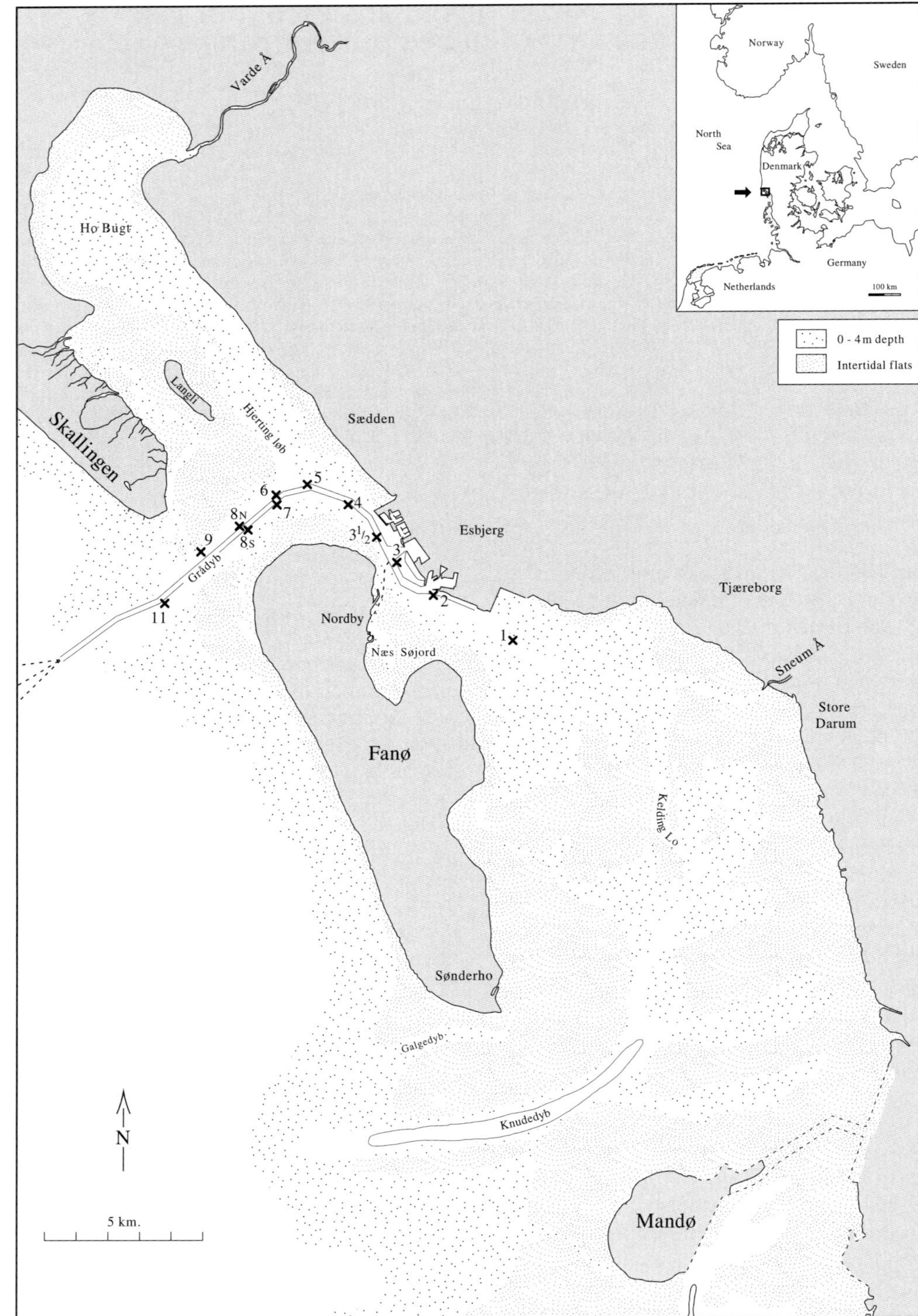

FIG. 1.—Location map of the study area, western coast of Denmark.

168 km². Of this approximately 37% are permanently covered by water, 45% are intertidal flats and 18% are saltmarsh situated above the mean high water level. The semidiurnal tidal range is frequently modified by wind tide, which, during storm surges, increases the high water level up to 3–4 m. According to a LANDSAT classification of the intertidal flats (Bartholdy and Folving, 1986) approximately 55% consist of sand flats, 13% of pure mudflats and 32% of mud- or mixed-flats. The tidal flats adjacent to the main channel are primarily sand flats. In contrast, in the inner parts of the tidal area, turbidity maxima and mussel banks build up mud flats along the channel margins and in sheltered zones in front of the salt marshes.

The tidal inlet is approximately 1000 m wide and 10 m deep at the entrance. The maximum current velocity (1 m/s–2 m/s) is strongest in the central part, flood-dominated in the inner parts, and ebb-dominated in the outer parts. The tidal excursion is on the order of 12 km–15 km.

The measured sedimentation rates in the accumulation areas are between 2 mg/l and 9 mm/yr (Bartholdy and Madsen, 1985), whereby the highest values were found in the salt marshes bordering the turbidity maximum in the northern part of Ho Bugt. The typical sedimentation rate in the salt marshes is 3 mm/yr, which is almost two to three times the present sealevel rise in this area of 1.1 mm/yr–1.2 mm/yr (Thomsen and Hansen, 1970, Aagard et al., 1995).

The freshwater discharge to Grådyb tidal area is almost exclusively supplied by two rivers, Varde Å in the northern part, and Sneum Å in the southern part. The outlet of the latter is situated at the tidal divide to Knude Dyb, which is the neighboring tidal area south of Grådyb. The mean freshwater supply from these rivers is very small compared to the tidal prism (<1%). The water in the main channel is well mixed. A model based on measurements of salinity in the tidal inlet, using the fresh water supply as a natural tracer (Ribe Amtsråd, 1988; Pejrup et al. 1993), shows that on average only 4–12% of the tidal prism is exchanged with North Sea water during a tidal period.

On the seaward side of the tidal inlet, the main channel cuts through a submerged ebb-tidal delta. This is constantly dredged in order to maintain a depth of 10.3 m in the navigation channel. Corresponding to the conditions (judged from maps) of similar undredged inlets south of Grådyb, the minimum depth in the channel passing the ebb tidal delta under natural conditions would be approximately 5 m.

The turbidity in the main channel varies in the typical way for a position between relatively clean water in the ocean and turbid water in the inner parts of the tidal area. At the southern border of the tidal area, a turbidity maximum is associated with the tidal divide to the south. As a consequence of this setting, the transport of fine-grained sediment is not only a result of tide and wave action, but also strongly affected by wind-induced currents, bringing turbid water from south to north, or relatively clean water from north to south.

METHODS

The current velocity and direction were recorded with Niskin winged current meters (model 6011 MKII), and the turbidity was measured with transmission infrared light turbidity meters (GMI model TU-150IR). Values were recorded every 5 minutes. Velocity was averaged over 40 discrete measurements spaced with 5 s intervals and turbidity was recorded as a mean of ten measurements with 1 s spacing. The meters were mounted on steel wires attached to concrete moorings on the sea floor and to buoyancy balls in the upper end. In general the current meters were placed 4 m above the bottom (close to the theoretical average mean current position) and the turbidity meters 1.5 m above this.

The turbidity signal (Tu) was transformed into Formacine Turbidity Units (FTU) on the basis of individual calibrations of each turbidity meter. Based on *in situ* calibrations (Bartholdy and Anthony, 1994) the FTU values were adjusted to concentration of suspended sediment (C) by:

$$C_{(mg/l)} = 1.06 \cdot Tu_{(FTU)} \quad (1)$$

Measuring the optical density of the water, the turbidity meter is by far more sensitive to the suspended fine-grained sediment than it is to suspended sand. The calibration of the turbidity signal was carried out with respect to this fact, and the turbidity data were interpreted relative to the concentration of fine-grained suspended sediment. In two data series (July 1992 and October 1993), control samples indicated that the lenses of the turbidity sensors were slightly covered by a surface film. This effect was corrected by means of a linear calibration function. In a few incidents, where the maximum range of the turbidity meter (180 mg/l) was exceeded, the concentrations were extrapolated. By running the truncated as well as the extrapolated values, it was in each case confirmed that this procedure did not change the direction of the net transport and had only marginal effects on the results. The net transport of suspended sediment, QS_n(kg$\times$m^{-2}/tidal period), was calculated as the net flux for each tidal period from low water to low water, by stepwise integration of the product between the velocity and the concentration of suspended sediment:

$$QS_n = \sum_{i=1}^{n} C_i \cdot U_i \cdot \Delta i \quad (2)$$

where Δi is the time interval (300 s) between the individual measurements, n is the number of measurements over a tidal period, C_i is the concentration in the ith interval (kg/m^3) and U_i is the corresponding current velocity (m/s), positive in the flood direction.

If the flood tidal excursion differs from the subsequent ebb tidal excursion, this tidal displacement comprises an advective component. In order to compensate for this, the net transport is corrected by the product between the transport weighted mean concentration, C_m (kg/m^3) and the tidal displacement Td (m):

$$QS_{cn} = QS_n - [C_m \cdot Td] \quad (3)$$

Where QS_{cn} is the corrected net transport (kgm^{-2}/tidal period) and C_m and Td are calculated as:

$$C_m = \sum_{i=1}^{n} C_i \cdot |U_i| \cdot \Delta i \Big/ \sum_{i=1}^{n} |U_i| \cdot \Delta i \quad (4)$$

and

$$Td = \sum_{i=1}^{n} U_i \cdot \Delta i \quad (5)$$

Wind speed, wind direction, and water level are recorded every 15 minutes by the harbor authorities in Esbjerg. Wind speed is measured at the harbor approximately 15 m above the ground and the tide gauge is situated in a harbor basin close to Station 3.

The grain-size distributions of the suspended sediment were analyzed by using a Braystoke SK 110 sedimentation tube sampler. It is a 1-m long tube (5 cm in diameter) with shutters in both ends and holds 2 l of sediment. After collecting the sample, the tube is placed in a vertical position and 10 discrete samples (0.2 l) are withdrawn from a tap in the lower end following a precalculated (temperature/salinity dependent) time scheme, based on an analysis of 1/2-phi fractions down to 5.5 phi (22 μm). The samples were stored in darkness and filtered through Millipore (0.45 μm) filters as quickly as possible.

RESULTS

Spatial Variation in the Tidal Current

Based on an analysis covering in general two spring cycles at each station (Fig. 2) results show the change in the tidal current conditions from the inner parts of the main channel (Station 1; 0 km) to the outer parts (Station 11; 14.8 km). The current velocity (Fig. 2A) is illustrated by its median value ($V_{50\%}$) and the value of the highest 1 percentile ($V_{1\%}$). The pattern is relatively clear, with the highest velocities in both ebb ($V_{1\%} \approx 1.5$ m/s; $V_{50\%} \approx 0.75$ m/s), and flood ($V_{1\%} \approx 1.2$ m/s; $V_{50\%} \approx 0.65$ m/s), located near 10 km (Station 6) in the central part of the T-shaped tidal area where the inlet splits into the two main tributaries behind the barriers. From Station 3 (4.7 km) and seaward, the ebb dominates over the flood. Further out (from Station 9; 12.4 km), both ebb and flood decrease in strength as a result of the increasing cross-sectional area across the ebb-tidal delta. A pattern showing an increased drop in the maximum flood current compared to the corresponding smaller drop in the maximum ebb current is clearly demonstrated. Landward from Station 3, the tidal curve asymmetry forces the flood current to dominate over the ebb current. As in the outer part, both ebb and flood drop in strength away from the central area. Here, the ebb current decreases faster than the flood current, giving the opposite asymmetry. The tidal currents described by the tidal excursion exhibit a similar pattern (Fig. 2B). In the central parts, the tidal excursion is between 12 km and 17 km. It drops to below 11 km in the inner parts, where the flood exceeds the ebb. In the outer parts, the flood current excursion diverges rapidly from that of the ebb, showing a drop below 9 km, whereas the ebb current excursion remains between 16 km and 17 km.

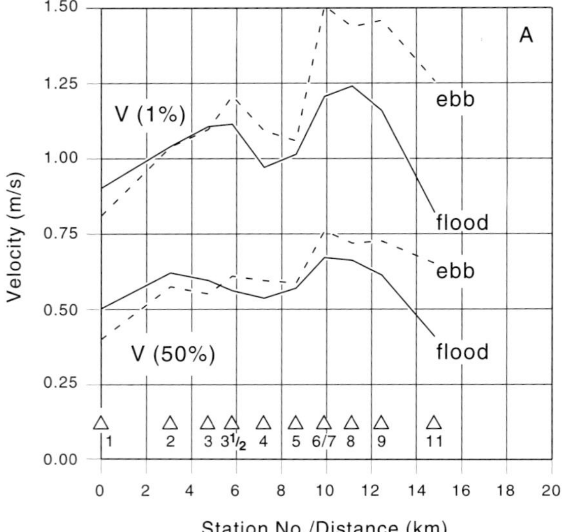

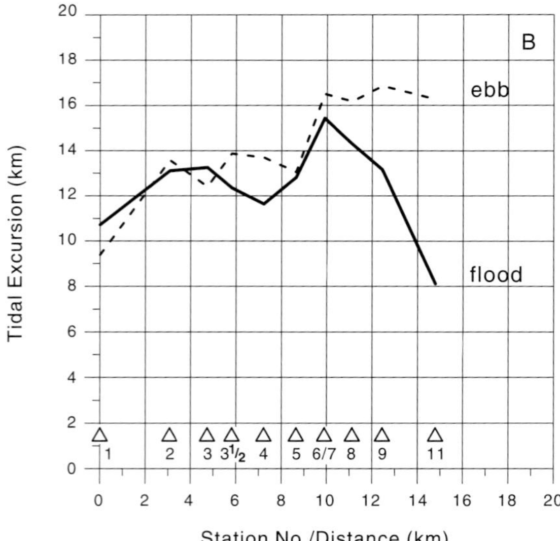

FIG. 2.—Tidal current variations in Grådyb from Station 1 at 0 km to Station 11 at 14.8 km (for location see Fig. 1). (A) The median current velocity and the current velocity exceeded 1% of the time. (B) The mean tidal excursion (no Station 10, it was never installed).

In the area close to Station 3 (4.7 km) the current strength and tidal excursion is of approximately the same magnitude in both flood and ebb. Thus, an evaluation of the fine-grained sediment budget from measurements in the main channel close to Station 3 should give the best results. Consequently, the following analysis of the tidal dynamics will focus on the results from Station 3.

Suspended Sediment

Suspended sediment transported in the Grådyb tidal area is composed of essentially two different types: sand and fine-grained material (<63 μm). The sand consists mainly of quartz grains (Bartholdy, 1983). The major clay minerals in the salt marsh deposits consist of illite (57%), followed by kaolinite (20%), chlorite (16%), and smectite (7%). The percentages are calculated as a mean of three samples from the salt marsh of the peninsula Skallingen (Deyu, 1987). The typical loss on ignition for the salt marsh clay is between 10% and 15%. The loss on ignition (500°C) of the suspended material is between 5% and 30% with the lower numbers strongly influenced by suspended sand (Bartholdy, 1993a). Fig. 3 shows the results of in situ median grain-size analysis carried out in 1992. These samples were collected during mixed (mostly calm) weather conditions. Half of them were collected during fortnightly visits to the measuring stations (1 m above the bottom and at the turbidity meter), and the other half from anchor stations (1 m above the bottom) during two tidal cycles (October 13, Station 9; and October 14, Station 8). The initial samples were collected during relatively rough weather conditions (wind velocity 10 – 15 m/s from west).

In Fig. 3A, the median grain sizes are plotted against the concentration for the whole sample, and in Fig. 3B only for the fraction finer than 63 μm. A comparison between these two distributions shows that only the fraction coarser than 63 μm is contributing to the proportionality between the median grain size and the concentration in Fig. 3A. It should be noted that the grain-size analysis procedure is less precise in the coarse fractions. The results coarser than 150 μm should therefore only be regarded as suggestive. The median grain-size of the fine-grained material is surprisingly stable. The mean value of the samples (excluding the October 13 values) is 26 μm with a standard deviation of 2 μm. It corresponds to a settling ve-

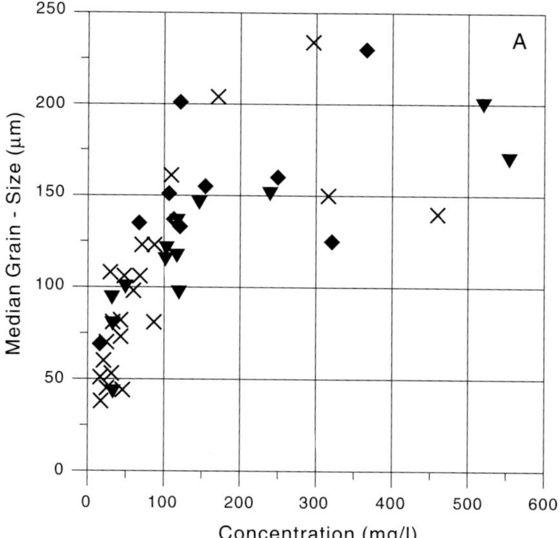

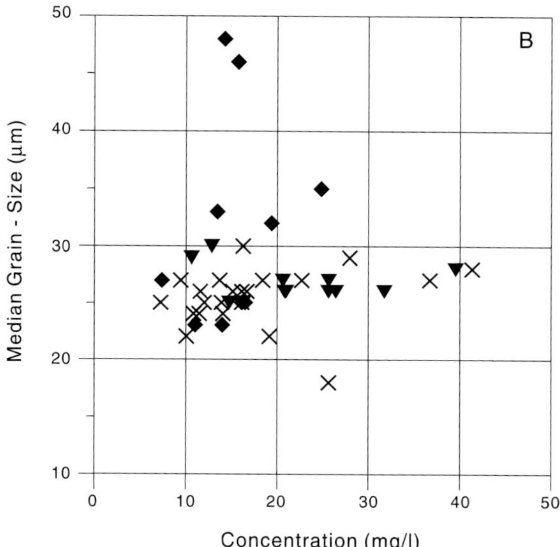

FIG. 3.—*In situ* median grain-sizes plotted as a function of concentration. X, from fortnightly visits to the measuring stations; diamonds, from Station 9, October 13, 1992; triangles, from Station 8, October 14, 1992. (A) For the whole sample. (B) Median diameters of the fine-grained material from the same samples as in (A). Only the parts finer than 63 µm are included in the calculations.

locity of $0.4 \cdot 10^{-3}$ m/s at 1°C and $0.6 \cdot 10^{-3}$ m/s at 20°C. When the data of relatively rough weather depart from this pattern with raised median grain sizes in the silt fraction, it is most likely a result of the input of material released by the raised dynamics. These increased values are not related to high concentrations in the silt fraction. The highest values are found in water entering the tidal area during flood and reach a maximum of 48 µm, which corresponds to a settling velocity of $1.2 \cdot 10^{-3}$ m/s at 1°C and $2.1 \cdot 10^{-3}$ m/s at 20°C.

If the observed change in the median grain-size with concentration (Fig. 3A) was related to flocculation, the particles finer than 63 µm would be affected. This is not the case, and it is therefore more probable that these particles originate from relatively stable older flocs like those described by Eisma et al. (1991a, 1991b).

As the material reaches the inner parts of the Wadden Sea it is included in the turbidity maxima, where high concentrations enable the formation of larger floc sizes (e.g. Pejrup, 1991).

Seasonal Variability of Sediment Export-Import

Transport of fine-grained sediments in the Grådyb tidal area is extremely dependent on weather conditions. During periods of fair weather it is not unusual to observe maximum concentrations in the main channel of less than 20 mg/l, whereas windy periods can be characterized by maximum concentrations of up to ten times this value. The southern part of Grådyb is an open system because only a tidal divide separates it from the Knudedyb tidal area to the south (Fig. 1). Fine-grained material tends to accumulate close to the watershed (Bartholdy and Folving, 1986). Thus, at Station 3 variations in the transport of fine-grained sediment are not only the result of tide and wave action, but also strongly affected by wind-induced currents (Anthony, 1995). Wind moves turbid water from south to north or relatively clean water from north to south. The transport conditions have to be monitored over a long period of time in order to give a comprehensive view of the transport dynamics. Also, it is important to distinguish between the real net transport and the transport produced by the advective component. The latter might cause the import of a large amount of sediment at the one end, while loosing a similar or even larger amount at the other. To correct for this advective component, the values for net sediment transport are calculated using a corrected net transport equation (Equation 3). The net transport is determined for every low water based on the product between the concentration of fine-grained sediment and the current velocity (Equation 2). The correction term is calculated as the tidal displacement (Equation 5) multiplied by the transport weighted mean concentration of the corresponding tidal period (Equation 4). Withdrawing this term from the net transport corrects for irregularities caused by local discontinuities. As shown in the following, it is also a useful approach for interpreting the overall transport conditions. Water lost through the turbidity maximum is most likely more turbid than water coming in from the North Sea. The correction term is therefore expected to generate values too low for situations where water is transported southward and values too high when the tidal displacement is to the north.

The measurements on which the following analyses are based cover 180 tidal periods from all seasons including all significant types of weather conditions. Four scenarios are chosen as typical: (A) summer with typical calm- to mixed-wind conditions (Fig. 4); (B) winter with a storm surge (Fig. 5); (C) spring with mixed- to stormy weather (Figs. 6 and 7); and (D) summer gale following a long period of calm weather (Fig. 8).

In the typical summer situation (Fig. 4), the concentration of fine-grained suspended material varies between 20 mg/l and 60 mg/l. The typical variation of turbidity for a measuring station situated seaward of a turbidity maximum is easily recognized. The concentration is low at high water and rises towards low water as a result of increased turbidity in the water coming from the inner parts of the tidal area. At low tide the concentration drops as a result of settling during slack water. A resuspension

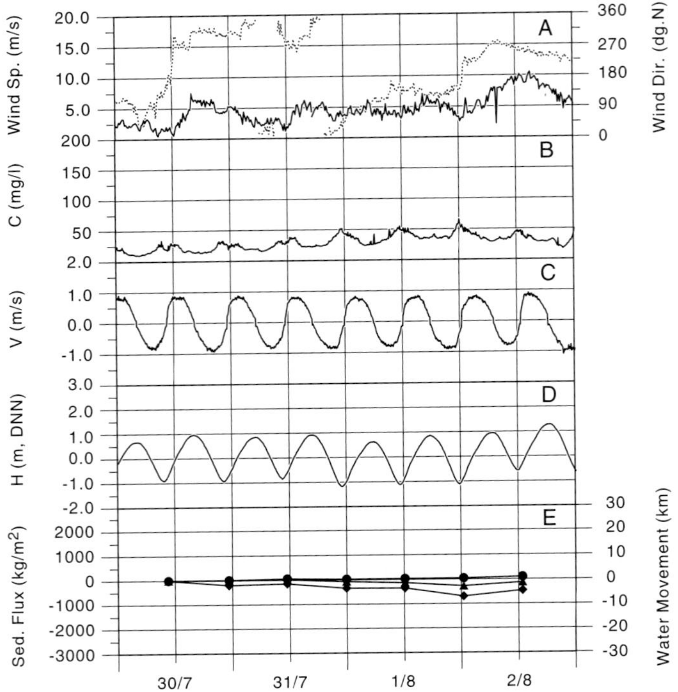

FIG. 4.—Wind-, tidal-, and transport dynamics at Station 3, July 30–August 2, 1992. (A) wind speed (full line) and wind direction (dotted line), (B) concentration of fine-grained material, (C) current speed, (D) water level, (E) tidal displacement (diamonds); net-flux of fine-grained sediment (triangles); corrected net-flux of fine-grained sediment (circles).

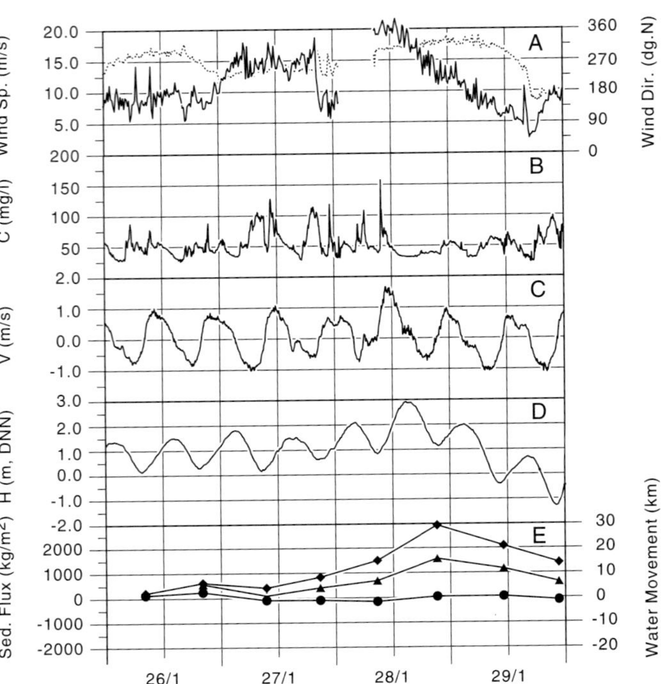

FIG. 5.—Wind-, tidal-, and transport dynamics at Station 3, January 26–29, 1994. (A) wind speed (full line) and wind direction (dotted line), (B) concentration of fine-grained material, (C) current speed, (D) water level, (E) tidal displacement (diamonds); net-flux of fine-grained sediment (triangles); corrected net-flux of fine-grained sediment (circles).

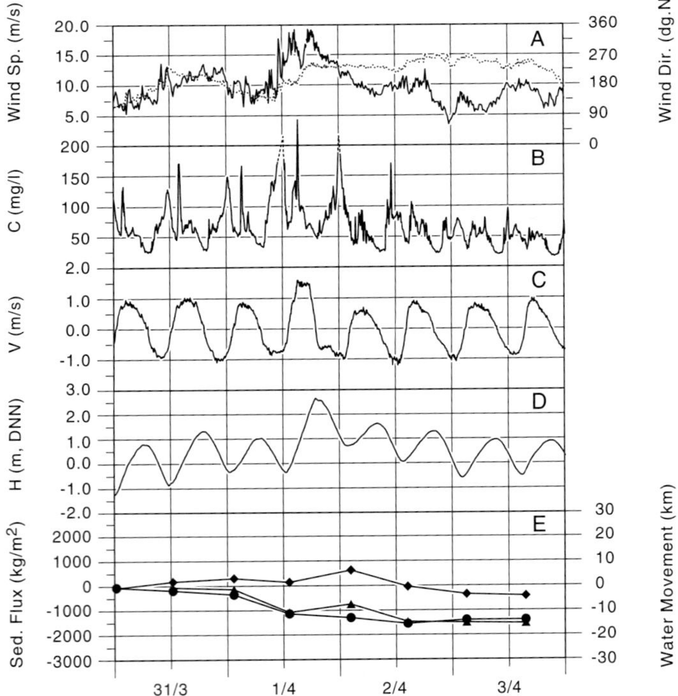

FIG. 6.—Wind-, tidal-, and transport dynamics at Station 3, March 31–April 3, 1994. (A) wind speed (full line) and wind direction (dotted line), (B) concentration of fine-grained material, (C) current speed, (D) water level, (E) tidal displacement (diamonds); net-flux of fine-grained sediment (triangles); corrected net-flux of fine-grained sediment (circles).

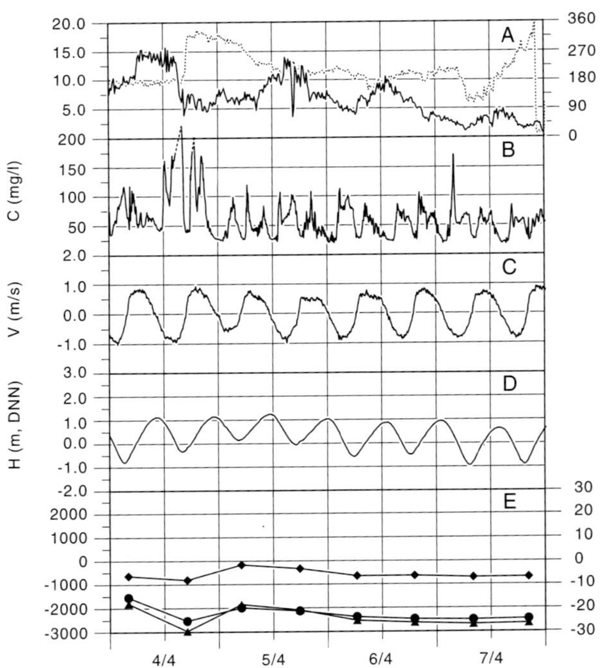

FIG. 7.—Wind-, tidal-, and transport dynamics at Station 3, April 4–7, 1994. (A) wind speed (full line) and wind direction (dotted line), (B) concentration of fine-grained material, (C) current speed, (D) water level, (E) tidal displacement (diamonds); net-flux of fine-grained sediment (triangles); corrected net-flux of fine-grained sediment (circles). The summation of tidal displacement and net fluxes are continued from Figure 6.

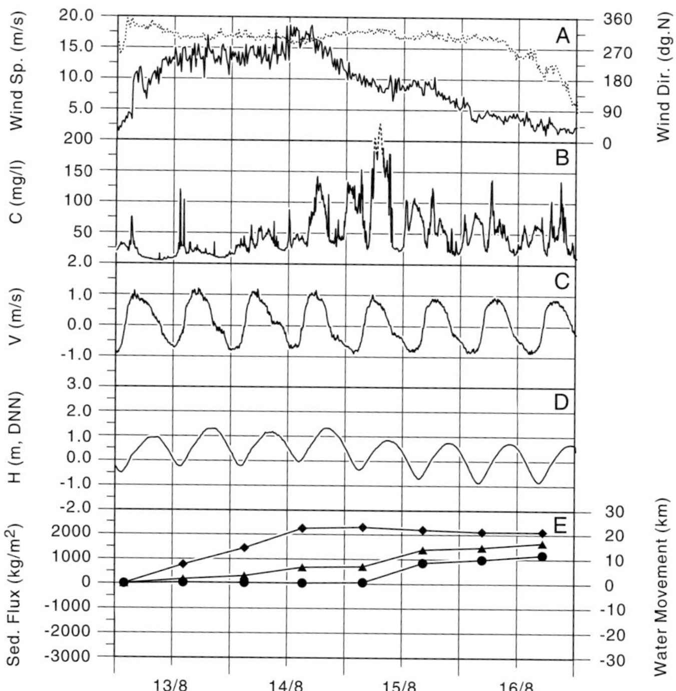

FIG. 8.—Wind-, tidal-, and transport dynamics at Station 3, August 13–16, 1994. (A) wind speed (full line) and wind direction (dotted line), (B) concentration of fine-grained material, (C) current speed, (D) water level, (E) tidal displacement (diamonds); net-flux of fine-grained sediment (triangles); corrected net-flux of fine-grained sediment (circles).

peak is formed at the turn of the tide, whereafter the concentration level drops towards the next high tide slack. Because of the general low concentrations in the water coming in from the North Sea, the resuspension peak following high tide is normally either absent or weak compared to that after low tide.

The 4-day period started with very calm wind conditions and concentration levels of fine-grained material around 25 mg/l, and ended with a fresh breeze from west to southwest with concentration levels of 20–60 mg/l. The water level dropped at the start of the period, and although it rose again towards the end, the net transport of water was directed outward. The variations in the concentration of fine-grained sediment were small. The outward-directed tidal displacement caused a small export of fine-grained sediment. The corrected net transport value is close to zero. The typical summer condition is dominated by a small circulation of fine-grained sediment and an approximate balance between export and import.

In contrast to summer, with relatively calm weather and a low transport rates, winter storm surges generate high concentrations and large transport rates (Fig. 5). The concentration of fine-grained sediment varied in this 4-day period between 20 mg/l and 150 mg/l. The wind came from westerly directions during the entire period, except for the last few hours when the direction changed to due south. At the beginning, the wind velocity was ≈10 m/s, whereafter it rose to 15 m/s. After a small drop, it reached a maximum level of 20 m/s before decreasing to 5 m/s over 24 hours until the end of the period.

The variation in the concentration of fine-grained sediment followed generally the same pattern as described above, although now accentuated with much higher concentrations strongly influenced by the advective component. Initially, the wind set up induced an import of a small amount of water from the north, whereafter a temporary change to a more southerly wind direction gave rise to a northward tidal displacement. This brought out turbid water from the turbidity maximum with maximum concentrations between 100 mg/l and 125 mg/l accompanied by large and rapid variations in the concentrations close to low tide. The high tide concentrations remained on the order of 20–40 mg/l. As the wind set up caused water to enter from the north, the turbidity maximum was pushed back and the concentration level dropped to around 30 mg/l. Towards the end of the period, the decreasing water level and corresponding discharge of water once again brought out turbid water from the inner parts of the tidal area, now reflecting the lower dynamics with concentration levels close to those from the start of the period.

For the duration of this incident, the ebb current maximum speed remained around 1 m/s and, thus, did not exceed that of the summer period. Only the maximum flood current reached extreme values (≈1.5 m/s) just before the highest high tide. This event was related to a concentration peak (>150 mg/l) that did not follow the pattern described above. The extremely high flood current was in this case associated with water levels higher than 2 m, and the peak is interpreted as derived from erosion in the more or less completely inundated tidal area. The low concentrations on the back part, before the velocity drops, indicate that it is a local source peak and not related to high sediment concentrations in the entering North Sea water.

As in the summer situation, the net transport of fine-grained sediment followed the tidal displacement, which after an increase to 30 km dropped to approximately half of this value. The total net transport of the observed period is approximately +600 kg/m². After correction due to the advective component it shows a small export of −100 kg/m². Because this correction procedure probably creates correction values that are too low when the tidal displacement is south-going, these results indicate that the combined effect of the storm generated a net export of fine-grained material from the study area. This indication of the effect of storms is confirmed by measurements in the mixed- to stormy-weather period with predominantly southwesterly winds (Figs. 6 and 7). This wind direction forced the tidal displacement to reverse, and thus, Station 3 was situated in the opposite end of the turbidity maximum. The variation pattern showed a substantial loss of sediment from the tidal area to the North Sea for both the net and the corrected net transport of fine-grained sediment. There is a tendency to regain lost material in the period directly after stormy events, but the overall trend suggests that stormy periods create a net loss of fine-grained sediment. It is well known that tidal channels have a tendency to be dominated by the ebb current, in contrast to the principally flood-dominated tidal flats on the sides of the tidal channel. When, as in this case, the tidal displacement follows that tendency, it is necessary to consider if such a pattern will alter the results. If the concentration of fine-grained sediment on the flats does not exceed the concentration measured in the channel (which is very likely, disregarding local effects), the effect of this circulation is adjusted by the correction term. If, on the other hand, the tidal displacement is primarily a result of water being pushed over the watershed from the south, the correction term is probably too high, as stated earlier. The correct value is, therefore, expected to range between the measured net transport and the corrected net transport values. Both values indicate a substantial loss of fine-grained sediment on the order of 2500 kg/m^2 for the 8-day period (Fig. 7).

The wind set up and high sediment concentration levels are the primary reasons for deposition of fine-grained material on the elevated salt marsh surfaces ($\approx$1 m DNN), and the salt marsh formation itself is a principal component of the positive net balance of the budget. These results indicate that the conditions necessary for deposition of fine-grained sediment on the salt marsh surfaces are the same conditions under which the system as a whole loses fine-grained material to the North Sea. This apparent inconsistency can only persist, if the transport system during other periods imports fine-grained sediment in such large amounts that an available store of fine-grained material can be maintained. This store consists primarily of mud flat deposits in the intertidal zone which, therefore, may provide depositional evidence of import events. In the sedimentary record, salt marsh deposits will reveal evidence of sedimentation related to stormy periods, distanced from the episodes when fine-grained sediment is imported (Fig. 8). A small summer gale with westerly to northwesterly wind followed a period of more than 1-1/2 months with very low wind speeds. In the beginning of the period, the variation in the concentration of fine-grained sediment followed the usual pattern. The gale produced little wind set up, but the tidal displacement was substantial (22 km, southerly). As a result the prominent flood tide peak produced in the fourth tidal period (Fig. 8) most likely originates from fine-grained material derived from the shelf outside of the tidal area. This substantial input settled during high water but resuspended in the following ebb period such that the net transport was only slightly affected. In the following flood period, the peak was even more prominent, with concentrations >180 mg/l (the limit of the turbidity meter). This time, the successive ebb current peak was weaker and the system was characterized by a significant net import, which continued in the following tidal periods despite a slightly out-going tidal displacement. The import continued, gradually lessening, for several days after the 4-day period.

The decrease in turbidity at high- and low tide is very pronounced during this sequence. Rapid settling indicates high fall velocities. The rate of concentration decrease can be used to calculate the settling velocity of the suspended sediment (Krone, 1962; cited in Amos and Mosher, 1985). After truncating occasional peaks during the concentration decrease, this procedure is used, as suggested by Amos and Mosher (1985) due to similar measurements in the Bay of Fundy. The median settling velocity related to the settling from the first five well-defined peaks in Fig. 8 is estimated at between 1×10^{-3} m/s and 3×10^{-3} m/s, corresponding to fall diameters between 35 μm and 61 μm. This is the same order of magnitude as found by placing the settling tube in the flood current of the relative rough weather period of October 13, 1992, and indicates that this material enters the Wadden Sea from the North Sea in the form of relatively large flocs.

There are no observations of the deposition of fine-grained material on the shelf outside this part of the Wadden Sea. Nevertheless, it can be speculated that long periods of calm weather can permit fine-grained material to settle on the shelf and increase the potential for high concentrations and fall-velocities during the subsequent windy period.

The total transport of all the 180 tidal periods surveyed at Station 3 (Fig. 9) demonstrates a positive net result of the tidal displacement that is caused by the slightly flood-dominated tendency at the measuring station (Fig. 2B). The mean concentrations (transport weighted) of the flood- and ebb periods are shown above the tidal displacement curve. In general, the windy periods induce concentrations above 40 mg/l with the lowest concentrations in the flood current. Exceptions to this pattern occur in Period 5, which is associated with the dynamics shown in Fig. 8, and to a smaller extent in Period 1. Also, this is a summer period, which, although it was less pronounced, followed a change from relatively calm to windier weather. The corrected net transport curve shows that long periods of small gains or losses of fine-grained sediment are followed by short periods of rapid loss or gain. The import in the beginning of Period 4 is associated with very low mean concentrations of suspended sediment (Cf and Ce), and is possibly due to a moderate spillage of dredge spoils from the navigation channel north of Station 3.

In general, the loss of fine-gained sediment is associated with stormy periods. The period of highest gain (Period 5) represents 25 tidal periods, of which three were responsible for one-third

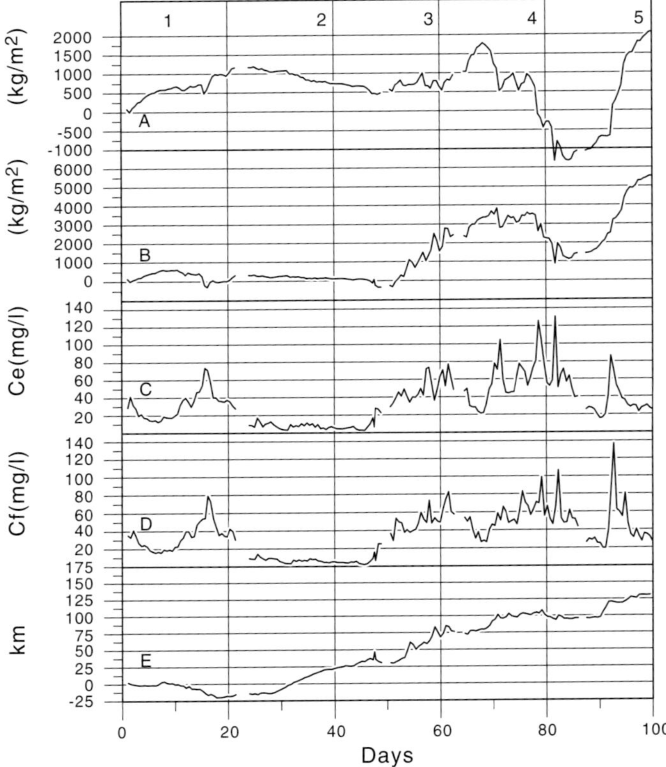

FIG. 9.—(A) Corrected net flux of fine-grained suspended sediment, (B) net flux of fine-grained suspended sediment, (C) transport weighted mean concentration of fine-grained suspended sediment—ebb (Ce), (D) transport weighted mean concentration of fine-grained suspended sediment—flood (Cf), (E) tidal displacement. The numbers above indicate the measuring periods (1) July 22–August 11, 1992; (2) October 18–November 12, 1993; (3) January 20–February 1, 1994; (4) March 18–April 4, 1994, and (5) August 10–22, 1994. The summed values are adjusted so that one period starts with the last value of the previous period.

of a total (corrected) net import of 3.1 t/m^2. Based on a rough estimate (mean conditions over the cross-section), and the import principle derived from Bartholdy and Madsen (1984), this import is equivalent to 25% of the yearly mean import from the North Sea.

CONCLUSIONS

In the typical T-shaped tidal area of Grådyb (mean tidal range $\approx$ 1.5 m), the maximum tidal currents are highest in both ebb ($V_{1\%} \approx$ 1.5 m/s; $V_{50\%} \approx$ 0.75 m/s) and flood ($V_{1\%} \approx$ 1.2 m/s; $V_{50\%} \approx$ 0.65 m/s) in the central part where the inlet splits up into two main tributaries behind the barriers. The ebb dominates over the flood in the outer parts and *vice versa*. In the central part, the tidal excursion is between 12 km and 17 km. It drops towards the inner parts where the flood exceeds the ebb. In the outer parts, the flood current excursion drops to below 9 km whereas the ebb current excursion maintains a value between 16 km and 17 km.

During calm and mixed weather conditions, the *in situ* median grain-size of the fine-grained material transported in the area is surprisingly stable with a mean value of 26 μm. Size increase during rough weather is not necessarily related to high concentrations in the silt fraction.

The transport of fine-grained sediment is extremely dependent on the weather conditions. In general, windy periods induce an increase of concentrations above 40 mg/l with mean values over the tidal period reaching a maximum of approximately 150 mg/l. In fair weather periods the typical mean concentration in the main channel is between 15 mg/l and 30 mg/l. The lowest measured mean concentration measured over a tidal period is 3.4 mg/l.

At the southern border of the tidal area, a turbidity maximum is located at the tidal divide to the south. As a consequence, the transport of fine-grained sediment is not only a result of tide- and wave action, but also strongly affected by wind-induced currents, bringing turbid water from south to north or relatively clean water from north to south.

Recordings of the transport of fine-grained sediment over 180 tidal periods, covering all seasons and significant weather types, showed that the important exchange of material between this part of the Wadden Sea and the North Sea is episodic. Long periods with low transport-rates and an insignificant small gain or loss of material are succeeded by short periods of rapid change.

The system is exporting sediment during stormy periods with wind set-up concentrated in the winter term, when large amounts of fine-grained material are mobilized and apparently lost through the exchange of locally turbid water with relatively clean water from the North Sea.

The wind set-up and high concentration of suspended sediment are the primary reasons for deposition of fine-grained material on the elevated salt marsh surfaces ($\approx$1 m DNN). Thus, these results indicate that salt marsh deposition occurs under the same conditions as those causing the system as a whole to lose fine-grained material to the North Sea. This apparent inconsistency can only persist in the long run if during other periods the system imports fine-grained sediment in amounts large enough to maintain an available store of fine-grained material. This store consists primarily of mud flat deposits in the intertidal zone which, therefore, hold the potential for depositional evidence of import events. In the sedimentary record, salt marsh deposits will reveal evidence of sedimentation related to stormy periods independent of the episodes in which fine-grained sediments are imported from the North Sea.

The system is importing during and after windy periods following long periods of calm weather conditions, which mostly appear in the summer. Such an event, of about two weeks, is assumed to be responsible for about 25% of the expected yearly net import. Sediment import was a result of high concentrations of fine-grained material (>180 mg/l), with relatively high settling velocities (1–3 mm/s), in the water entering from the North Sea. This is speculated to be a result of the settling of fine-grained sediment on the shelf during long periods of calm weather, which thus builds the potential for high concentrations and fall-velocities during the subsequent windy period.

ACKNOWLEDGMENTS

This paper is based on an investigation carried out for the harbor authorities in Esbjerg, Denmark. We are pleased to express our gratitude to Hennning Nørgaard and Erik Brenneche in "the Castle" for their never-failing assistance and cooperation, and to Peter Kempf and his crew in the boat house on the pier for field assistance and good humor even in the dark and stormy hours. Several students at the Institute of Geography, University of Copenhagen, participated enthusiastically in the fieldwork, for which we are grateful, as we also are to Kirsten Simonsen and Heini Larsen at the Skalling Laboratorium in Esbjerg for good operation.

REFERENCES

AAGARD, T., NIELSEN, N., AND NIELSEN, J., 1995, Skallingen—Origin and Evolution of a Barrier Spit: Meddelelser fra Skalling-Laboratoriet, v. XXXV, 85 p.

AMOS, C. L., AND MOSHER, D. C., 1985, Erosion and deposition of fine-grained sediments from the Bay of Fundy: Sedimentology, v. 32, p. 815–832.

ANTHONY, D., 1995, Variasions mønsteret i transporten af finkornet suspenderet sediment i den sydlige del af Grådyb, det danske vadehav: Unpublished M.S. Thesis, Institute of Geography, University of Copenhagen, 83 p.

BARTHOLDY, J., 1983, Recent Sedimentologi i Hobo Dybs Tidevandsområde: Unpublished Ph.D. Dissertation, Institute of Geography, University of Copenhagen, 217 p.

BARTHOLDY, J., 1993a, Miljømæssig vurdering af uddybning af Grådyb: Delrapport 7, Feltresultater. Dynamiske målinger, manuelle registreringer. Kyst-Havnesamarbejdets Forlag, Lemvig, 45 p.

BARTHOLDY, J., 1993b, Miljømæssig vurdering af uddybning af Grådyb: Delrapport 6, Feltresultater. Dynamiske målinger, selvregistrerende udstyr. Kyst-Havnesamarbejdets Forlag, Lemvig, 119 p.

BARTHOLDY, J., AND MADSEN, P. P., 1985, Accumulation of fine-grained material in a Danish tidal area: Marine Geology, v. 67, p. 121–137.

BARTHOLDY, J., AND FOLVING, S., 1986, Sediment classification and surface type mapping in the Danish Wadden Sea by remote sensing: Netherlands Journal of Sea Research, v. 20, p. 337–345.

BARTHOLDY, J., AND ANTHONY, D., 1994, Monitering af turbiditet i forbindelse med uddybningen af Grådyb: Institute of Geography, University of Copenhagen, unpublished report to the harbor authorities in Esbjerg, 92 p.

BARTHOLDY, J., AND PEJRUP, M., 1994, Holocene Evolution of the Danish Wadden Sea: Senckenbergiana Maritima, v. 24, p. 187–209.

DEYU, Z., 1987, Sedimentological investigation of fine-grained marine sediments, with particular reference of features for recognition: Unpublished Theses, Department of Geology, Aarhus University, Denmark, 148 p.

DYER, K., 1994, Estuarine sediment transport and deposition, *in* Pay, K., ed., Sediment Transport and Depositional Processes: London, Blackwell, p. 193–218.

EISMA, D., 1981, Supply and Deposition of Suspended matter in the North Sea, in Nio, S. D., Shuttenhelm, R. T. E., and van Weering, eds., Sedimentation in the North Sea Basin: London, Blackwell, p. 415–428.

EISMA, D., AND KALF, J., 1987, Dispersal, concentration and deposition of suspended matter in the North Sea: Journal of Geological Society London, v. 144, p. 415–428.

EISMA, D., BERNARD, P., CADÉE, G. C., ITTEKKOT, V., KALF, J., LAANE, R., MARTIN, J. M., MOOK, W. G., VAN PUT, A., AND SCHUMACHER, T., 1991a, Suspended-matter particle size in some West-European Estuaries; Part 1, Particle size distribution: Netherlands Journal of Sea Reasearch, v. 28, p. 193–214.

EISMA, D., BERNARD, P., CADÉE, G. C., ITTEKKOT, V., KALF, J., LAANE, R., MARTIN, J. M., MOOK, W. G., VAN PUT, A., AND SCHUMACHER, T., 1991b, Suspended-matter particle size in some West-European Estuaries; Part 1, A review on floc formation and break-up: Netherlands Journal of Sea Reasearch, v. 28, p. 215–220.

KE, X., EVANS, G., AND COLLINS, M. B., 1996, Hydrodynamics and sediment dynamics of The Wash embayment, eastern England: Sedimentology, v. 43, p. 157–174.

KRONE, R. B., 1962, Flume studies of the transport of sediments in estuarial shoaling processes: Hydraulic Engineering Laboratory and Sanitary Engineering Research Laboratory, University of California, Berkeley.

MCCAVE, I. N., 1973, Mud in the North Sea, *in* Goldberg, E. D., ed., North Sea Science: Cambridge, Massachusetts, Massachusetts Institute of Technology Press, p. 75–100.

PEJRUP, M., 1991, The influence of flocculation on cohesive sediment transport in a microtidal estuary, *in* Smith, D. G., Reinson, G. E., Zaitlin, B. A., and Rahmani, R. A., eds., Clastic Tidal Sedimentology: Calgary, Canadian Society of Petroleum Geologists Memoir 16, p. 283–290.

PEJRUP, M., BARTHOLDY, J., AND JENSEN, A., 1993, Supply and exchange of Water and Nutrients in the Grådyb Tidal Area, Denmark: Estuarine, Coastal and Shelf Science, v. 36, p. 221–234.

POSTMA, H., 1961, Transport and accumulation of suspended matter in the Dutch Wadden Sea: Netherlands Journal of Sea Research, v. 1, p. 148–190.

POSTMA, H., 1967, Sediment Transport and Sedimentation in the Estuarine Environment, in Lauf, H., ed., Estuaries, American Association for the Advancement of Science, p. 186–221.

POSTMA, H., 1981, Exchange of Material Between the North Sea and the Wadden Sea: Marine Geology, v 40, p. 199–213.

RIBE AMTSRÅD, 1988, Vandudskiftningsmodel og næringssaltsbalance for Grådybs tidevandsområde: Report to the council of Ribe, ISBN 87-7342-446-3, 73 p.

SHERIDAN, M. F., WHOLETZ, K. H., AND DEHN, J., 1987, Discrimination of grain-size subpopulations in pyroclastic deposits: Geology, v. 15, p.367–370.

SÜNDERMANN, J., 1994, Suspended particulate matter in the North Sea: Field observations and model simulations, *in* Charnock, H., Dyer, K. R., Huthnance, J. M., Liss, P. S., Simpson, J. H. and Tett, P. B., eds., Understanding the North Sea System: London, Chapman & Hall for The Royal Society, p. 45–52.

THOMSEN, H., AND HANSEN, B., 1970, Middelvandstand og dens ændring ved de danske kyster: Det Danske Meteoroliske Institut Meddelelser, v. 23, 24 p.

VAN ES, F. B., 1977, A preliminary carbon budget for a part of the Ems Estuary: The Dollard. Helgol. Wiss. Meeresunters., v. 30, p. 283–293.

VAN STRAATEN, L. M. J. U., AND KUENEN, Ph. H., 1958, Tidal action as a cause for clay accumulation: Journal of Sedimentary Petrology, v. 28, p. 406–413.

VARIABILITY OF FINE-GRAINED SUSPENDED SEDIMENT CONCENTRATIONS IN A BACK-BARRIER TIDAL CHANNEL DURING CALM WEATHER CONDITIONS

DENNIS ANTHONY* AND JESPER BARTHOLDY

University of Copenhagen, Institute of Geography, Oester Voldgade 10, DK-1350 Copenhagen K.

ABSTRACT: Concentration of fine-grained suspended sediment in a microtidal estuarine environment (northern part of the European Wadden Sea; mean tidal range = 1.5 m) is examined in order to clarify how this is influenced by flood/ebb, diurnal inequality and neap/spring during calm weather conditions. The diurnal inequality was reflected in the fine-grained suspended sediment concentration as an alternation of the maximum and minimum concentrations caused by the variation in advection as well as an alternation in the resuspension strength. The onset of the ebb current generally caused higher resuspension than the onset of the flood current, which presumably was due to the fact that a larger portion of the tidal flats was affected by the ebb current than by the flood current because of the difference in water level. The variability of the concentration of fine-grained suspended sediment over a neap/spring/neap cycle is found to be controlled primarily by the changes in the mean water level. A decrease in mean water level caused a rise in the concentration level because of a seaward extension of the turbidity maximum. An increase in mean water level could also cause a rise in the concentration level, presumably because of exposure of the tidal flats to the current. Because of the relatively low tidal range, however, the resulting long-term tide is strongly modified by the wind effects even in relatively calm periods.

INTRODUCTION

Short-term dynamics of fine-grained suspended sediment in estuarine environments have been measured, described and interpreted by several authors in the past. As a consequence the basic mechanisms of advection, settling and resuspension are generally well understood (see, e.g., Dyer, 1994, and Eisma, 1993). Earlier work dealing with this topic in greater detail has been published by Gelfenbaum (1983) and Nichols (1986). Because of pronounced wind-tide interactions and strong bioturbation, the dynamics of suspended sediment in the northern part of the Wadden Sea cannot be directly related to any signatures in the sedimentary record. On the other hand, relatively clean astronomically induced variations may be useful in elucidating relationships between hydrodynamics and transport of fine-grained sediment, and thereby significantly add to the understanding of the overall processes controlling this depositional system.

The objective of this paper, thus, is to examine potential effects of astronomically induced variations in suspended sediment concentrations in a microtidal estuarine environment of the Danish Wadden Sea. The paper focuses on the significance of the semidiurnal tide, the diurnal inequality and the neap-spring-neap influence on the fine-grained suspended sediment concentration.

The data sets presented and discussed in this paper were collected in connection with an environmental survey carried out for the harbor authorities of Esbjerg, Denmark. Long time-series of current velocity and turbidity were collected in the nearby tidal area prior to a planned deepening of the navigation channel. A two-week period in autumn 1993, characterized by extraordinary low wind speed, was chosen for examination of the astronomical influence on the concentration of fine-grained suspended sediment.

STUDY AREA

The tidal area landwards of the barrier-type peninsula Skallingen and the northern part of the barrier island Fanø is connected to the North Sea through the Grådyb tidal channel, which is the northernmost inlet of the European Wadden Sea (Fig. 1). According to the definition of Perrilo (1995), the area is a secondary estuary and can be classified as a restricted coastal lagoon. Landwards of the barrier islands are extensive marsh lands that are bordered by tidal flats. These contain a high content of fine-grained sediment—especially in the inner parts of the estuary.

The tidal range is approximately 1.5 m, and the tidal prism at the entrance is approximately 140×10^6 m^3. The access channel through the ebb tidal delta is continuously dredged in order to maintain a minimum depth of 10.3 m below mean spring low tide for navigational purposes. A detailed description of the area is given in Bartholdy and Pejrup (1994).

METHOD

Tidal currents and turbidity were measured continuously every 5 minutes with a winged current meter (General Oceanics 6011 MkII) and an infrared-light based turbidity meter (Geological and Marine Instrumentation, TU-150IR), respectively. The instruments were mounted on a mooring with the current meters placed 4 m above the bottom (close to the theoretical depth-averaged mean current position) and the turbidity meter approximately 1.5 m above the current meter. The turbidity readings were converted into suspended sediment concentration on the basis of *in situ* calibrations (Bartholdy and Anthony, 1994). This was done by calibrating the different turbidity meters to a standardized unit (Formazine Turbidity Units), from where the concentration is calculated by a linear relationship to the calibration samples ($r = 0.89$). At one- to two-week intervals the lenses of the turbidity meter were cleaned and the difference in readings immediately before and after this procedure were noted as the maximum defilement of the lenses. The concentration level was then corrected for this error by subtracting a linear function, starting from zero at the beginning of the data set and ending with the maximum defilement of the lenses.

Measuring the optical density of the water, the turbidity meter is far more sensitive to fine-grained suspended sediment than it is to suspended sand. The turbidity data are therefore interpreted as being directly related to the concentration of fine-grained suspended sediment (FGSS).

RESULTS

The data set presented in this study was collected at Station 3 (Fig. 1) and describes a full neap-spring-neap cycle from Oc-

*Present address: Geological Survey of Denmark and Greenland, Thoravej 8, DK-2400 Copenhagen NV.

Tidalites: Processes and Products, SEPM Special Publication No. 61

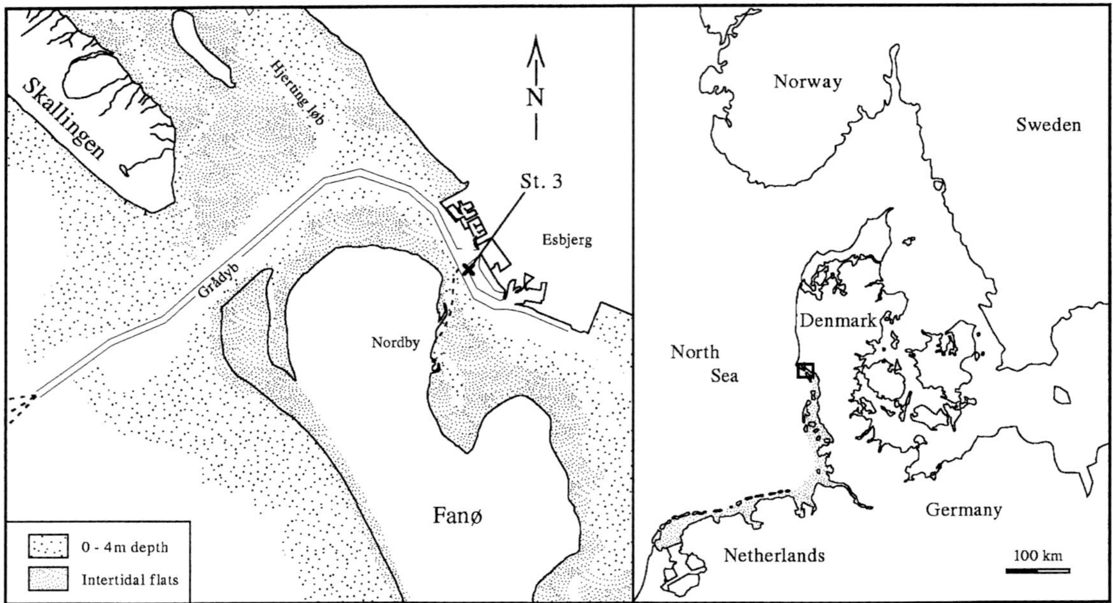

FIG. 1.—Location of the study area in the northern part of the Danish Wadden Sea. The measurements from both the mooring station and the anchor station are from Station 3 in the Grådyb tidal channel.

tober 22 to November 11, 1993. Apart from isolated resuspension peaks with concentrations up to 32 mg l^{-1}, the concentration varied between 2 and 17 mg l^{-1}, which is very low compared to a more typical range of 15–100 mg l^{-1} (Bartholdy, 1993, and Bartholdy and Anthony, 1994). It is assumed that the reason for such low turbidity readings were the extraordinary low wind speeds recorded immediately prior to, and during, the study period.

As evident from Figure 2F, the concentration at high-water slacks (HWS) was consistently lower than that of low-water slacks (LWS). From this it can be inferred that the measuring station was located at the seaward side of a turbidity maximum. Superimposed on this advective pattern—which produced an oscillatory variation in concentration over time because of the gradient in the concentration of FGSS through the estuary—are the effects of settling and resuspension. This feature produces the characteristic two-peaked signature during a tidal cycle (see also Jago et al., 1994).

In the following, this data set is used to interpret the variations caused by the diurnal inequality and neap-spring conditions. In order to evaluate the representativity of the data, and to describe the conditions related to the shortest astronomically induced variation, that is, the flood-ebb cycle, results from an anchor station at Station 3, with measurements at different levels above the bottom, were examined first.

Vertical Distribution of Fine-grained Suspended Sediment

Long-term measurements of salinity and temperature at Station 3 have shown that there is virtually no stratification due to salinity or temperature differences. Because of settling and resuspension, however, the FGSS becomes stratified in the water column as shown in Fig. 3. The diagram reflects the concentration at five levels above the bottom, covering the entire settling and resuspension period surrounding a low water slack. The profiles were measured from an anchor station near Station 3 using a turbidity meter connected to a personal computer. Each value represents an average over approximately 1 minute of continuous measurements at each level. The current velocity was calculated from the tidal-gauge data recorded at the nearby harbor by means of a simple empirical model (Anthony, 1995). Wind conditions were rather calm (about 5 m/s), so that the model is assumed to predict at least 95% of the current velocity variation.

When the current velocity falls below about 0.2 m/s, stratification in the concentration of FGSS develops due to settling. The resulting gradient increases towards slack. When the current velocity reaches about 0.5 m/s after current reversal, resuspension occurs and continues until the maximum velocity (about 0.7 m/s in this case) is reached. The resuspension amplifies the vertical concentration gradient, and the resuspended sediment is observed to gradually mix upwards into the water column.

In the course of the long time series, concentration was only measured at one level, that is, close to 5 m above bottom (m ab). Resuspension obviously occurs over a longer period of time than measured at the fixed level because of the time it takes for the resuspended sediment to be mixed into the water column. It is, however, assumed that the measured resuspension at the 5 m ab-level, in general, is proportional to the amount of resuspended sediment in the entire water column even though small resuspension events may be strongly diluted before they reach this level. When calculating the depth-integrated (and velocity-weighted) cumulative sediment flux for the present period, using first all five levels and thereafter only the 5 m ab-level, the results turned out to be almost identical (151.3 kg/m^2 and 149.5 kg/m^2). With this in mind, the effect of the diurnal inequality on the concentration of the FGSS could then be examined.

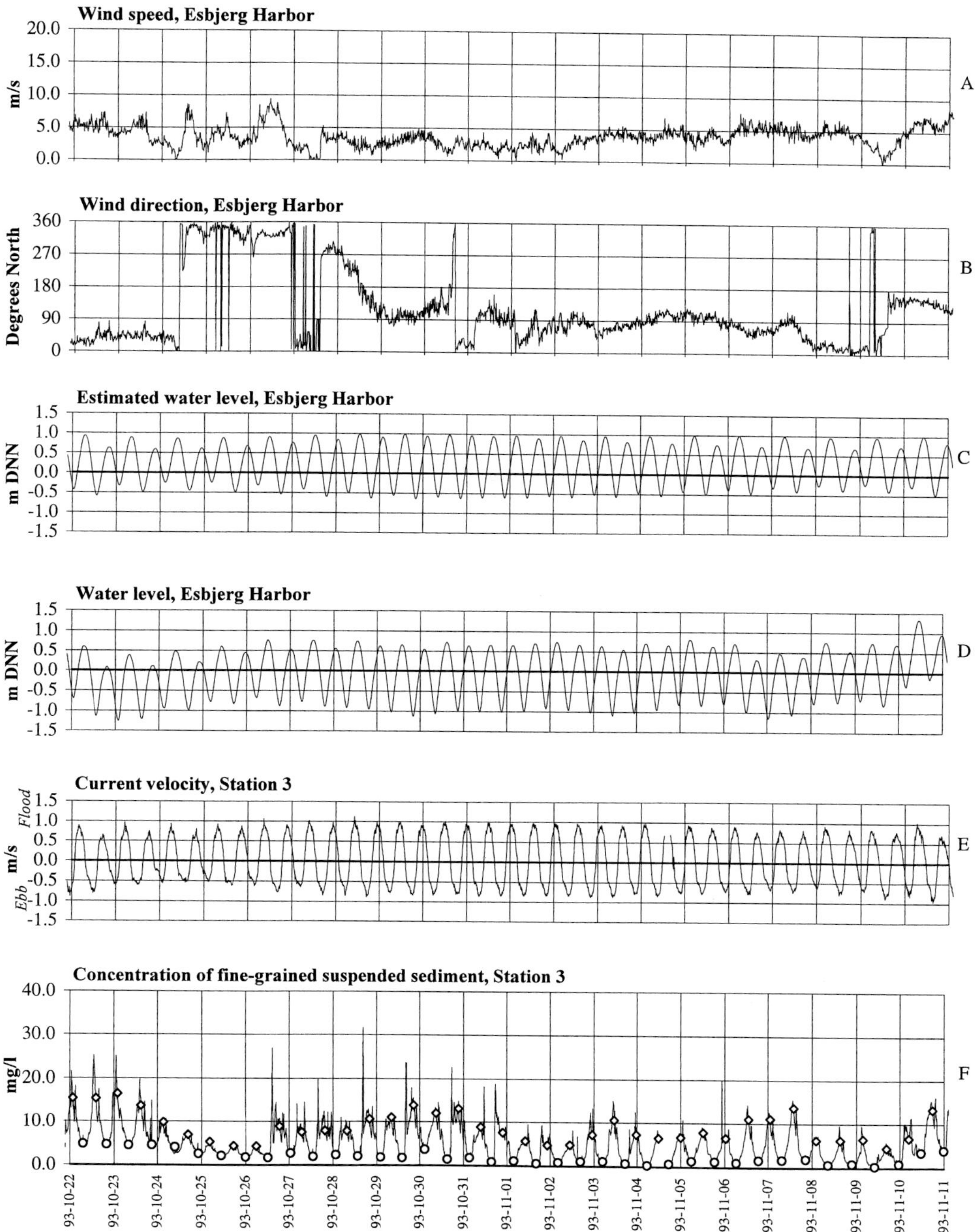

FIG. 2.—Dynamic parameters during a neap-spring-neap cycle at Station 3 in the Grådyb tidal channel. Wind speed (A) and wind direction (B) were measured 15.5 m above ground. Water level (D) is measured in a wave-dampened well in one of the basins in Esbjerg Harbor. The estimated water level (C) was calculated by the Danish Meterological Institute. Data on current velocity (E) and concentration of fine-grained suspended sediment (F) were collected at a mooring station (Station 3) in the Grådyb tidal channel. The circles and diamonds in F indicate the concentration at high and low water slacks, respectively.

Diurnal Inequality

As seen in Fig. 2, a distinct diurnal inequality is found in some periods. Fig. 4 shows one of these periods—covering seven tidal cycles in greater detail. The diurnal inequality alternatively produces a dominant and a subordinate current (strong and weak), and the subordinate and the dominant current components will produce relatively small and large advection-based variations in the concentration of FGSS, respectively. Fig. 4 begins with a subordinate tidal cycle (a flood and an ebb period with a relatively weak maximum velocity). The subsequent tidal cycles alternate between relatively high and relatively low maximum current velocities. The corresponding concentration of FGSS shows a pattern reflecting the variable degree of advection caused by the diurnal inequality. It is reflected in a lower concentration at HWS in the dominant cycle as compared to the subordinate cycle and a corresponding higher concentration around LWS in the dominant cycle.

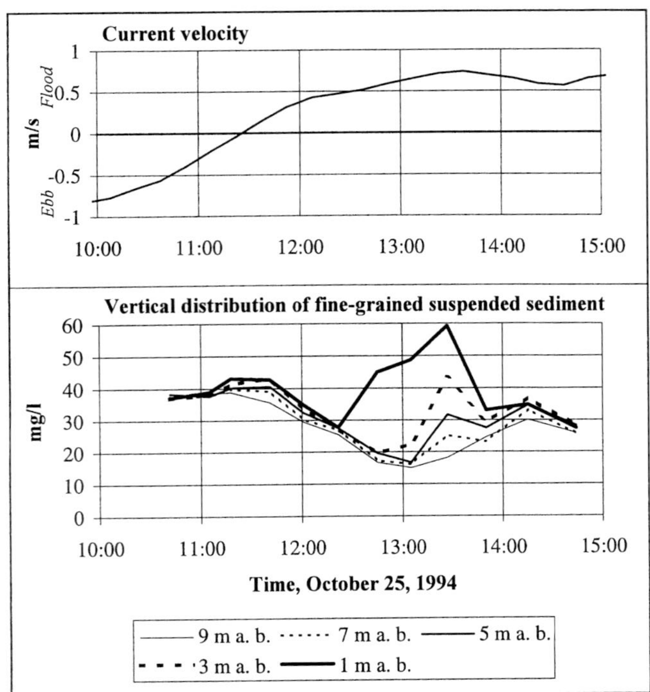

FIG. 3.—Vertical distribution of fine-grained suspended sediment and mean current velocity during a low water slack period at Station 3 in the Grådyb tidal channel (m ab = m above bottom; the mean depth at Station 3 was approximately 10 m).

Moreover, it is likely, that the alternating maximum currents should result in alternating levels of settling and especially resuspension, the amount of resuspension being dependent on the shear stress (Mehta, 1989) and, hence, the current velocity. Resuspension produces concentration peaks during both the flood and the ebb, being consistently more pronounced during the dominant ebb as compared to the subordinate ebb. No clear pattern emerges during the flood periods. It is noteworthy that the highest resuspension values are recorded during the dominant ebb current, because this current follows the lowest slack concentrations from the preceeding dominant flood current, which provides less material for resuspension. Whereas the variation in the maximum current strength is not very pronounced, and the resuspension is completed before the maximum velocity is reached, it cannot be the maximum current velocity, which is the controlling factor. A possible explanation may be that the higher water level during the dominant tidal periods allows the ebb current to cause stronger erosion in higher intertidal areas close to the channel margins. If this were the case, then the diurnal inequality could be responsible for a rhythmic net transport directed towards the open North Sea. This interpretation is, in fact, supported by the calculated net transport of FGSS (Fig. 4D). The excess of FGSS in the dominant ebb period results in a net loss of approximately 0.2 kg/m^2 over tile period (through the cross-section), even though the net water displacement over the same period is directed landward (Fig. 4C). It has also been considered whether the observed pattern could be a consequence of anthropogenic influences, such as the dumping of fine-grained sediment dredged from the harbor basins. Any such effect, however, is considered to be negligible because of the high repeatability of tile pattern observed and because dumping of dredge spoils is only allowed at specific sites and at times of the tidal cycle that have the least influence on the amount of fine-grained suspended sediment passing through the Grådyb channel close to the harbor.

Concentration of Fine-grained Suspended Sediment During a Neap-Spring Cycle

Fig. 2 shows a continuous time series of wind speeds (A), wind direction (B), theoretical and actual water level for Es-

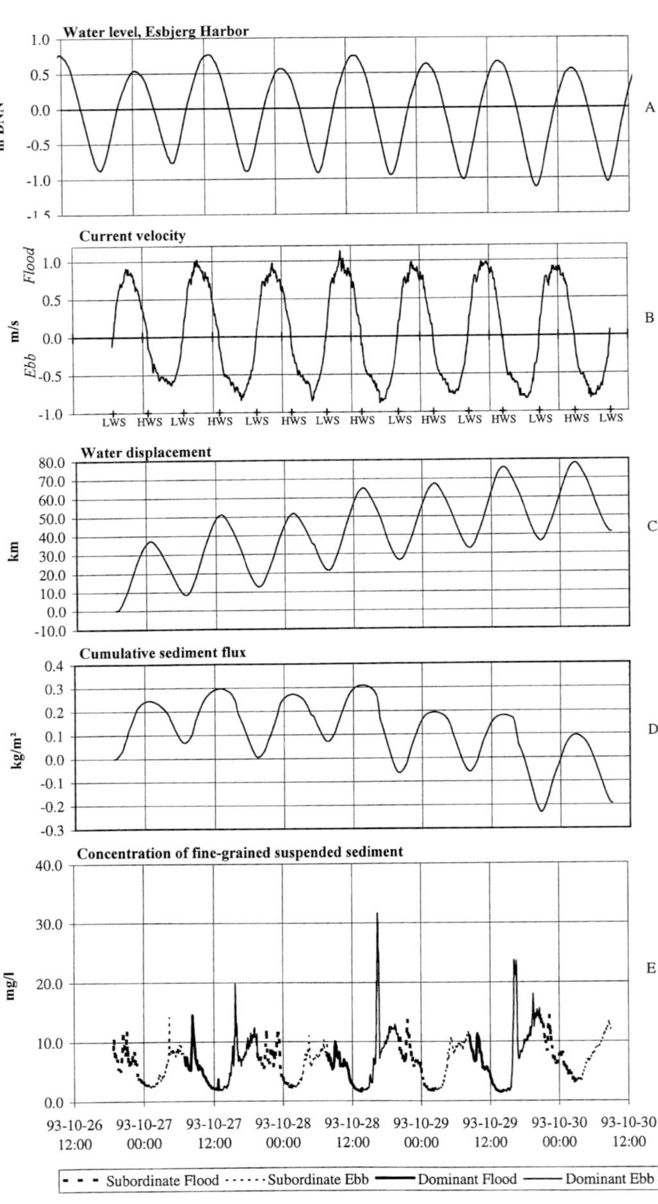

FIG. 4.—Water level (A), current velocity (B), water displacement (C), cumulative sediment flux (D), and concentration of fine-grained suspended sediment (E) during a period with distinct diurnal inequality (LWS = low water slack, HWS = high water slack; the designations subordinate and dominant refer to the relative strength of the current velocity in the course of the diurnal inequality).

bjerg Harbor (C and D), current velocity (E) and concentration of fine-grained sediment (F) for October 22—November 11, 1993. The wind conditions for the period are generally rather calm. At the beginning of the period the wind speed fluctuates from 0 to 9 m/s, with an average around 4–5 m/s. The corresponding wind directions for most of the period are from a northerly or easterly direction.

During the study period the astronomical phases were responsible for a distinct diurnal inequality in the water level around neap-tide (October 24 and November 7, 1993). By contrast, the diurnal inequality is diminished at spring-tide (October 30, 1993) because of the phase-transition in the tidal components. The dominant component of the semidiurnal tide reaches spring-size tidal heights a few days after the neap-tide, which lasts until a few days before the subsequent neap.

The estimated and the measured water levels are rather similar, though there are some disturbances in the measured water levels, for example at the beginning of the period, which coincides with a few days of northeasterly winds at speeds of 4–7 m/s. During the subsequent extended period of low wind speed and easterly wind directions, the measured water levels are generally lower than the estimated water levels. This is due to the fact that the model calculations for the estimated water level assume a westerly wind direction, which is statistically expected for this time of the year.

The description and interpretation of the concentration of FGSS begins just before neap tide and continues until shortly after the subsequent neap tide.

October 22, 00:00 hr to October 26, 12:00 hr, 1994: This period crosses a neap-tide cycle. The maximum current velocity generally declines towards neap tide on October 24, whereafter it gradually increases. The wind is on the order of 0–8 m/s, but mostly around 5 m/s, and the direction is northeast, shifting to northwest. Water levels are generally low, with a mean around −0.5 m Danish Normal Datum (DNN) over the first two days, but reaches normal levels over the following 2½ days. Current velocities generally decrease towards neap but increase after neap towards spring.

The FGSS is relatively high during the first two days, with both the HWS and the LWS concentration reaching the highest values for the entire period ($\approx$5 mg l^{-1} and $\approx$15 mg l^{-1} respectively). Over the following 2½ days both the HWS and the LWS concentrations, as well as the difference between them, decrease gradually.

The high concentrations during the first two days versus the low concentrations over the following 2½ days are primarily due to the differences in water level. The easterly winds have drawn some water away from the area, thereby extending the turbidity maximum in a seaward direction, an effect clearly influencing the measurements at Station 3. The turbidity maximum is located close to the tidal divide behind the barrier island Fanø, but because it consists of a more or less mobile suspension, the lowering of the mean water level can, from a continuity point of view, be considered as a "flattening" of the turbidity maximum. When the water returned to the area, relatively non-turbid North Sea water entered the system and "pushed" the turbidity maximum back to place. The reason no resuspension peaks were observered at the end of the period is probably due to a combination of the lack of suspendable sediment and the generally low maximum current velocity which has a relatively low erosive potential and is less efficient in mixing the resuspended sediment upwards into the water column.

October 26, 12:00 hr to November 2, 12:00 hr: This period extends from shortly after neap-tide to just after spring tide. On the first day, maximum wind speeds registered from the northwest. During the following days, the wind speeds were below 5 m/s, starting with a westerly direction, but later shifting to an easterly direction. The maximum current velocity of the subordinate component did not increase much towards spring tide, whereas that of the subordinate component increased considerably. At the same time the concentration increased abruptly compared to the previous period. In fact, both the HWS and the LWS concentrations increased above the levels observed towards the end of the previous period, although slight decreases registered towards the end of the period. Strong resuspension occurred over approximately the first five days.

The increase in FGSS is coincident with maximum wind speed, high water level and strong current velocity. Because the increase in concentration actually begins with a resuspension peak shortly before the maximum current speed is reached, the current cannot be the primary controlling factor. As in the previous case, the most plausible explanation would be that the higher water level now allows resuspension of fine-grained sediment on the tidal flats. The decrease in concentration at the end of the period is probably due to exhaustion of the fine-grained deposits because neither the water level nor the current velocity have changed significantly.

November 2, 12:00 hr to November 9, 00:00 hr: This period begins shortly after spring tide and ends at the following neap tide. The wind speed is approximately 4 m/s, blowing from the east at the beginning of the period, and increasing slightly as it shifts towards the north. The mean water level drops slightly twice, on November 3 and November 6–7. The current velocity generally decreased towards neap. Notably, the concentration of FGSS distinctly increased during the periods of water level drop, implying the same mechanism as found in the first period—namely a flattening of the turbidity maximum during a lowering of mean water levels.

November 9, 00:00 hr to November 11, 00:00 hr: This period commences just after neap-tide. The wind speed is very low on the first day, but increases to approximately 8 m/s from the southeast on the following day. Over these two days, the mean water level increased approximately 0.5 m, and the current velocity increased in general. The concentration of FGSS is rather low during the first day, and there is very little resuspension. On the following day, the concentration increases distinctly, presumably due to the concomitant increase in water level.

Over the period discussed in this paper, the variation pattern of the FGSS does not appear to be directly linked to the neap-spring-neap cycle, but rather to the relative height of the mean tide. The latter is influenced by both the astronomical tide and effects of wind. Ideally, the water levels would be expected to reflect the neap- and spring-tide conditions. However, in this case the wind-induced sea level variations clearly overrode the neap-spring variation, notwithstanding that the period was characterized by extraordinarily calm conditions. The above mentioned variation patterns indicate that under calm conditions there is a general export of fine-grained sediment from the inner part of the system. This is, however, coincident with an import during much different weather conditions (Bartholdy and Anthony, this volume).

SUMMARY AND CONCLUSION

In this study, the variation pattern of fine-grained suspended sediment in the Grådyb main channel of the Danish Wadden Sea was investigated during calm weather. The data set is based on measurements from an anchor station and from self-recording equipment at a mooring station over a neap-spring-neap cycle.

The vertical distribution of fine-grained suspended sediment is examined. Settling through the water column becomes apparent when the mean current velocity falls below approximately 0.2 m/s. After current reversals, resuspension occurs when the mean current velocity exceeds approximately 0.5 m/s.

Examination of the concentration pattern during a period with distinct diurnal inequality reveals stronger resuspension during the ebb than during the flood phase, and the resuspension during the dominant ebb phase is consistently larger than during the subordinate tide. The main controlling factor is the varying tidal height, which more strongly exposes the fine-grained tidal flats to resuspension at the onset of the ebb current following the dominant flood. This alternation of the concentration would presumably give rise to alternating depositions. Because of the very low concentrations and the slight net export of suspended sediment, deposition would only be local and very small in scale.

The observed variations in the concentration of fine-grained suspended sediment over a neap-spring-neap period were evidently strongly linked to even small fluctuations in the mean water level. During a temporary lowering of the mean water level, the concentration level at Station 3 was raised because the turbidity maximum in the inner part of the system was extended in a seaward direction. When, on the other hand, the mean water level is temporarily raised, for example by onshore winds, a larger portion of the upper-fine-grained tidal flats becomes exposed to the current. This results in an increase in the resuspension, and thereby the concentration, of fine-grained suspended sediment. During steady dynamic conditions and with no fluctuations in the mean water level, it would seem that the availability of potentially resuspended fine-grained sediment becomes rapidly exhausted. These mechanisms indicate that some of the fine-grained suspended sediment is being exported to the North Sea during calm weather conditions.

Finally, it ought to be emphasized that the above description is only valid in a microtidal environment during calm wind conditions. The net export is probably due to the very low concentrations, which get very quickly enriched in fine-grained suspended sediment as the water flows across fine-grained tidal flats.

ACKNOWLEDGMENTS

We thank the harbor authorities at Esbjerg Harbor for wind and water level data, the crew of the working vessel named Grådyb for assistance and good seamanship during the field work, the Danish Meterological Institute for the theoretical water level data, Anita Folly for comments on the manuscript and Kent Pørksen for the drawing of the study area.

REFERENCES

ANTHONY, D., 1995, A simple model for current speed in tidal channels: Danish Journal of Geography, v. 95, p. 12–18.

BARTHOLDY, J., 1993, Miljømæssig vurdering af uddybning af Grådyb. Feltresultater. Dynamiske målinger, manuelle registreringer. Delrapport nr.7. Technical Report to the Authorities at Esbjerg Harbour: Institute of Geography, University of Copenhagen, 45 p.

BARTHOLDY, J., AND ANTHONY, D., 1994, Monitering af turbiditet i forbindelse med uddybning af Grådyb. Technical Report to the Authorities at Esbjerg Harbour: Institute of Geography, University of Copenhagen. 92 p.

BARTHOLDY, J., AND PEJRUP, M., 1994, Holocene evolution of the Danish Wadden Sea: Senckenbergiana maritima, v. 24, p.187–209.

DYER, K. R., 1994, Estuarine sediment transport and deposition, *in* Pye, K., ed., Sediment Transport and Depositional Processes: Oxford, Blackwell Scientific Publications, p. 193–218.

EISMA, D., 1993, Suspended Matter in the Aquatic Environment: Berlin, Springer Verlag, 315 p.

GELFENBAUM, G., 1983, Suspended-sediment response to semidiurnal and fortnightly tidal variations in a mesotidal estuary: Columbia River, U.S.A.: Marine Geology, 52, p. 39–57.

JAGO, C. F., BALE, A. J., GREEN, M. O., HOWARTH, M. J., JONES, S. E., MCCAVE, I. N., MILLWARD, G. E., MORRIS, A. W., ROWDEN, A. A., AND WILLIAMS, J. J., 1994, Resuspension processes and seston dynamics, southern North Sea, *in* Charnock, H., Dyer, K. R., Huthnance, J. M., Liss, P. S., Simpson, J. H., and Tett, P. B., eds., Understanding the North Sea System: London, The Royal Society, Chapman and Hall, p. 97–113.

MEHTA, A. J., 1989, On estuarine cohesive sediment suspension behavior: Journal of Geophysical Research, 94 (C10), p. 14,303–14,314.

NICHOLS, M. M., 1986, Effects of fine sediment resuspension in estuaries, *in* Mehta A. J., ed., Estuarine Cohesive Sediment Dynamics: Berlin, Springer Verlag, p. 5–42.

PERILLO, G. M. E., 1995, Definitions and geomorphologic classification of estuaries, *in* Perillo, G. M. E., ed., Geomorphology and Sedimentology of Estuaries: Amsterdam, Elsevier, p. 17–47.

HIERARCHY OF CONTROLS ON CYCLIC RHYTHMITE DEPOSITION: CARBONIFEROUS BASINS OF EASTERN AND MID-CONTINENTAL U.S.A.

A. W. ARCHER

Department of Geology, Kansas State University, Manhattan, Kansas 66506, USA

ABSTRACT: Cyclic rhythmites (tidal rhythmites) provide fine-scale resolution of deposition within specific types of tidal systems. Rhythmites of this kind have been reported from many sites throughout geologic time; nonetheless, they are particularly widespread and common within the Carboniferous basins of the eastern and mid-continental United States. These rhythmites have great potential for providing information regarding the local, regional, and global controls on their deposition. Widespread occurrence of cyclic rhythmites within these basins outlines a hierarchy of parameters that controlled deposition. These controls include: (1) local and regional depositional setting, (2) elevated paleotidal ranges, (3) sequence stratigraphic and basinal conditions, (4) a unique paleoceanographic configuration, and (5) specific conditions of astrophysical alignments.

INTRODUCTION

Recognition of Cyclic Rhythmites

The study area for this analysis includes the Central Appalachian, Eastern Interior, and Western Interior basins (Fig. 1). Earlier work on cyclic rhythmites was first performed in the Eastern Interior Basin. In this area, Kuecher (1983) was the first to clearly delineate the tidal origin of the cyclic rhythmites. Subsequently, Kvale et al. (1989) presented detailed analysis and delineated the neap-spring cycles, as well as other tidal periodicities, and concluded that the paleotidal deposition was controlled by a mixed, predominantly semidiurnal system. The common association of the cyclic rhythmites to major economic goals was subsequently noted by Kuecher et al. (1990), Kvale and Archer (1990), and Archer and Kvale (1993). Prior to the recognition of the tidal controls on sedimentation, many of these facies had been considered to have formed within terrestrial paleoenvironments (Archer and Maples, 1984; Maples and Archer, 1987).

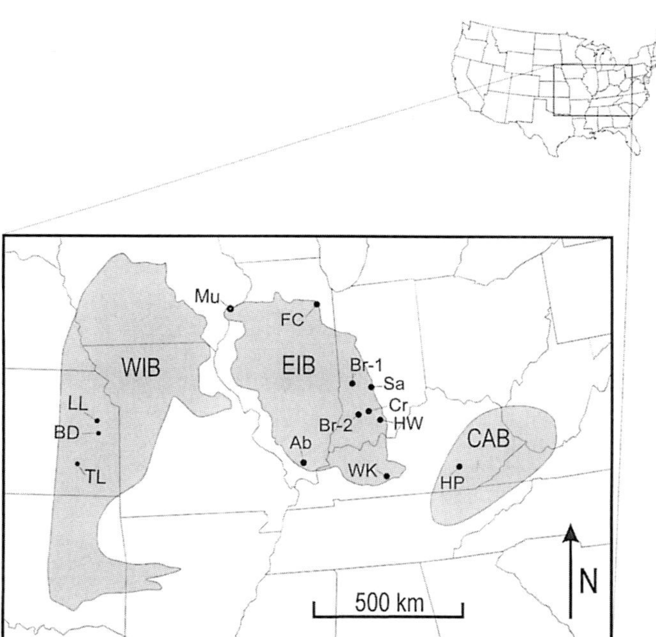

FIG. 1.—Within the eastern and mid-continental U.S., basins containing cyclic rhythmites include: Central Appalachian Basin (CAB), Eastern Interior Basin (EIB), and the Western Interior Basin (WIB). In the CAB, Greb and Archer (1995) described the Hazel Patch (HP) sandstone rhythmites. In the EIB, rhythmite occurrence includes sites: (Mu) Muscatine, Iowa (Archer et al., 1992); (FQ) Francis Creek Shale (Kuecher, 1983; Kuecher et al., 1990); (Ab) Abbott Sandstone in southern Illinois (Kvale and Archer, 1991); (Br-1, 2) Brazil Formation in Indiana (Kvale and Archer, 1990); (Sa) Salem Limestone (Brown et al., 1990); (Cr) Crane-area study (Kvale and Barnhill (1994); (HW) Hindostan Whetstone (Kvale et al., 1989; Kvale and Archer, 1991); (WK) in western Kentucky (Greb and Archer, this volume). In the WIB, rhythmite sites include: (LL) Lonestar Lake spillway (Archer, 1991, 1994); (BD) Buildex quarry (Lanier et al., 1993); and (TL) Toronto Lake area (Archer et al., 1995b).

Other examples of Carboniferous cyclic rhythmites were subsequently reported from the mid-continental U.S. in the western Interior Basin (Archer, 1991; Feldman et al., 1993; Lanier et al., 1993; Archer et al., 1994a, b; Archer and Feldman, 1995). Similar rhythmites also occur within the Central Appalachian Basin (Martino and Sanderson, 1993; Greb and Archer, 1995).

The ability to recognize tidal controls within cyclic rhythmites was greatly aided by modern analogs. Visser (1980) and Allen (1981) presented compelling evidence that tidal periodicities could be encoded within a series of foresets. Consequently, similar neap-spring cycles were noted in vertically accreted laminae. From the upper reaches of the macrotidal Bay of Fundy, Nova Scotia, Canada, Dalrymple and Makino (1989) and Dalrymple et al. (1991) described cyclic rhythmites. Tessier et al. (1989) and Tessier (1993) described similar styles of sedimentation within the macrotidal Bay of Mont Saint Michel along the northwestern coast of France. Within tidal channels in The Netherlands, Van den Berg (1981) and Roep (1991) presented details about mesotidal cyclic rhythmites.

The many other examples of non-Carboniferous cyclic rhythmites are beyond the intended scope of this discussion. Important works on Precambrian rhythmites include Williams (1991) and Chan et al. (1994). Summaries and analyses of other examples of deposits containing tidal cyclicities are presented by Nio and Yang (1989) and Archer (1996a).

Terminology

For both modern and ancient types the term cyclic rhythmites (Greb and Archer, 1995; Archer, 1995) is preferred herein. A number of other terms have been used to describe cyclic rhythmites; the widely used term tidal rhythmites is problematic for two reasons. First, there is an unfortunate mixture of descriptive (rhythmites) and genetic (tidal) terminology. Secondly, the term rhythmite has dichotomous implications and can be interpreted in two ways: (1) as the repetitive, graded laminae, or (2) as the periodic thickness variations of groups of laminae. Another somewhat older but similar term, rhythmic tidalites, has sim-

Tidalites: Processes and Products, SEPM Special Publication No. 61

ilar problems. Nonetheless, all these terms (cyclic rhythmites, tidal rhythmites, and rhythmic tidalites) are essentially synonymous.

PRINCIPAL TIDAL PERIODICITIES

Neap-spring Cycles

Neap-spring periods are the characteristic cycles that define cyclic rhythmites. The cycles can be related directly to semimonthly lunar periods. Modeling of cyclic rhythmites, based on a range of modern macrotidal diurnal to semidiurnal tidal-station data, indicates that various combinations of flood- or ebb-related flow can produce such rhythmites (Archer, 1995). Archer et al. (1991), Kvale et al. (1995), and Archer (1996a) have discussed a number of other periods from rhythmites; however, these will be discussed under the topic of astronomical alignments.

Origin of Laminae

Cyclic rhythmites exhibit a variety of laminae types and there is a spectrum of intergradational forms. Discussion of two contrasting endpoints, however, can account for much of the variability. Archer et al. (1994a) defined these endpoints as silty and heterolithic rhythmites. Silty rhythmites were formed from fine-sand to silt-rich muds and consist of planar, normally graded laminae or thin beds. The upper parts of the silty laminae are gradational to thin, laterally persistent, muddy drapes. Within centimeter-scale silty rhythmites, incipient to well-developed climbing-ripple lamination can occur at the base. In very thick ($\geq$10 cm) silty rhythmite beds, a zone of crude lamination and water-escape structures occurs below the zone of climbing-ripple lamination (Lanier et al., 1993). These structures indicate that very rapid sedimentation, dominated by suspension settling, produced these silty rhythmites. The presence of upright trees, encased within silty rhythmites, further confirms the occurrence of rapid rates of deposition (Kvale et al., 1989; Kuecher et al., 1990; Kvale and Archer, 1991; Lanier et al., 1993).

The other endpoint of rhythmite types, heterolithic rhythmites, consists of finely interlaminated sand-mud layers. These exhibit a variety of tidal bedding including lenticular, wavy, and flaser bedding (Reineck and Wunderlich, 1968). In addition, there are muddy rhythmites with very thin (a few sand-grain diameters) sand streaks (termed pinstripe bedding by Klein, 1977). These forms of sand-mud rhythmites can be defined as heterolithic rhythmites.

Multi-element Rhythmites

Particularly within silty rhythmites, some examples exhibit prominent thick-thin pairing and more complex relationships of lithologically identical laminae (Fig. 2). It is important not to conceptually confuse such a pairing with a sand-mud couplet, such as produced during current flow followed by a stillstand (see Reineck and Wunderlich, 1968). Archer et al. (1995) referred to this thick-thin pairing as a two-element rhythmite. The pairing can be formed by: (1) subordinate-dominant relationships in reversing flows, or (2) systems that have a higher- and a lower-high tide (a diurnal inequality, see de Boer, 1989). In the first case, higher velocities during flood tides (the so-called dominant in this case) may result in significantly thicker laminae than those produced during lower-velocity ebb tides (the subordinate). Of course, the converse is possible, that is, that the ebb flow could have higher current speeds than the flood tides. Regardless of whether flood or ebb is dominant, this type of reversing-flow deposition could result in two-element rhythmites.

Despite the possible scenario above, details of the depositional fabric of Carboniferous two-element rhythmites described here indicates that they occurred primarily during the dominant flow with very little to no deposition during the subordinate flow (Archer et al., 1995). The principal evidence for this conclusion is that there is not a linear relationship between the thicknesses of the thicker laminae as compared to the thinner laminae. Thus, the thick-thin pairing appears related to a significant diurnal inequality within the tidal system (de Boer et al., 1989). This inequality refers to a difference of heights of successive high tides and is controlled by changes in lunar declination (the tropical periods, see Kvale et al., 1989; Archer et al., 1991; Kvale et al., 1995). Various statistical and mathematical techniques can be applied to rhythmite-thickness series in order to test for the presence of such a diurnal inequality. These will not be summarized here, but the interested reader is referred to de Boer et al. (1989), Nio and Yang (1989), and Archer et al. (1995).

These two-element rhythmites appear to have formed within flood-dominant, inland tidal systems and more complex, three- and four-element rhythmites have also been reported (Archer et al., 1995). The four-element forms (Fig. 2) have recorded both flood and ebb deposition for the higher and lower high tides in a paleotidal system that had a strong diurnal inequality. The three-element forms (Fig. 2) are similar, but ebb deposition during the lower high tide did not produce appreciable sedimentation.

ANALYSES AND MODELING OF CYCLIC RHYTHMITES

Mathematical Analysis

Tidal heights can be very closely approximated by harmonic series; this is the basis of modern tidal predictions and thus is also the basis for analyses and modeling of rhythmites. Tides can be decomposed into harmonic components with varying amplitudes, periods, and phases. In a similar manner, thickness series of rhythmites can be decomposed using Fourier or similar analyses. Use of such mathematical tools can aid in the reconstruction of paleotidal systems from data obtained from cyclic rhythmites.

Tidal systems are intrinsically cyclic and the parameters of such cycles are well understood. Thus, it is particularly easy to analyze the formation of and model the generation of cyclic rhythmites. Unlike many sedimentary systems, which contain considerable stochastic components, the modeling of cyclic rhythmites is quite deterministic. This means that direct numerical, rather than probabilistic, analyses can be implemented.

Sequences of laminae-thickness data although not actual time series, are amenable to scrutiny using standard and commonly used time-series techniques. The simplest approach is to plot rhythmite-laminae thicknesses as a bar chart. When depicted in this manner, rhythmites with well-formed neap-spring cycles exhibit obvious sinusoidal trends. Another relatively simple

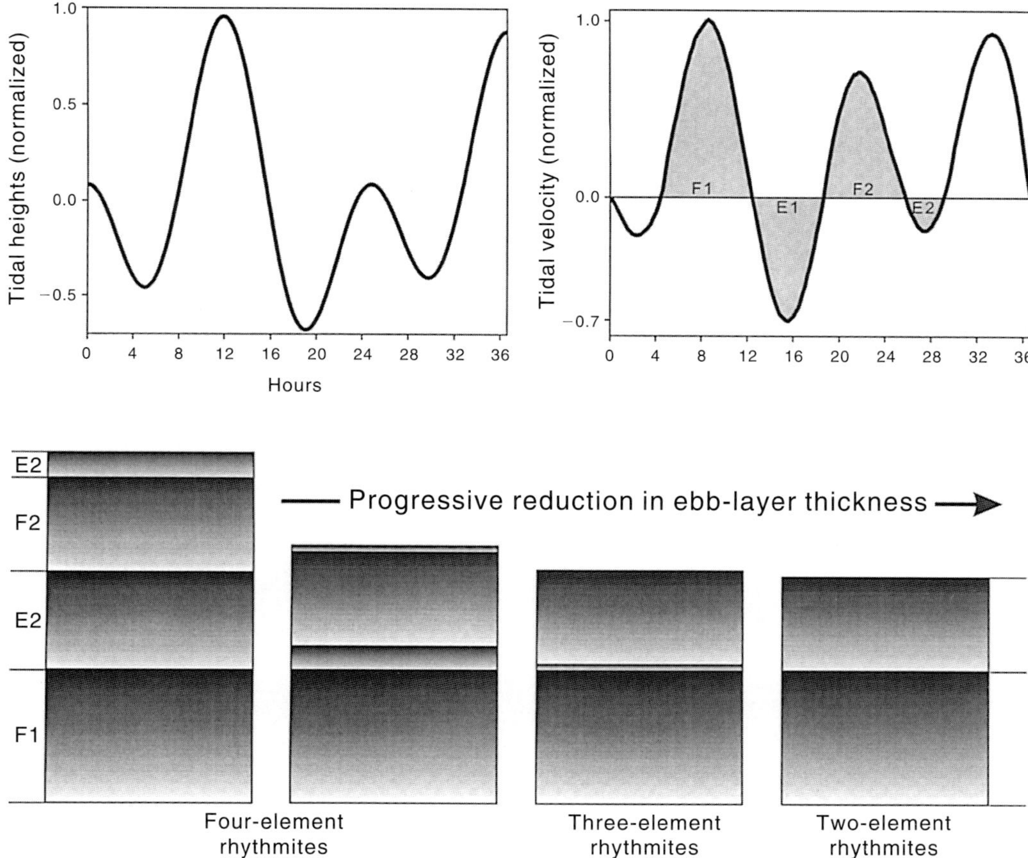

FIG. 2.—Production of multi-element rhythmites within a semidiurnal tidal regime. Complete preservation of all flood and ebb events results in four-element rhythmites, whereas progressive reduction in amount of sedimentation during ebb phases results in three- and ultimately two-element rhythmites.

technique involves autocorrelation, in which the thickness series is compared directly to itself with a progressive degree of offsets. Martino and Sanderson (1993) used this approach with some success. Rhythmites can contain a variety of closely overlapping periodicities, however, and autocorrelation is not readily capable of resolving this finer-scale variability.

More sophisticated analyses can be made using various techniques of spectral estimation, which extracts the periodicities and amplitudes of wave-like variation within the data. Press et al. (1988) discuss maximum-entropy methods (MEM) and fast-Fourier transforms (FFT) that can resolve overlapping periodicities. Archer (1996a) used these methods to extract meaningful periodicities from relatively short sequences of cyclic rhythmite data.

Modeling

Because of the deterministic nature of tidal cyclicities, it is relatively easy to model the potential for cyclic rhythmite development. Earlier attempts at modeling relied on predicted high- and low-water information, such as that available from the National Oceanic and Atmospheric Administration (1990) (Fig. 3A). Such tidal-height information can be used to directly simulate the relative amount of sediment accumulation that would potentially be deposited during high tides. The simplest approach is merely to proportionally equate sedimentation to tidal height, or

$$S_i \approx h_i$$

where S is the thickness of an individual lamina and h is the height of a high tide. Archer (1991) and Archer and Clark (1992) applied this function over a series of high tides, $h_i, \ldots, h_n$, and generated a simple approximation of a cyclic rhythmite (Fig. 3C). A slightly more sophisticated approach is to use the changes in tidal height. For example, to model rhythmite development by flood tides, the difference between the preceding low water to the following high water, or

$$S_i \approx h_i - h_{i-1}$$

This approach results in more clearly defined cycles because it accentuates the thicknesses during spring tides and reduces the thicknesses during neap tides (Fig. 3D). This technique was used to model rhythmites in Archer et al. (1991).

Modern depositional settings capable of producing cyclic rhythmites, and inland tidal systems in general, tend to have linear flow that is parallel to the channel axis. Thus, the flow is generally directed into or out of the depositional system and the complexities of circumrotary tidal systems are eliminated. Linear flow allows a close approximation to current speed by taking the change in water levels over time (Fig. 3B). Furthermore, the cube of these current speeds allows a direct approximation to sedimentation, thus

$$S_i \approx \left(\frac{h_i - h_{i-1}}{t_i - t_{i-1}}\right)^3$$

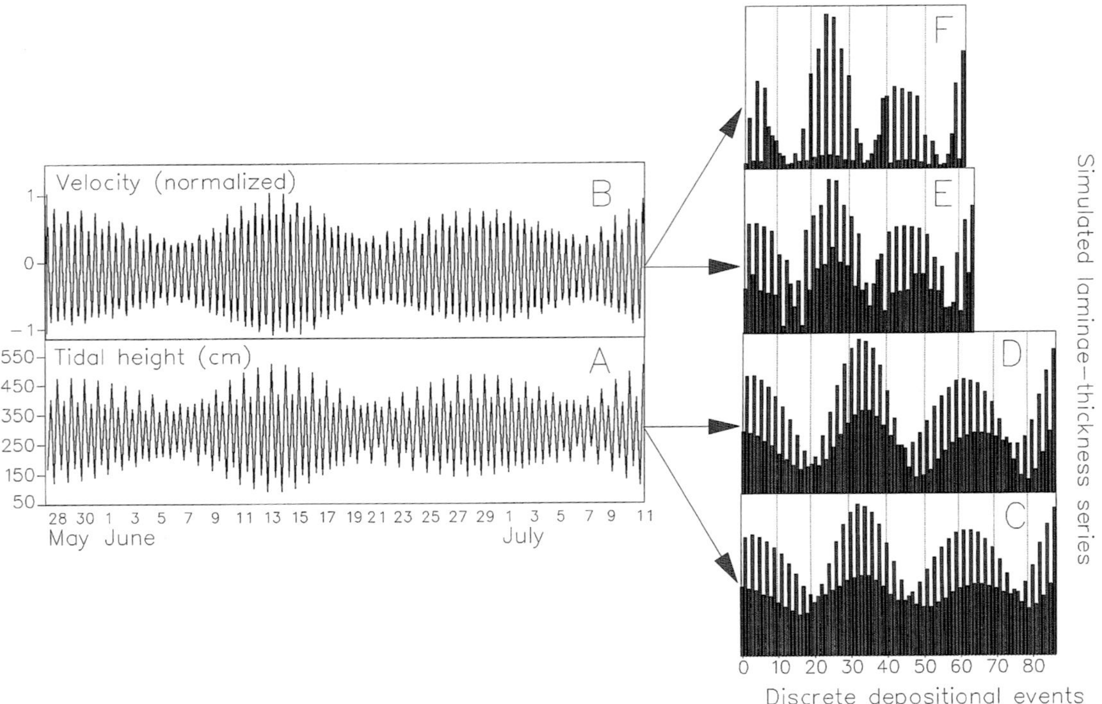

FIG. 3.—Various aspects of rhythmite modeling based on: (A) predictive tidal-height data and (B) conversion of data to simulated velocities. Data is based on predictive high-low tidal predictions (NOAA, 1990) for Ch'ang Chiang Approach, China (see Archer, 1995). Rhythmite series can be modeling based upon: (C) height of tides, (D) changes in tidal heights, (E) relative sedimentation related to velocity, or (F) relative sedimentation related to relative position within the intertidal zone.

where t refers to time of the specific tidal height h. This type of modeling can be combined with a velocity threshold, that is, sedimentation would not occur until current speeds above a specific threshold were achieved (Fig. 3E). Archer (1995) applied this type of modeling to a range of diurnal to semidiurnal macrotidal systems. Finally, this last approach can be modified by allowing the threshold to be the relative position within the tidal frame. At high intertidal positions, this modeling approach yields simulated rhythmites in which the diurnal inequality results in very prominent thick-thin lamina pairing (Fig. 3F).

One of the fundamental weaknesses of such modeling lies in the assumption that sedimentation is temporally related to current speeds where, in fact, there is commonly some degree of lag between current speeds and turbidity in such systems. Obviously, more sophisticated modeling could be undertaken, however, the four simple approaches described above all yield rhythmite-thickness series that closely correlate to those derived from modern and ancient rhythmites.

HIERARCHY OF CONTROLS ON CYCLIC RHYTHMITES

Of all the various controlling parameters, sufficiently dynamic tidal sedimentation must have occurred or else cyclic rhythmites could not have formed. For production and preservation of cyclic rhythmites, the local sedimentational dynamics (such as current speeds) must have exceeded thresholds for deposition but simultaneously must have been below erosive levels.

These constraints need to have been maintained not only throughout individual flood-ebb cycles, but also throughout most of the neap-spring cycles. Thus, nearly continuous sedimentation and extended non-erosive periods were required to deposit and preserve cyclic rhythmites.

The following section discusses the various controls, ranging from smaller- to larger-scale. This ordering is from the perspective of the actual site of deposition. Thus, small scale refers to localized factors. We will proceed from these up to regional and subsequently to global-scale phenomena and controls. This hierarchy of controls include: (1) depositional setting, (2) effects of tidal range, (3) sequence stratigraphic controls, (4) paleoclimatic effects, (5) paleoceanographic controls, and (6) effects of astronomical alignments. Although there is some overlap within some of these categories, this ordering can provide a useful perspective for discussion of the various levels of controlling factors.

Depositional Setting

One of the most important local factors for shallow-water rhythmites is the relative placement of the depositional system within the tidal zonation. Sites in the uppermost intertidal zone will only be flooded by the highest spring tides, thus deposition of cyclic rhythmites cannot occur at such high levels. At lower levels, more complete cycles can be formed. In this context, Dalrymple et al. (1991) and Tessier (1993) discuss modern rhythmites and Archer (1994) discusses ancient counterparts. In essence, the number of events within neap-spring cycles becomes progressively reduced, or truncated, at depositional sites located progressively higher in the tidal frame (Fig. 4). Such constraints, however, do not directly affect subtidal deposits unless they are directly coupled with sediment production in the intertidal zone.

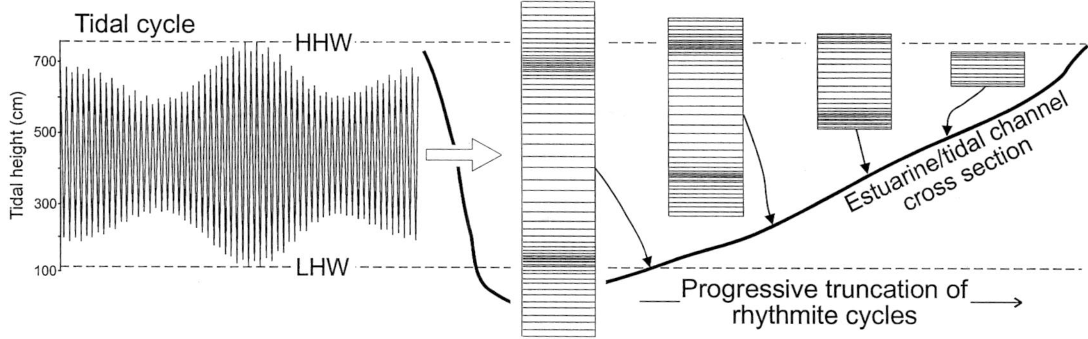

FIG. 4.—Truncation, or loss of discrete laminae, as related to relative position within the intertidal zone. At relatively higher levels, progressively fewer tidal events result in the production of a recognizable lamina. HHW, high high water; LHW, low high water.

Another important aspect of the depositional setting is that sufficient accommodation space is needed to receive the rapidly forming cyclic rhythmites. For intertidal-zone rhythmites, the relative amount of available space relates to the tidal range; thus macrotidal systems have significantly more depositional space available than microtidal systems. For intertidal to shallow subtidal settings, channels commonly provide accommodation space whereby either accreting bars are developed within migrating channels or abandoned channels are filled with rhythmites (see Greb and Archer, 1995).

Cyclic rhythmites can only occur within settings in which tidal sedimentation totally dominates the depositional regime. Other depositional processes, such as storms, waves, and fluvial influx, must either be eliminated or so overwhelmed by the tidal controls so as to have insignificant effects on sedimentation. A great number of Carboniferous rhythmites appear to have formed within significantly inland, tidally influenced settings. This included freshwater to brackish, macrotidal estuarine settings that were formed within bedrock-confined valleys, or incised valley fills (IVFs) (Archer et al., 1994a, b), this topic is discussed further under sequence stratigraphy. Regarding modern analogs, the depositional settings were apparently similar to fluvio-tidal bars within modern macrotidal estuaries (Dalrymple et al. 1991, 1992; Tessier et al., 1995; Johnson et al., 1996). Some cyclic rhythmites exhibit yearly cycles that are probably related to seasonal variations in fluvial sediment input (Kvale et al., 1994; Archer et al., 1994b).

Another common depositional setting for Carboniferous rhythmites includes tidal channels. Brown et al. (1990) report occurrences of cyclic rhythmites in a carbonate-dominated system. Similar occurrences in clastic-dominated coastal settings have been reported (Kvale and Archer, 1990; Greb and Archer, 1995), and Roep (1991) presents a particularly well documented modern analog for such channel-constrained rhythmites.

Smith et al. (1990) describe subtidal, deeper-water cyclic rhythmites produced via a so-called tidal drawdown model. In such a setting the high sediment concentrations, related to emergence and submergence of intertidal flats, can result in small-scale, density-driven underflows. Such a model could explain the occurrence of ancient examples of apparently deeper-water cyclic rhythmites such as those described by Williams (1991).

Tidal Range

For well-formed and relatively complete cycles to be produced, sedimentation must have been dynamic over a significant portion of the neap-spring cycle. Some paleotidal systems would have only resulted in laminae production during a few of the most energetic spring tides. Only during such periods would there have been sufficient sediment flux to produce recognizable laminae. Greb and Archer (1995) discuss that this type of system would be capable only of the production of noncyclic rhythmites.

It is apparent that high sediment flux during each tidal cycle is a basic requirement needed for the deposition of cyclic rhythmites. To date, the most completely described modern analogs occur in settings with elevated tidal ranges. Macrotidal examples include the Bay of Mont Saint Michel in northwestern France. Cyclic rhythmites occur within fluvio-estuarine zones as well as in the non-estuarine inner portions of the bay that are protected from wave and storm energies (Tessier et al., 1989, 1995; Tessier, 1993). Much of the coarser sediment in the rhythmites is biogenic and consists of silt- and sand-sized, abraded particles of shells. Thus, Mont Saint Michel provides an intriguing analog for the various carbonate-rich cyclic rhythmites of the Carboniferous (Brown et al., 1990; Archer and Feldman, 1994).

For siliciclastic-dominated systems, cyclic rhythmites have been documented to be forming in the upper reaches of the macrotidal Bay of Fundy, Nova Scotia, Canada (Dalrymple and Makino, 1989; Dalrymple et al., 1991). These rhythmites appear to provide a useful analog for many of the intertidal, Carboniferous rhythmites including the silty and heterolithic forms.

Within the intertidal, macrotidal, modern analogs discussed above, a number of factors serve to greatly reduce or largely eliminate any significant degree of biogenic reworking of the cyclic rhythmites. Such factors include: (1) very high rates of sediment flux that are detrimental to filter-feeding activities, (2) rapid and extreme salinity fluctuations, which can range from nearly normal marine through brackish, and even freshwater, and (3) extended periods of exposure during low tides.

Within mesotidal systems, rhythmites with well-formed neap-spring cycles have also been documented (Roep, 1991). Such examples appear to be more channel confined, and in such settings, the sediment flux approaches levels that are more extensively developed within macrotidal systems. Given the relatively recent recognition of cyclic rhythmites, however, much more work is needed in a variety of modern analogs before we will completely understand the range of conditions in which neap-spring cycles can be preserved.

Sequence Stratigraphic Controls

The various cyclic rhythmites discussed herein appear to have formed within portions of the lowstand (LST) and transgressive systems tracts (TST). During the development of these systems tracts specific types of tidally dominated environments were developed that involved localized, very high rates of sedimentation. Kvale and Archer (1990, 1991) and Archer and Kvale (1993) presented many examples of cyclic rhythmites that were closely associated with coal beds. Some of the more important settings within the LST include lowstand deltaic systems and incised valley fills (IVFs). In some settings, the upper parts of the IVFs might involve TST lithofacies, however IVFs appear to be a useful model for much of the rhythmite-bearing sections discussed herein (Archer et al., 1994b; Feldman et al., 1995) (Fig. 5). Cyclic rhythmites appear to be particularly related to IVFs and related estuarine systems. Greb and Archer (1995) and Archer and Greb (1995) report that such relationships are common in the Central Appalachian Basin. Kvale and Barnhill (1994) report similar occurrences in the Eastern Interior coal basin.

Conversely, a tidally influenced, deltaic system appears to explain certain, primarily subtidal rhythmite occurrences (see Archer and Clark, 1992; Feldman et al., 1992, 1993). This component of cyclic rhythmite deposition is less well understood, however, the high sediment flux required within such settings can probably be explained in part by the tidal drawdown model of Smith et al. (1990).

Basinal Controls

Structural aspects of the depositional setting exerted a number of important controls, primarily in regard to basin geometry and water depths. Development of a specific geometry would have enhanced the potential for a significant degree of tidal amplification within the most inland parts of the basin. These effects are possible both within foreland and cratonic (epeirogenic) basins. In the eastern U.S., continental collision and uplift of the Appalachian orogen created the large, trough-like Appalachian Basin. This basin and related cratonic basins created a seaway that extended along much of the eastern half of North America. During times of cyclic rhythmite development, this seaway appears to have been closed to the north but open to the southwest, where it opened ultimately into the global paleo ocean. It is generally difficult to estimate maximum water depths for much of the stratigraphic section. Nonetheless, the cyclic rhythmites indicate a strong degree of tidal resonance, and this would have required sufficient depths to allow a progressive wave to oscillate and resonate with the open oceanic tides.

In a similar manner, a significant degree of resonance must have occurred within the cratonic basins of the Eastern Interior and Western Interior Basins. For interior basins, such as the Illinois Basin, it is important to separate the geometry of the overall filling sequence from what must have been the paleobathymetry that existed at a specific deposition interval. In other words, the structural and sedimentological aspects may be considerably different; although the thickest basin fill occurs in southern Illinois, the basin did not experience centripetal drainage during deposition. In fact, paleocurrent analyses indicate a generally southerly trend for much of the history of basin filling. Thus, the depositional basin opened seaward to the southwest. In addition, based on the pervasive occurrence of Carboniferous rhythmites, the basinal shorelines must have had a flaring, or at least linear geometry, in this direction. Similar to the Central Appalachian Basin discussed above, basinal depths must have been sufficient to allow oscillation and resonance. Based upon numerous rhythmite records, tidal amplification was particularly well developed even into the most inland reaches in west-central Indiana.

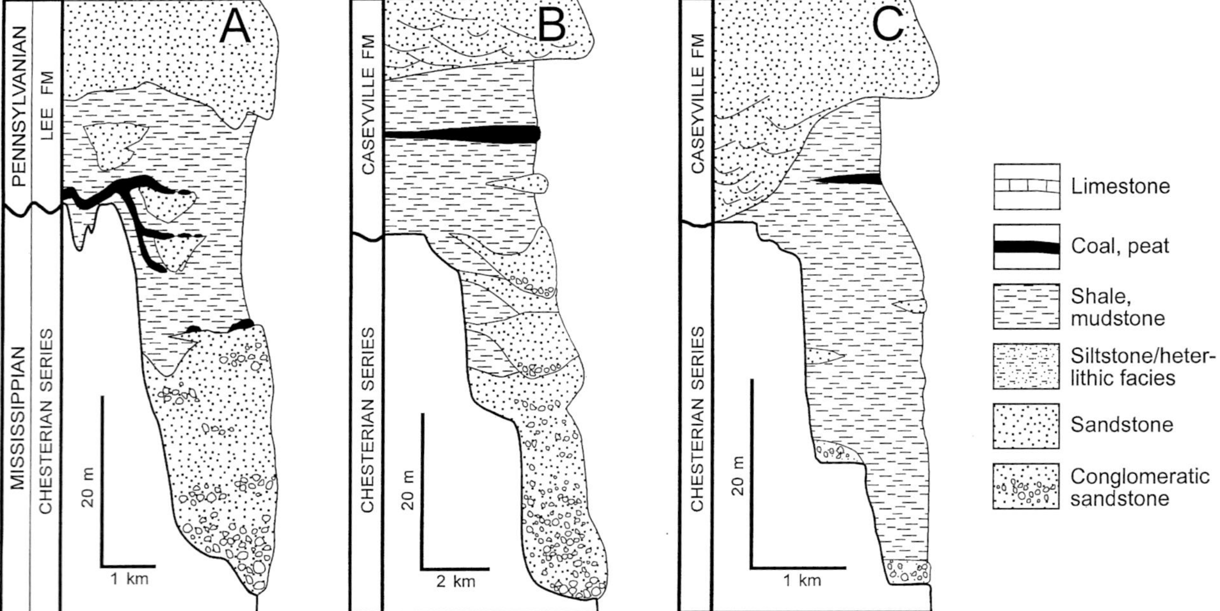

FIG. 5.—Vertical sequences in Pennsylvanian incised valley fills (IVFs): (A) Livingston Conglomerate in the Central Appalachian Basin; (B) Rochester IVF in Eastern Interior Basin (Illinois Basin); and Madisonville IVF in the Eastern Interior Basin (Illinois Basin) (adapted from Archer and Greb, 1995).

In the Western Interior coal basin, which principally includes parts of Kansas, Missouri, and Oklahoma, similar resonance must have occurred. As compared to the more eastern basins, the rhythmites in this basin do not commonly exhibit a semidiurnal signal; thus, two-element rhythmites are relatively rare. Although work is ongoing, this suggests that this basinal paleotide exhibited more of a diurnal signature. If this preliminary conclusion holds up as more data are acquired, it would indicate that diurnal resonances must have predominated. This is potentially important because the period of such resonances, having periods twice as long as semidiurnal, can be developed in systems of approximately twice the length as those required for semidiurnal resonance.

Paleoclimatic Effects

Regional.—

Many of the Carboniferous cyclic rhythmites were formed in near-equatorial settings and within a paleoclimatic regime that was presumably dominated by a strong degree of chemical weathering. Comparison of latitudinal controls on sediment types along modern coasts would suggest that the Carboniferous coastal settings were very muddy (Fig. 6). In addition to their low-paleolatitude position, many of the Carboniferous rhythmites were apparently formed in near-coastal (paralic) settings. Thus, maritime influences would further enhance the wetness of the paleoclimates. Presence of wet paleoclimates is further substantiated by the common occurrence of cyclic rhythmites that directly overlie coals (Kvale and Archer, 1990; Archer et al., 1994a).

Other paleoclimatic indicators are related to fluvial sizes and modes of deposition. The large-scale nature of the fluvial systems associated with the incised-valley systems and rhythmites also indicates that wet paleoclimates predominated during the Carboniferous. Based upon estimates of paleodrainage-basin area and comparisons to modern tropical rivers, the potential for Amazon-scale rivers existed particularly in the eastern U.S. during the Upper Carboniferous (Archer and Greb, 1995). At times of sea-level lowstands, these rivers deeply incised an integrated drainage network into older, underlying strata. During subsequent transgressions, these incised valleys were flooded to become estuaries.

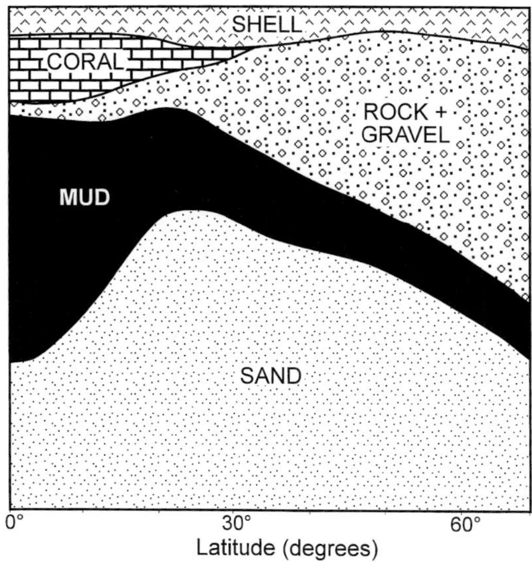

FIG. 6.—Relative proportions of various types of coastal and inner shelfal sediments as related to latitude. At the low latitudes inferred for many Carboniferous systems, mud-dominant facies would have been prominent (adapted from Archer and Feldman, 1995, originally from Hayes, 1967).

Global.—

During the late Paleozoic, and particularly throughout much of the Carboniferous, large-scale continental glaciations were occurring in the southern hemisphere. These Gondwanan glaciations apparently involved considerable ice volumes. Thus, episodes of glaciation and deglaciation had major influences on global sea level. This potential glacio-eustatic flux is not easy to estimate, however, approximations can be made based upon relative area of ice coverage. This leads to estimates of 45–190 m of glacio-eustatic flux, depending upon various ice sheet estimates (Crowley and Baum, 1991). Although estimates based on area might be too high, Archer et al. (1994b) present data from incised-valley sequences in Kansas (U.S.) indicating that eustatic flux ranged from at least 30–40 m.

These glacial cycles may have been controlled by paleo-Milankovitch orbital parameters, and Heckel (1986) suggested cycle lengths to have been in the ≈100 ka range. Within the Eastern and Western Interior Basins, the Carboniferous stratigraphic succession is notably cyclic (the so-called cyclothems). These cycles appear to relate more-or-less directly to the ongoing glacio-eustatic oscillations (see Heckel, 1986). Of particular interest here is the potential for repetitive development of incised valleys and subsequent filling episodes. This potential was apparently quite high during formation of the Carboniferous rhythmites and, as discussed in other sections, accentuated the development of tidally dominated, high-tidal range depositional environments.

Paleoceanographic Constraints

During the Carboniferous, and continuing through the end of the Paleozoic, ongoing continental accretion resulted in the formation of the supercontinent, Pangea. As the continents combined, a very large global paleocean, Panthalassa, was formed. Because Panthalassa was a unified body of water it could have resonated more freely with the various tractive forces of the moon and sun. Related to this potential for unconstrained resonance, modeling of Panthalassa tides suggests that open-ocean tides may have been 50% higher than modern equivalents. Tidal amplification upon the continental shelf and within inland settings would have created widespread macrotidal systems along the coasts of Pangea, especially those coasts directly adjacent to Panthalassa. Tidal amplification would have been particularly prevalent within structural embayments, such as the Appalachian and Interior coal basins of the eastern U.S., that had essentially direct connections to the open Pangean coast.

In addition to higher-than-normal tides, the extent of Panthalassa could have resulted in nonastronomical, basin-specific resonances. Unusual periodicities in rhythmites from northeastern Kansas have been attributed to intrabasinal resonance, or seiche, of the global paleocean (Fig. 7). Because of the large diameter of Panthalassa, this seiche had a period of about 50 hours (Archer, 1996b).

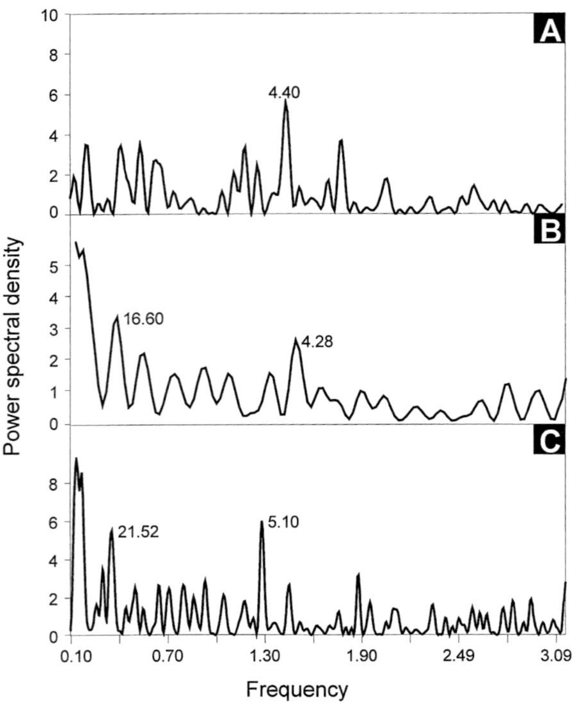

FIG. 7.—Harmonic analyses of rhythmites of the Douglas Group (latest Carboniferous) of Kansas (U.S.). Shown are periodicities within: (A) large-scale cross-bedding of the predominantly fluvial facies, (B) large-scale planar foresets of the fluvio-estuarine transitional facies, and (C) small-scale foresets from barforms in the estuarine facies. Neap-spring cycles are lacking in A, but occur in B and C. Nontidal harmonics, ranging from four to five bedforms per cycle, occur in all facies and have been interpreted as representative of a paleoceanographic seiche (Archer, 1996b).

Despite the relatively common occurrence of Carboniferous cyclic rhythmites in the eastern U.S., they are far less commonly reported in Europe (for exceptions, see Read, 1992; Aitkenhead and Riley, 1996). This may merely be related to sampling and as the recognition of such facies becomes more commonplace, many examples may be reported. Conversely, it may relate to paleoceanographic controls. Some paleogeographic reconstructions, which admittedly are tentative, indicate a smaller, isolated ocean on the eastern side of Pangea. This ocean, termed the Eastern Ocean (Archer, 1996b), was much smaller and relatively isolated from Panthalassa (Fig. 8).

Astronomical Alignments and Tidal Effects

Ultimately, the tidal periods that generate cyclic rhythmites need to be related to extraterrestrial forces. These forces can be principally related to the gravitational effects of the moon and sun on the earth's oceans. Alignment of the moon and sun result in higher, or spring, tides. The maximal heights of these spring tides commonly vary between one series of springs to the next series of spring tides, which occur approximately two weeks later. This variability has been termed a fortnightly inequality and is related to temporal convergence of perigee (closest approach of the moon to the earth) with new or full moon (Archer, 1996a). Many ancient rhythmite series exhibit very well developed fortnightly inequalities and indicate that deposition occurred during perigean-spring tides (Tessier et al., 1989; Archer, 1996a). Exceptionally close alignments of perigee and new or full moon occur at a variety of longer-term cycles, which include 8-, 18-, and 1600-year cycles (see Nio and Yang, 1989; Archer, 1996a) (Fig. 9). Thus, many rhythmites may have been

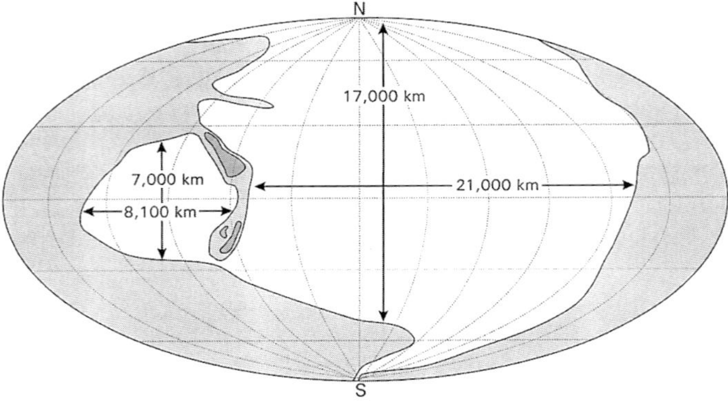

FIG. 8.—Pangean paleogeography for the Late Carboniferous (306 Ma) adapted from Scotese and Golonka (1992). This depiction has been longitudinally rotated 180° from the original to emphasize the lateral extent of the global paleocean Panthalassa.

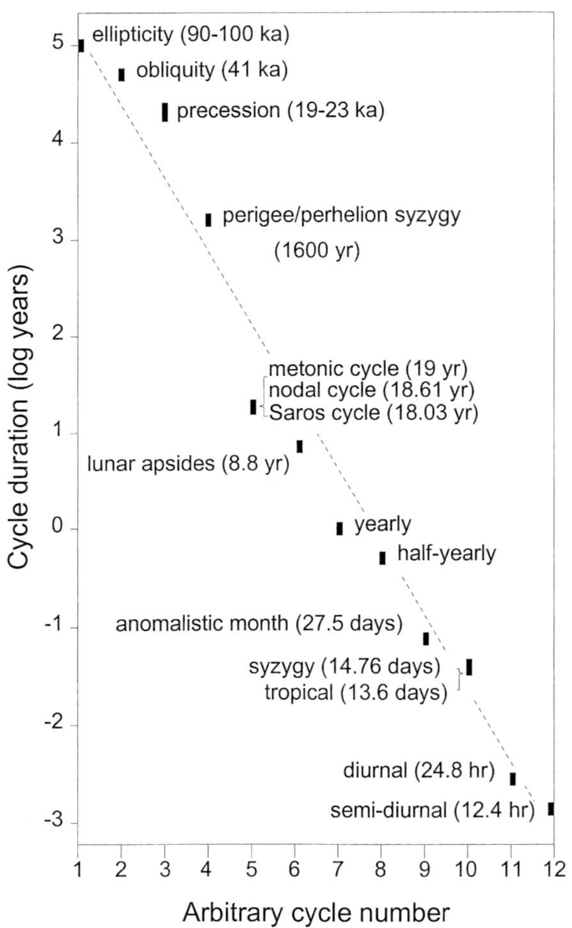

FIG. 9.—Range of periodicities expressed by earth-moon-sun orbital system include longer-term Milankovitch-scale parameters. The periods more relevant to tidal processes are in a variety of 18–19-year periods. Cyclic rhythmites can commonly express annual down through diurnal and semidiurnal periods.

deposited during periods of exceptionally high, astronomical tides that relate to such alignments, because during such periods, the higher than normal tides would have had an increased potential for rapid sedimentation and rhythmite production.

DISCUSSION

Of the various factors that affected cyclic rhythmite deposition during the Carboniferous, it is relatively apparent that macrotidal conditions appeared to dominate in many areas. This elevated tidal range was directly related to several factors described above; their combined effect produced high tidal ranges and dynamic tidal sedimentation at the various sites of deposition. The configuration of the global paleocean, Panthalassa, apparently led to a strong resonance with the astronomical tide-raising forces. The resultant higher, open-oceanic tides, as suggested by modeling, produced common macrotidal conditions along the coasts of Pangea, or at least along those coasts along the margins of Panthalassa.

Embayed coastlines, which were the byproduct of tectonism and continental convergence, led to increased coastal and inland tidal amplification. At various times during the Carboniferous, both foreland and intracratonic embayments must have evolved very effective resonant geometries. This would have included portions of the Central Appalachian, Eastern Interior, and the Western Interior Basins.

Paleoclimates played a considerable role; there were dramatic and relatively short-term ($\approx$100 ka) sea-level fluctuations related to Gondwanan glaciations. During the lowstands, large, wet-seasonal to tropical rivers incised large-scale river valleys. These incised valleys served as sediment sinks during ensuing transgressions. Valley-fill sequences provided the accommodation and preservation space for deposition of the macrotidal rhythmites. Deposition of rhythmites within incised valleys, within abandoned fluvio-estuarine channels or tidal channels, and deposition overlying compacting peats all increased the preservation potential for the cyclic rhythmites.

Finally, specific alignments of astronomical tide-producing forces may have occasionally increased the tidal heights. For example, the occurrence of apparently perigean spring tides in cyclic rhythmites indicates that this could be an important, secondary factor in the commonality of Carboniferous rhythmites.

CONCLUSIONS

Because of the various controls cited above, cyclic rhythmites are particularly common in basins within the eastern and mid-continental U.S. This commonality is related to a number of overlapping factors, which at a larger scale relate to the unique global paleogeographic and paleoceanographic conditions of the late Paleozoic. On a smaller scale, wet tropical conditions, large-scale fluvial and estuarine systems, and specific depositional environments enhanced the production and preservation of cyclic rhythmites. Ongoing glacio-eustacy, and its controls on sequence stratigraphy, also played an important role. The convergence of these many factors led to a tremendous potential for rhythmite deposition, which may have been relatively unique in earth history.

ACKNOWLEDGMENTS

Components of research on Carboniferous depositional settings were funded by the National Science Foundation; this has included work on Lagerstätten and controls on fossil preservation (EAR-9018079) and coring in the Douglas Group, Kansas, and comparisons to modern analogs (EAR-9405123). Additional work on the Douglas Group was funded by the Kansas Department of Transportation (Grant K-TRAN: KSU-91-6). Partial field expenses were provided by the Kansas, Indiana, and Kentucky state geological surveys. Various grants from agencies at Kansas State University funded specific activities related to this research—particularly the Center for Research in Computer-Controlled Automation (CRCCA Project 90E041), Faculty Development Awards, and Bureau of Grants and Research.

REFERENCES

Allen, J. R. L., 1981, Lower Cretaceous tides revealed by crossbed sets with mud drapes: Sedimentology, v. 289, p. 579–581.

Aitkenhead, N., and Riley, N. J., 1996, Kinderscoutian and Marsdenian successions in the Bradup and Hag Farm boreholes, near Ilkley, west Yorkshire: Proceedings of the Yorkshire Geological Society, v. 51, p. 115–125.

Archer, A. W., 1991, Modeling of tidal rhythmites using modern tidal periodicities and implications for short-term sedimentation rates, *in* Franseen, E. K., Watney, W. L. Kendall, C. G. St. C. and Ross, W. C., eds., Sedimentary Modeling: Computer Simulations for Improved Parameter Definition: Kansas Geological Survey Bulletin, v. 233, p. 185–194.

Archer, A. W., 1994, Extraction of sedimentological information via computer-based image analyses of gray shales in Carboniferous coal-bearing sections of Indiana and Kansas, USA: Mathematical Geology, v. 26, p. 47–65.

Archer, A. W., 1995, Modeling of tidal rhythmites based on a range of diurnal to semidiurnal tidal-station data: Marine Geology, v. 123, p. 1–10.

Archer, A. W., 1996a, Reliability of lunar orbital periods extracted from ancient cyclic tidal rhythmites: Earth and Planetary Science Letters, v. 141, p. 1–10.

Archer, A. W., 1996b, Panthalassa: Paleotidal resonance and a global paleocean seiche: Paleoceanography, v. 11, p. 625–632.

Archer, A. W., and Clark, G. R., II, 1992, Depositional environment of the *Dunbarella* beds: an exercise in paleoecology and sediment cyclicity, *in* Zideck, J., ed., Geology and Paleontology of the Kinney Brick Pit Quarry, Late Pennsylvanian, central New Mexico: New Mexico Bureau of Mines and Mineral Resources Bulletin, v. 138, p. 27–36.

Archer, A. W., and Feldman, H. R., 1994, Tidal influence in Carboniferous fine-grained limestones [Rhythmites tidales dans les calcaires caboniferes aux Etats-Unis]: Geobios, M.S., no. 16, p. 283–291.

Archer, A. W., and Feldman, H. R., 1995, Incised valleys and estuarine facies of the Douglas Group (Virgilian): Implications for similar Pennsylvanian sequences in the U.S. Mid-Continent, *in* Hyne, N., ed., Sequence Stratigraphy of the Mid-Continent: Tulsa Geological Society, p. 119–140.

Archer, A. W., Feldman, H. R., Kvale, E. P., and Lanier, W. P., 1994a, Pennsylvanian (Upper Carboniferous) fluvio- to tidal-estuarine coal-bearing systems: Delineation of facies transitions based upon physical and biogenic sedimentary structures: Palaeogeography, Palaeoclimatology, Palaeoccology, v. 106, p. 171–185.

Archer, A. W., and Greb, S. F., 1995, An Amazon-scale drainage system in the early Pennsylvanian of Central North America: Journal of Geology, v. 103, p. 611–628.

Archer, A. W., Kuecher, G. J., and Kvale, E. P., 1995, The role of tidal-velocity asymmetries in the deposition of silty tidal rhythmites (Carboniferous, Eastern Interior Coal Basin): Journal of Sedimentary Research, v. A65, p. 408–416.

Archer, A. W., and Kvale, E. P., 1993, Origin of gray-shale lithofacies ("clastic wedges") in U.S. midcontinental coal measures (Pennsylvanian): an alternative explanation, *in* Cobb, J. C., and Cecil, B., eds., Modern and Ancient Coal-Forming Environments: Boulder, Geological Society of America Special Paper, v. 286, p. 181–192.

Archer, A. W., Kvale, E. P., and Johnson, H. R., 1991, Analysis of modern equatorial tidal periodicities as a test of information encoded in tidal rhythmites, *in* Smith, D. G., Reinson, G. E., Zaitlain, B. A., and Rahmani, R. A., eds., Clastic Tidal Sedimentology: Canadian Society of Petroleum Geologists Memoir, v. 16, p. 189–196.

ARCHER, A. W., KVALE, E. P., AND LUDVIGSON, G. A., 1992, Comparison between tidal sediments (Morrowan) deposited on the eastern and western margins of the Illinois Basin: Geological Society of America, Abstracts with Programs, v. 24, p. 2.

ARCHER, A. W., LANIER, W. P., AND FELDMAN, H. R., 1994b, Fluvio-estuarine transitions within incised paleovalley fills, Douglas Group (Stephanian; Upper Carboniferous) of Kansas, U.S.A, *in* Dalrymple, R. W., Boyd, R., and Zaitlin, B. A., eds., Incised Valley Fill Systems: Tulsa, Society for Sedimentary Geology Special Paper, v. 51, p. 175–190.

ARCHER, A. W., AND MAPLES, C. G., 1984, Pennsylvanian nonmarine trace-fossil assemblages: Southwestern Indiana: Journal of Paleontology, v. 58, p. 448–466.

BROWN, M. A., ARCHER, A. W., AND KVALE, E. P., 1990, Neap-spring tidal cyclicity in laminated carbonate channel-fill deposits and its implications: Salem Limestone (Mississippian), south-central Indiana, U.S.A: Journal of Sedimentary Petrology, v. 60, p. 152–159.

CHAN, M. A., KVALE, E. P., ARCHER, A. W., AND SONETT, C. P., 1994, Oldest direct evidence of lunar-solar tidal forcing encoded in sedimentary rhythmites, Proterozoic Big Cottonwood Formation, central Utah: Geology, v. 22, p. 791–794.

CROWLEY, T. J., AND BAUM, S. K., 1991, Estimating Carboniferous sea-level fluctuations from Gondwanan ice extent: Geology, v. 19, p. 975–977.

DALRYMPLE, R. W., AND MAKINO, Y., 1989, Description and genesis of tidal bedding the Cobequid Bay-Salmon River estuary, Bay of Fundy, Canada, *in* Taira, A., and Masuda, F., eds., Sedimentary Facies of the Active Plate Margin: Tokyo, Terra Scientific, p. 151–177.

DALRYMPLE, R. W., MAKINO, Y., AND ZAITLIN, B. A., 1991, Temporal and spatial patterns of rhythmite deposition on mud flats in the macrotidal Cobequid Bay-Salmon River estuary, Bay of Fundy, Canada, *in* Smith, D. G., Reinson, G. E., Zaitlin, B. A., and Rahmani, R. A., eds., Clastic Tidal Sedimentology: Canadian Society of Petroleum Geologists Memoir, v. 16, p. 137–160.

DALRYMPLE, R. W., ZAITLIN, B. A., AND BOYD R., 1992, A conceptual model of estuarine sedimentation: Journal Sedimentary Petrology, v. 62, p. 1130–1146.

DE BOER, P. L., OOST, A. P., AND VISSER, M. J., 1989, The diurnal inequality of the tide as a parameter for recognizing tidal influence: Journal of Sedimentary Petrology, v. 59, p. 912–921.

FELDMAN, H. R., ARCHER, A. W., KVALE, E. P., CUNNINGHAM, C. R., MAPLES, C. G., AND WEST, R. R., 1993, A tidal model for Carboniferous conservat Lagerstatten formation: Palaois, v. 8, p. 485–498.

FELDMAN, H. R., ARCHER, A. W., MAPLES, C. G., AND WEST, R. R., 1992, The Kinney Brick Company Quarry: Preliminary analysis using an estuarine model, *in* Zideck, J., ed., Geology and Paleontology of the Kinney Brick Pit Quarry, Late Pennsylvanian, central New Mexico: New Mexico Bureau of Mines and Mineral Resources Bulletin, v. 138, p. 21–26.

FELDMAN, H. R., GIBLING, M. R., ARCHER, A. W., WIGHTMAN, W. G., AND LANIER, W. P., 1995, Stratigraphic architecture of the Tonganoxie Paleovalley Fill (Lower Virgilian) in Northeastern Kansas: American Association of Petroleum Geologists Bulletin, v. 79, p. 1019–1043.

GREB, S. F., AND ARCHER, A. W., 1995, Rhythmic sedimentation in a mixed tide and wave deposit, Hazel Patch Sandstone (Pennsylvanian), Eastern Kentucky Coal Field: Journal of Sedimentary Research, v. B65, p. 93–106.

HAYES, M. O., 1967, Relationship between coastal climate and bottom sediment type on the inner continental shelf: Marine Geology, v. 5, p. 111–132.

HECKEL, P. H., 1986, Sea-level curve for Pennsylvanian eustatic marine transgressive-regressive depositional cycles along midcontinent outcrop belt, North America: Geology, v. 14, p. 330–334.

JOHNSON, T. W., GILL, J. D., AND ARCHER, A. W., 1996, Modern analog of cyclical tidal rhythmites, Bay of Fundy, Canada: Kansas Academy of Science, Abstracts, v. 15, p. 27.

KLEIN, G. DE-V., 1977, Clastic Tidal Facies: Champaign, Illinois, Continuing Education, 149 p.

KUECHER, G. J., 1983, Rhythmic sedimentation and stratigraphy of the Middle Pennsylvanian Francis Creek Shale near Braidwood, Illinois: Unpublished M.S. Thesis, Northeastern Illinois University, Chicago, Illinois, 143 p.

KUECHER, G. J., WOODLAND, B. G., AND BROADHURST, F. M., 1990, Evidence of deposition from individual tides and of tidal cycles from the Francis Creek Shale (host rock to the Mazon Creek Biota), Westphalian D (Pennsylvanian), northeastern Illinois: Sedimentary Geology, v. 68, p. 211–221.

KVALE, E. P., AND ARCHER, A. W., 1990, Tidal deposits associated with low-sulfur coals, Brazil Formation (Lower Pennsylvanian), Indiana: Journal of Sedimentary Petrology, v. 60, p. 563–574.

KVALE, E. P., AND ARCHER, A. W., 1991, Characteristics of two Pennsylvanian-age semidiurnal tidal deposits in the Illinois Basin, U.S.A, *in* Smith, D. G., Reinson, G. E., Zaitlain, B. A., and Rahmani, R. A., eds., Clastic Tidal Sedimentology: Canadian Society of Petroleum Geologists Memoir, v. 16, p. 179–188.

KVALE, E. P., ARCHER, A. W., AND JOHNSON, H. R., 1989, Daily, monthly, and yearly tidal cycles within laminated siltstones (Mansfield Formation: Pennsylvanian) of Indiana: Geology, v. 17, p. 365–368.

KVALE, E. P., AND BARNHILL, M. L., 1994, Evolution of lower Pennsylvanian estuarine facies within paleovalleys, Illinois Basin, Indiana, *in* Dalrymple, R. W., Boyd, R., and Zaitlin, B. A., eds., Incised-Valley Systems: Origin and Sequence Stratigraphy: Tulsa, Society for Sedimentary Geology Special Paper, v. 51, p. 191–207.

KVALE, E. P., CUTRIGHT, J., BILODEAU, D., ARCHER, A. W., JOHNSON, H. R., AND PICKETT, B., 1995, Analysis of modern tides and implications for ancient tidalites: Continental Shelf Research, v. 15, p. 1921–1943.

KVALE, E. P., FRASER, G. S., ARCHER, A. W., ZAWISTOSKI, A., KEMP, N., AND MCGOUGH, P., 1994, Evidence of seasonal precipitation in Pennsylvanian sediments of the Illinois Basin: Geology, v. 22, p. 331–334.

LANIER, W. P., FELDMAN, H. R., AND ARCHER, A. W., 1993, Tidally modulated sedimentation in a fluvial to estuarine transition, Douglas Group, Missourian-Virgilian, Kansas: Journal of Sedimentary Petrology, v. 63, p. 860–873.

MAPLES, C. G., AND ARCHER, A. W., 1987, Trace-fossil holotypes from the freshwater Hindostan Whetstone Beds (Late Carboniferous) of Indiana: Journal of Paleontology, v. 61, p. 890–897.

MARTINO, R. L., AND SANDERSON, D. D., 1993, Fourier and autocorrelation analysis of estuarine tidal rhythmites, lower Breathitt Formation (Pennsylvanian), eastern Kentucky, USA: Journal of Sedimentary Petrology, v. 63, p. 105–119.

NIO, S. D., AND YANG, C. S., 1989, Recognition of tidally-influenced facies and environments: Short Course Note Series #01, International Geoservices BV, Reaal 5, 2353 Leiderdorp, The Netherlands.

National Oceanic and Atmospheric Administration, 1990, Tide Tables, 1991, High and Low Water Predictions, Central and Western Pacific Ocean and Indian Ocean: Riverdale, NOAA, 391 p.

PRESS, W. H., FLANNERY, B. P., TEUKOLSKY, S. A., AND VETTERLING, W. T., 1988, Numerical recipes in C, the art of scientific computing: New York, Cambridge University Press, 735 p.

READ, W. A., 1992, Evidence of tidal influences in the Arnsbergian rhythmites in the Kincardine Basin: Scottish Journal of Geology, v. 28, p. 135–142.

REINECK, H.-E., AND WUNDERLICH, F., 1968, Classification and origin of flaser and lenticular bedding: Sedimentology, v. 11, p. 99–104.

ROEP, TH. B., 1991, Neap-spring cycles in a subrecent tidal channel fill (3665 BP) at Schoorldam, NW Netherlands: Sedimentary Geology, v. 71, p. 213–230.

SCOTESE, C. R., AND GOLONKA, J., 1992, PALEOMAP Project, Department of Geology, University of Texas at Arlington, 20 maps.

SMITH, N. D., PHILLIPS, A. C., AND POWELL, R. D., 1990, Tidal drawdown: A mechanism for producing cyclic sediment laminations in glaciomarine deltas: Geology, v. 18, p. 1013–1018.

TESSIER, B., 1993, Upper intertidal rhythmites in the Mont-Saint-Michel Bay (NW France): Perspectives for paleoreconstruction: Marine Geology, v. 110, p. 355–367.

TESSIER, B., ARCHER, A., LANIER, W. P., AND FELDMAN, H. R., 1995, Comparison of ancient tidal rhythmites (Carboniferous of Kansas and Indiana, USA) with modern analogues (the Bay of Mont-Saint-Michel, France): Oxford, International Association of Sedimentologists Special Publication, v. 24, p. 259–271.

TESSIER, B., MONTFORT, Y., GIGOT, P., AND LARSONNEUR, C., 1989, Enregistrement des cycles tidaux en accretion verticale, adaption d'un outil de traitement mathmatique: Examples en baie du Mont-Saint-Michel and dans la molasse marine miocene du bassin de Digne. Bull. Geol. Soc. France, v. 8, no. 5, p. 1029–1041.

VAN DEN BERG, J. H., 1981, Rhythmic seasonal layering in a mesotidal channel fill sequence, Oosterschelde Mouth, the Netherlands: Oxford,International Association Sedimentology Special Publication, v. 5, p. 147–159.

VISSER, M. J., 1980, Neap-spring cycles reflected in Holocene subtidal large-scale bedform deposits: a preliminary note: Geology, v. 8, p. 543–546.

WILLIAMS, G. E., 1991, Upper Proterozoic tidal rhythmites, South Australia: Sedimentary features, deposition, and implications for the earth's paleorotation, *in* Smith, D. G., Reinson, G. E., Zaitlain, B. A., and Rahmani, R. A. eds., Clastic Tidal Sedimentology: Canadian Society of Petroleum Geologists Memoir, v. 16, p. 161–177.

TIDAL CYCLES: ANNUAL VERSUS SEMI-LUNAR RECORDS

BERNADETTE TESSIER
Université de Lille 1, Laboratoire de Sédimentologie et Géodynamique, URA 719 CNRS. 59 655 Villeneuve d'Ascq, France

ABSTRACT: Marsh deposits frequently consist of a succession of packages interpreted to be annual cycle records. Annual sedimentary cycles are usually recognized because of a contrasted seasonal differentiation of sedimentation. In the inner estuary of the Bay of Mont-Saint-Michel, detailed lamination analysis of some marsh packages that are not heavily disturbed by root traces demonstrates that they can be made of planar silty-mud couplets that thicken and thin systematically. In terms of facies and lamina thickness evolution, these sedimentary cycles are similar to neap-spring-neap tidal rhythmites. However, the number of couplets they contain is inconsistent with the number of tides that are able to reach the supratidal domain during a single fortnightly cycle, but it fits the average number of tides that inundate the marshes during one year. The problem with the occurrence of such annual records arises from their possible misinterpretation as neap-spring-neap cycle records in ancient tidal facies. This confusion could lead to erroneous conclusions about sedimentation rate, tidal regime, environmental context, and orbital parameters, conclusions that are usually inferred from the analysis of neap-spring-neap tidal rhythmites.

INTRODUCTION

Tidal rhythmites constitute a common facies in inner estuarine environments although they are known in delta-front settings as well. Present-day estuarine examples have been described in detail in the Cobequid Bay-Salmon River estuary (Dalrymple and Makino, 1989; Dalrymple et al., 1991) and in the Bay of Mont-Saint-Michel Estuary (Tessier, 1993), and many ancient examples interpreted as estuarine facies have been analyzed (Rahmani, 1988; Kvale and Archer, 1989, Tessier and Gigot, 1989; Brown et al., 1990; Kuecher et al., 1990; Kvale and Archer, 1990, Read, 1992; Lanier et al., 1993; Martino and Sanderson, 1993; Archer et al., 1994; Archer and Feldman, 1994; Chan et al., 1994; Deynoux et al., 1993; Kvale et al., 1994; Greb and Archer, 1995). In all these studies, tidal rhythmites are defined as vertically accreted tidal beddings that record semi-diurnal, diurnal and neap-spring-neap cycles (Dalrymple, 1992; Tessier et al., 1995). In rock records, tidal rhythmites are recognized on the basis of two main criteria: 1) the vertical cyclic evolution of bed/lamina thickness related to tidal range evolution, and 2) the number of elementary sedimentary units recorded within each fortnightly cycle record (i.e., 14-day-long neap-spring-neap cycle). Detailed analyses of these tidal rhythmites in terms of facies and cyclicity evolution may provide some useful information estimating sedimentation rates, for reconstructing palaeoenvironment and basin morphology, and in some cases, pointing out variations of orbital parameters through time such as Earth's paleorotation and lunar orbit (Williams, 1989, 1991; Chan et al., 1994; Sonett et al., 1996).

As mentioned previously, the recognition of ancient tidal rhythmites and the interpretation inferred from this recognition are based on the number of sedimentary units contained in each neap-spring-neap cycle record identified. Interpretations consist of comparing this number of beds with the ideal values of 14 or 28 related to the number of tides per neap-spring-neap cycle in a diurnal (one tide per day) or semi-diurnal (two tides per day) system, respectively. However, interpretations are made keeping in mind that 1) these values have changed over geological time and, 2) records may be incomplete with respect to these ideal values. Incomplete records may occur in subtidal environments due to possible nondeposition during low energy neap tides, or incomplete records characterize upper intertidal settings because the highest topographic levels in the tidal frame are not inundated by neap tides. This last remark is especially well illustrated when observing the tidal rhythmites present in the innermost part of the Cobequid Bay-Salmon River estuary and the Mont-Saint-Michel Bay estuary, where only incomplete upper intertidal neap-spring-neap cycle records are recognized (Fig. 1A).

This paper discusses the occurrence of sedimentary cycles (Fig. 1B) that frequently constitute the marsh facies of the inner estuary of the Bay of Mont-Saint-Michel and closely simulate neap-spring-neap tidal rhythmites in terms of facies and lamina thickness evolution. However, the number of beds these cycles contain is inconsistent with the number of tides that are able to reach the marsh elevation during one fortnightly cycle. Therefore, these marsh cycles cannot be interpreted as neap-spring-neap tidal rhythmites. The paper focuses on the description of these sedimentary cycles and discusses possible confusion with neap-spring-neap cycle records and the consequences of such misinterpretation in rock record analysis.

SETTING

The Bay of Mont-Saint-Michel, located in the western English Channel between Brittany and the Cotentin Peninsula (Fig. 2), represents one of the more remarkable examples of a coastal macrotidal environment (Larsonneur, 1989). The tidal range reaches up to 15.3 m, and the tidal regime is semi-diurnal with a slight-diurnal inequality.

In the eastern part of the Bay, two small rivers, the Sée and the Sélune, give rise to a wide estuarine system. These rivers are characterized by a low mean annual discharge, that is, the Sélune: 20 m^3/s; the Sée: 8 m^3/s. Their sediment input is negligible (Larsonneur, 1989).

Morphologically, the evolution of the estuary from its head to its mouth fits quite well the model developed by Dalrymple et al. (1992) in the Cobequid Bay-Salmon River estuary (Bay of Fundy, Canada). The inner part of the system consists of a single channel (the channel of the Sée in the north and the channel of the Sélune in the south) evolving seaward from a straight to meandering and then again straight pattern (Fig. 2). According to Dalrymple et al. (1992), this straight-meandering-straight channel portion represents the fluvio-tidal transition. Following the outer straight portion, the two channels coalesce and give rise to a wide braided domain characterized by a multiple channel-and-shoal system. This braided domain constitutes most of the outer estuary—no longitudinal tidal bars are developed in the mouth system (Lanier and Tessier, this volume).

Along the transitional zone of the Mont-Saint-Michel estuary, upper intertidal and supratidal domains are characterized

Tidalites: Processes and Products, SEPM Special Publication No. 61

FIG. 1.—Sedimentary cycles observable in the inner estuary of Mont-Saint-Michel. (A) Cycles in upper intertidal facies are interpreted as neap-spring-neap cycle records (in Tessier, 1993). n: neap, s: spring. Coin is 2.8 cm. (B) Cycles in marsh facies interpreted as annual cycle records. Pencil is 14 cm. White dashes indicate the limit between successive neap-spring-neap cycles on (A), and annual cycles on (B).

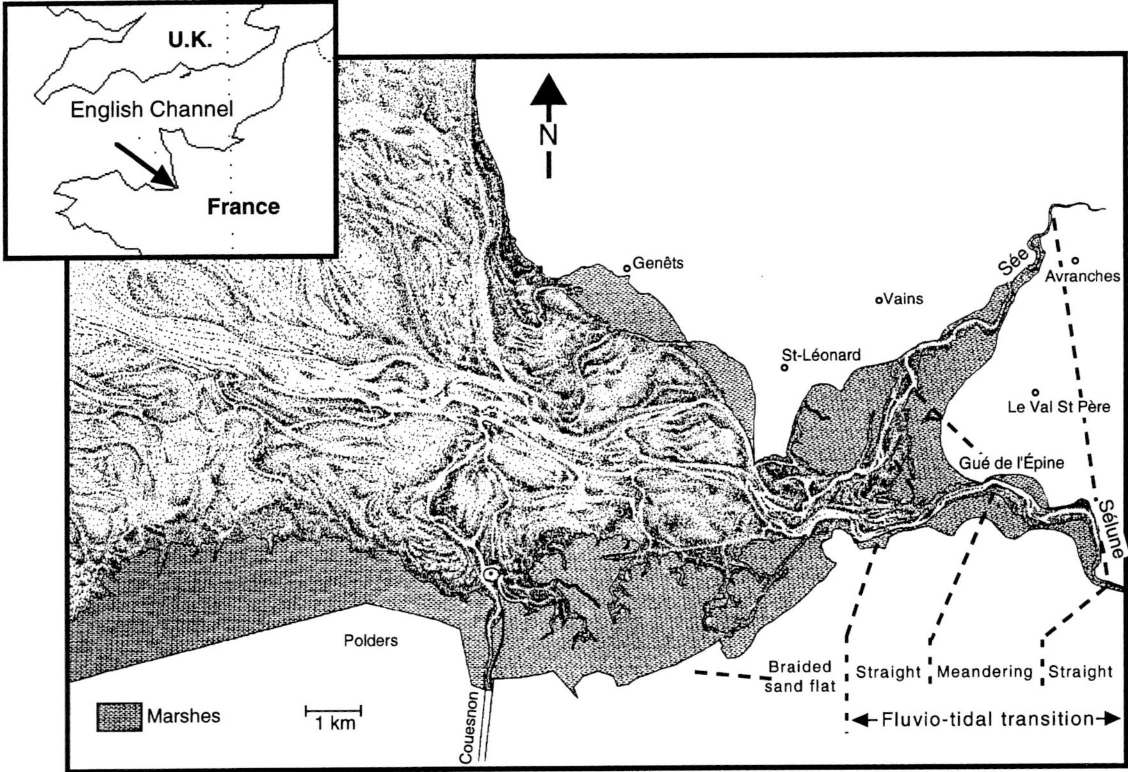

FIG. 2.—Morphosedimentary map of the Mont-Saint-Michel estuary (Larsonneur, 1989, drawn after a satellite image from June 1986). Nomenclature used for channel shape description is after Dalrymple et al., 1992.

by particular sediments locally called tangues. The tangues are composed of fine-grained, silt-dominated sediment that contains as much as 60% of marine carbonates (shell debris).

NEAP-SPRING-NEAP AND ANNUAL CYCLE RECORDS

Laterally to upper intertidal areas, marshes, vegetated mainly by halophite plants such as *Spartina, Salicornia, Sueda* and *Obione,* develop extensively along the transitional zone. Along some erosional rims of channel meanders, clean and thick (3–4 m-high) cross-sections allow the analysis of facies successions from sand-dominated channel and sand flat facies to mud-dominated marsh facies.

The lower sand-dominated part of the succession frequently displays ripple cross-bedding and upper plane beds. Tidal rhythmites with neap-spring-neap cycle records are not recognized in these facies that reflect the high energy of deposition in and close to the channel.

Well-developed neap-spring-neap cycle records are localized in the middle part of the succession (e.g., Fig. 1A) and constitute the most characteristic upper intertidal flat facies of the transitional zone. They are composed of couplets that range from a few millimeters to a few centimeters in thickness. Each couplet is made of a silty planar or rippled bed capped by a muddy bed. The couplets are stacked 10–12 to form decimetric thick cycles in which they thicken and thin progressively. Fig. 3 illustrates couplet thickness analysis of an example of neap-spring-neap rhythmites. The average number of couplets per cycle (T = 10 on Fig. 3B) matches the number of high tides able to reach the upper intertidal domain during a single fortnightly cycle (see Tessier, 1993, for further description of these tidal rhythmites in the Mont-Saint-Michel estuary).

Above the upper intertidal facies made of neap-spring-neap rhythmites, the fine-grained, silt-dominated marsh facies appear quite abruptly. These marsh facies commonly display a vertical succession of centimetric- to decimetric- thick packages (Fig. 4). In the uppermost part of this succession, roots heavily disturb the bedding of these packages. In the lower part, bedding may be best preserved so that detailed observations and bed thickness analysis are possible. The bedding consists of millimetric- to centimetric- thick couplets of silty and muddy planar beds (Fig. 1B). In many cases, successive couplets within each package appear to thicken and thin progressively upwards. An analysis of thickness evolution performed on an example of such marsh packages is illustrated in Fig. 5. It is clear that the thickness evolution displays a cyclic pattern in which cycles of two orders appear: first order cycles that contain, on average, 31 couplets, and second order cycles containing about 16 couplets (Fig. 5).

With respect to their bedding characteristics, that is, successive couplets and vertical thickness evolution, these marsh sedimentary cycles are comparable to neap-spring-neap rhythmites. However, as demonstrated in Fig. 5, it is obvious that the number of couplets contained in both first-order and second-order cycles is higher than that of high tides able to reach the marsh elevation during a single neap-spring-neap cycle. As mentioned previously, neap-spring-neap sedimentary records observed in the upper intertidal facies below the marsh deposits contain specifically about 10–12 couplets. Therefore, the marsh packages cannot be interpreted as neap-spring-neap rhythmites.

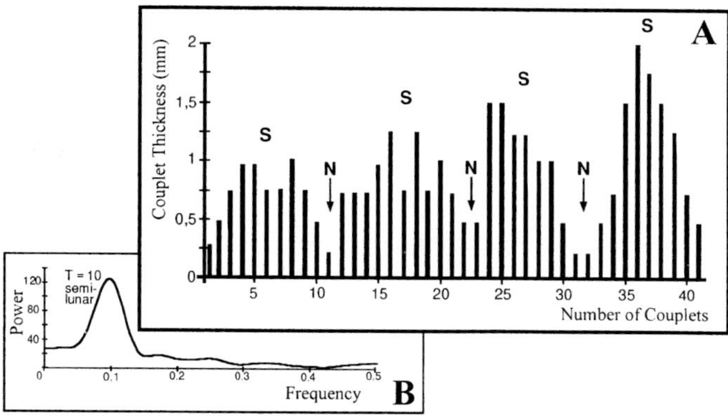

FIG. 3.—Example of couplet thickness analysis in millimeter-scale neap-spring-neap cycle records in the Mont-Saint-Michel estuary. (A) Field data. n: neap, s: spring. (B) Power spectra (FFT).

FIG. 4.—Marsh facies succession in the Mont-Saint-Michel estuary with typical stacked packages. The white line denotes the sharp limit between intertidal facies with neap-spring-neap rhythmites and marsh facies with annual rhythmites. Scale bar in centimeters.

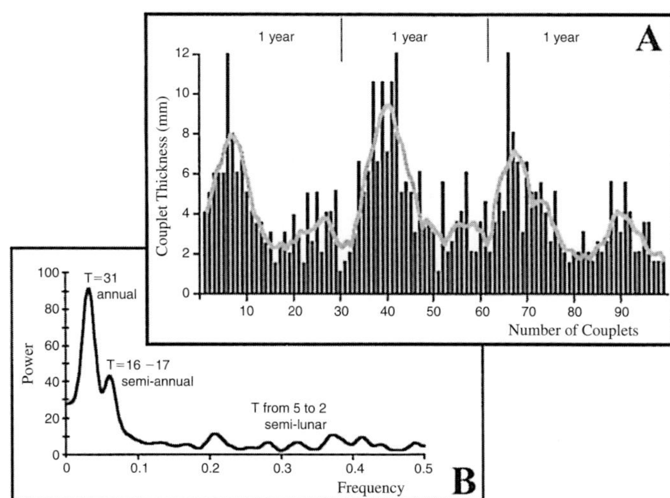

FIG. 5.—Example of couplet thickness analysis in marsh annual cycle records such as those illustrated on figures 1B and 4. (A) Field data. (B) Power spectra (FFT).

Studies have been performed in the Bay of Mont-Saint-Michel to estimate the annual rate of tidal inundation in the different domains of the tidal frame (Phlipponneau, 1956; Madec, 1991; J.-P. Auffret, personal communication, 1995). In the supratidal area, that is, on the marshes, this annual rate is evaluated to 3–5% (Fig. 6). Analysis of tidal range over a period of 15 years (1980–1994) demonstrates that the annual number of tides whose range is high enough for marshes to be flooded is comprised most of the time between 30 and 45, with extreme values of 8 and 50, and an average of 30. The number of couplets counted in the first-order marsh cycles fits this average value. Therefore, it is assumed that these sedimentary cycles represent the record of annual cycles during which are recorded only the higher high tides that are able to flood the marshes. The first-order cycles (T = 31) correspond to the general trend of tidal energy evolution over one year and the second order cycles (T = 16) represent the two equinoxial periods (Fig. 5). This evolution pattern is illustrated in Fig. 7, which shows, as an example, the tidal range evolution during 1984 and 1985, considering only the ranges that allow marsh flooding.

These annual cycles observed in marsh facies should be considered to be composed, in fact, of an amalgamation of several very incomplete neap-spring-neap cycles, according to the principle of cut-off discussed by Archer (1991). Besides, this fortnightly component is probably represented by the periods that vary from two to five in the example of annual records illustrated in Fig. 5. In Fig. 7, the neap-spring-neap component corresponds to almost the same range of periods. However, tides usually have a very low energy when they reach the marshes. Marsh flooding is of short duration and occurs arround slack water stage. These conditions do not favor high sedimentation rates, so couplets that compose the marsh packages remain fairly thin. Neap-spring-neap cycle records are thus hardly discernable, and the annual and semi-annual evolution of tidal energy are the only components to be significantly recorded in these marsh tidal rhythmites.

Finally, one may think that the passage from neap-spring-neap rhythmites to annual rhythmites should be progressive, resulting from the deposition of neap-spring-neap records that are increasingly incomplete. It is frequently observed, on the contrary, that this passage from intertidal facies to marsh facies is sharp (Fig. 4), which could be explained in part by the fact that the rate of tidal inundation is not strictly linear. The rate appears to decrease more rapidly once marsh elevation is reached, and this evolution leads to a relatively abrupt, rather than progressive, decrease in the number of tides able to inundate the supratidal area (Fig. 6).

DISCUSSION

Annual cycles have been already recognized in marsh facies of estuarine environments. One of the most striking examples is described by Allen (1990) in the Severn estuary. Another example is described by Borrego et al. (1995) in a Spanish estuary. Annual cycles associated with upper intertidal facies are described by Van den Berg (1981) in the Oosterschelde estuary and have also been observed in the Bay of Mont-Saint-Michel (in Larsonneur, 1989, 1994) and in the Cobequid Bay-Salmon River estuary (Dalrymple et al., 1991). Ancient examples are also reported (e.g., Hamberg, 1991). In both cases, marsh and upper intertidal facies, annual cycles are recognized because of a differentiation between summer/autumn sedimentation and winter/spring sedimentation: winter/spring deposits are generally characterized by high-energy, sand-dominated facies, whereas summer/autumn deposits consist of burrowed, low-energy, mud-dominated facies. This differentiation is not the characteristic feature of the annual cycles described herein. Almost all tidal events recorded in the marsh packages illustrated in Fig. 1B and Fig. 5 are similar in terms of lithology and bedding. However, the annual rhythmites discussed in this paper belong to the lower part of the marsh facies succession. In the upper part, marsh packages become increasingly thinner and rooted (Fig. 4) so that their recognition as annual rhythmites based on lamina thickness and lamina number analysis is no longer possible. The annual origin of these upper packages is

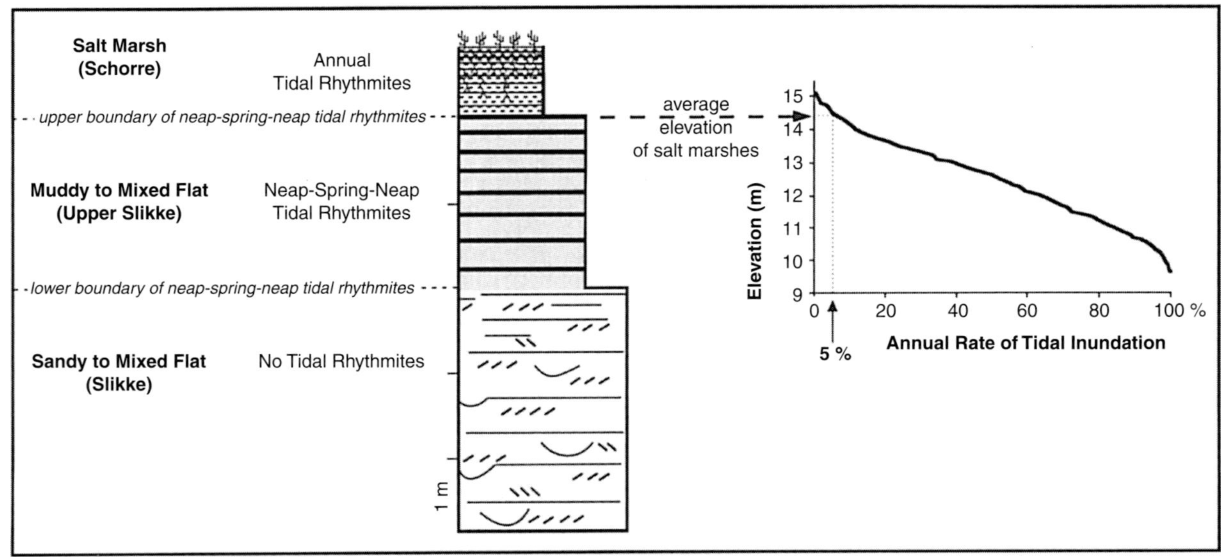

FIG. 6.—Annual rate of tidal inundation in the Bay of Mont-Saint-Michel. On the marshes, this rate is less than 5%. The average marsh elevation coincides with a slightly quicker rate decrease, explaining in part why the passage from neap-spring-neap to annual rhythmites is sharp rather than progressive.

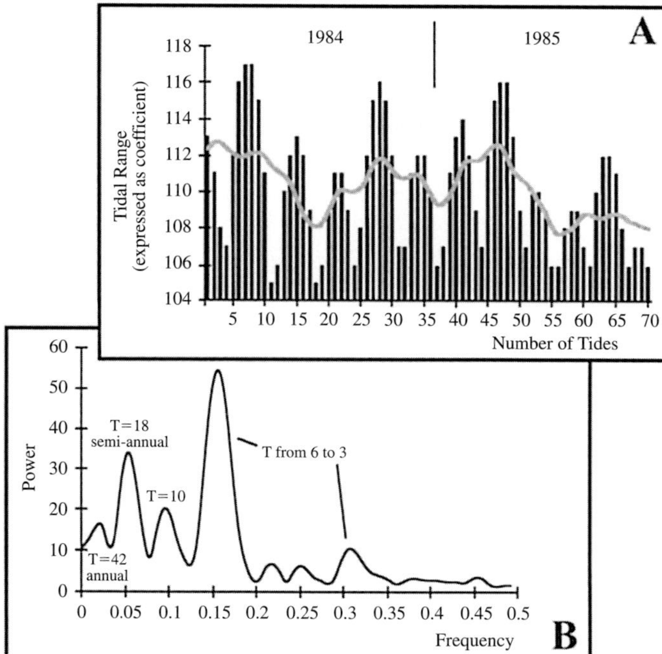

FIG. 7.—(A) Evolution of the range of the tides that have potentially flooded the marsh domains during the years 1984 and 1985 in the Mont-Saint-Michel estuary. (B) Power spectra (FFT). The periods from 10 to 3 correspond to the fortnightly cycles which are more or less amalgamated.

supported by the seasonal differentiation that characterizes their general constitution. They are formed by a succession of silt-dominated laminae and mud-dominated, heavily rooted laminae, the latter reflecting soil development that occurs during summer.

The occurrence of annual rhythmites such as those illustrated in Fig. 5 points out the problem that may arise in recognizing and interpreting tidal rhythmites. Neap-spring-neap rhythmites and annual rhythmites are characterized by a same feature: a cyclic bed/lamina thickness evolution. Moreover, as it has been demonstrated through the example shown in Fig. 5, annual cycle record (i.e., on Fig. 5, first-order cycle: T = 31) can be composed by a thick-thin pair of semi-annual cycle records (i.e., on Fig. 5, second-order cycle: T = 16). This cycle combination pointed out in annual rhythmites mimics closely the combination between the fortnightly component and the lunar component (i.e., high spring [perigee]/low spring [apogee]) that is frequently recorded in neap-spring-neap rhythmites (e.g., Tessier and Gigot, 1989; Archer, 1996). As a consequence, the confusion between neap-spring-neap rhythmites, especially those made of parallel to pinstriped silty to muddy laminated facies (e.g., Kvale and Archer, 1990) and annual rhythmites is highly probable in ancient sedimentary records. This confusion is particularly conceivable in formations with no evidence of roots (because of age and/or climatic constraints) that would allow identification of marsh deposits.

Detailed analysis of neap-spring-neap rhythmites in rock records may provide useful information about: 1) sedimentation rates, 2) paleotidal regime (semi-diurnal or diurnal) inferred from the number of events recorded per neap-spring-neap cycle, 3) basin morphology inferred from the latter information—the prevailing oscillation of a given tidal component (e.g., diurnal rather that semi-diurnal) depends, among other factors, on the natural period of the basin, that is, on the basin size and shape, 4) palaeoenvironment of deposition by considering the completeness of the neap-spring-neap cycle record with respect to the position in the tidal frame, 5) orbital parameter variation through geological time. In case of long (multi-year) and continuous preserved record of neap-spring-neap cycles, rhythmite characteristics such as the number of fortnightly cycles per year are used as criteria for orbital extraction.

Misinterpretation of annual rhythmites as neap-spring-neap rhythmites will lead to erroneous conclusions about all of these aspects.

CONCLUSION

This paper presents evidence that tidal rhythmites preserved in estuarine deposits can be composed either by neap-spring-neap cycle records or annual cycle records. In modern settings such as the Mont-Saint-Michel Estuary, annual tidal rhythmites exclusively characterize marsh facies and can be easily distinguished from neap-spring-neap rhythmites that are best preserved in upper intertidal deposits. In ancient formations, the distinction between annual and semi-lunar rhythmites may be more problematic owing to the fact that both are characterized by the same feature, that is, a cyclic evolution of bed/lamina thickness. With respect to this feature, especially to the number of tidal events recorded per year, the modern annual rhythmites described herein have their proper specificities and have, therefore, to be considered as just an example of annual rhythmites. The main purpose of this paper is to draw attention to the possibility that cyclic tidal bedding that seems to be composed of semi-diurnal, diurnal, semi-lunar and lunar components can be of annual origin as well. This last remark has to be considered carefully, especially when these tidal components are identified and used for orbital extraction.

ACKNOWLEDGMENTS

I thank Allen W. Archer (Kansas State University) for providing a critical review of an earlier version of this paper. The AnalySeries software (Paillard et al., 1996) was used for spectral analysis. It includes the ARAN software for Spectral analysis developed at Brown University for the SPECMAP program and communicated by W. Prell.

REFERENCES

ALLEN, J. R. L., 1990, Salt-marsh growth and stratification: A numerical model with special reference to the Severn Estuary, southwest Britain: Marine Geology, v. 95, p. 77–96.

ARCHER, A. W., 1996, Reliability of lunar orbital periods extracted from ancient cyclic tidal rhythmites: Earth and Planetary Science Letters, v. 141, p. 1–10.

ARCHER, A. W., 1991, Modeling of tidal rhythmites using tidal periodicities and implications for short-term sedimentation rates, *in* Franseen, E. K., et al., eds., Sedimentary Modeling: Computer simulations and methods for improved parameter definition: Kansas Geological Survey Bulletin, v. 233, p. 185–194.

ARCHER, A. W., AND FELDMAN, H. R., 1994, Tidal rhythmites in fine-grained carboniferous limestones, U.S.A.: Geobios, v. 16, p. 275–281.

ARCHER, A. W., FELDMAN, H. R., KVALE, E. P., AND LANIER, W. P., 1994, Comparison of drier- to wetter-interval estuarine roof facies in the Eastern and Western Interior coal basins, USA.: Palaeogeography, Palaeoclimatology, Palaeoecology, v. 106, p. 171–185.

Borrego, J., Morales, J. A., and Pendon, J. G., 1995, Holocene estuarine facies along the mesotidal coast of Huelva, south western Spain, *in* Flemming, B. W., and Bartolomä, A., eds., Tidal Signatures in Modern and Ancient Sediments: International Association of Sedimentologists Special Publication, v. 24, p. 151–170.

Brown, M. A., Archer, A. W., and Kvale, E. P., 1990, Neap-spring tidal cyclicity in laminated carbonate channel fill deposits and its implications: Salem Limestone (Mississipian), South central Indiana, USA.: Journal of Sedimentary Petrology, v. 60, p. 152–159.

Chan, M. A., Kvale, E. P., Archer, A. W., and Sonett, C. P., 1994, Oldest direct evidence of lunar-solar tidal forcing encoded in sedimentary rhythmites, Proterozoic Big Cottonwood Formation, central Utah: Geology, v. 22, p. 791–794.

Dalrymple, R. W., 1992, Tidal depositional systems, *in* Walker, R. G., and James, N. P., eds., Facies models: Response to sea level change: St John's, Newfounland, Geological Association of Canada, p. 195–218.

Dalrymple, R. W., and Makino, Y., 1989, Description and genesis of tidal bedding in the Cobequid Bay-Salmon River estuary, Bay of Fundy, Canada, *in* Taira, A., and Masuda, F., eds., Sedimentary Facies in the Active Plate Margin: Tokyo, Terra Scientific Publishing Company, p. 151–177.

Dalrymple, R. W., Makino, Y., and Zaitlin, B. A., 1991, Temporal and spatial patterns of rhythmite deposition on mud flats in the macrotidal Cobequid Bay-Salmon River, Bay of Fundy, Canada, *in* Smith, D. G., Reinson, G. E., Zaitlin, B. A. and Rahmani, R. A., eds., Clastic Tidal Sedimentology: Calgary, Canadian Society of Petroleum Geologists Memoir 16, p. 137–160.

Dalrymple, R. W., Zaitlin, B. A., and Boyd, R., 1992, Estuarine Facies models: Conceptual basis and stratigraphic implications: Journal of Sedimentary Petrology, v. 62, p. 1130–1146.

Deynoux, M., Duringer, P., Ramzy, K., and Villeneuve, M., 1993, Laterally and vertically accreted tidal deposits in the upper Proterozoic Madina-Kouta Basin, southeastern Senegal, West Africa: Sedimentary Geology, v. 84, p. 179–188.

Greb, S. F., and Archer, A. W., 1995, Rhythmic sedimentation in a mixed tide and wave deposit, Hazel Patch Sandstone (Pennsylvanian), Eastern Kentucky coal field: Journal of Sedimentary Research, v. B65, p. 96–106.

Hamberg, L., 1991, Tidal seasonal cycles in a lower Cambrian Shallow marine sandstone (Hardeberga Fm.), Scania, Southern Sweden, *in* Smith, D. G., Reinson, G. E., Zaitlin, B. A., and Rahmani, R. A., eds., Clastic Tidal Sedimentology: Calgary, Canadian Society of Petroleum Geologists Memoir 16, p. 255–274.

Kuecher, G. J., Woodland, B. G., and Broadhurst, F. M., 1990, Evidence of deposition from individual tides and tidal cycles from the Francis Creek Shale (host rock to the Mazon Creek Biota), Westphalian D (Pennsylvanian), northeastern Illinois: Sedimentary Geology, v. 68, p. 211–221.

Kvale, E. P., and Archer, A. W., 1989, Daily, monthly, and yearly tidal cycles within laminated siltstones of the Mansfield Formation (Pennsylvanian) of Indiana: Geology, v. 17, p. 365–368.

Kvale, E. P., and Archer, A. W., 1990, Tidal deposits associated with low sulfur coals, Brazil Formation (Lower Pennsylvanian), Indiana: Journal of Sedimentary Petrology, v. 60. p. 563–574.

Kvale, E. P., Fraser, G. S., Archer, A. W., Zawistoski, A., Kemp, N., and McGough, P., 1994, Evidence of seasonal precipitation in Pennsylvanian sediments of the Illinois Basin: Geology, v. 22, p. 331–334.

Lanier, W. P., Feldman, H. R., and Archer, A. W., 1993, Tidal sedimentation from a fluvial to estuarine transition, Douglas Group, Missourian-Virgilian, Kansas: Journal of Sedimentary Petrology, v. 63, p. 860–873.

Larsonneur, C., 1989, La baie du Mont-Saint-Michel: Bulletin de l'Institut Géologique du Bassin d'Aquitaine, v. 46, p. 5–74.

Larsonneur, C., 1994, The Bay of Mont-Saint-Michel: A sedimentation model in a temperate macrotidal environment: Senckenbergiana maritima, v. 24, p. 3–63.

Madec, V., 1991, Un modèle spatial des taux d'émersion appliqué au domaine intertidal de la Baie du Mont-Saint-Michel: Oceanologica Acta, v. 14, p. 525–529.

Martino, R. L., and Sanderson, D. D., 1993, Fourier and autocorrelation analysis of estuarine tidal rhythmites, lower Breathitt Formation (Pennsylvanian), Eastern Kentucky, USA.: Journal of Sedimentary Petrology, v. 63, p. 105–119.

Paillard, D., Labeyrie, L., and Yiou, P., 1996, Analyseries 1.0: A Macintosh software for the analysis of geographical time-series, Eos, Transactions, American Geophysical Union, v. 77, p. 379.

Phlipponneau, M., 1956, La Baie du Mont-Saint-Michel: étude de morphologie littorale: Société Géologique et Minéralogique de Bretagne, Memoir 11, p. 1–125.

Rahmani, R. A., 1988, Estuarine tidal channel and nearshore sedimentation of a late Cretaceous epicontinental sea, Drumheller, Alberta, Canada, *in* de Boer, P. L., van Gelder, A., and Nio, S. D., eds., Tide-influenced Sedimentary Environments and Facies: Dordrecht, Sedimentary and Petroleum Geology, Reidel Publishing Company, p. 433–471.

Read, W. A., 1992, Evidence of tidal influences in the Arnsbergian rhythmites in the Kincardine Basin: Scottish Journal of Geology, v. 28, p. 135–142.

Sonett, C. P., Kvale, E. P., Zakharian, A., Chan, M. A., and Demko, T. M., 1996, Late Proterozoic and Paleozoic tides, retreat of the moon, and rotation of the Earth: Science, v. 273, p. 100–104.

Tessier, B., 1993, Upper intertidal rhythmites in the Mont-Saint-Michel Bay (NW France): Perspectives for paleoreconstruction: Marine Geology, v. 110, p. 355–367.

Tessier, B., and Gigot, P., 1989, A vertical record of different tidal cyclicities. An example from the Miocene marine molasse of Digne (Haute Provence, France): Sedimentology, v. 36, p. 767–776.

Tessier, B., Archer, A. W., Lanier, W. P., and Feldman, H. R., 1995, Comparison of modern analogues (the Bay of Mont-Saint-Michel) with ancient tidal rhythmites (Carboniferous of Kansas and Indiana, U.S.A.), *in* Flemming, B. W., and Bartolomä, A., eds., Tidal Signatures in Modern and Ancient Sediments: International Association of Sedimentologists Special Publication, v. 24, p. 259–271.

Van den Berg, J., 1981, Rhythmic seasonal layering in a mesotidal channel fill sequence, Oosterschelde mouth, The Netherlands: Oxford, International Association of Sedimentologists Special Publication, v. 5, p. 147–159.

Williams, G. E., 1989, Late Precambrian tidal rhythmites in South Australia and the story of the Earth's rotation: Journal of the Geological Society, London, v. 146, p. 97–111.

Williams, G. E., 1991, Upper Proterozoic tidal rhythmites, South Australia: Sedimentary features, deposition and implications for the Earth's paleorotation, *in* Smith, D. G., Reinson, G. E., Zaitlin, B. A., and Rahmani, R. A., eds., Clastic Tidal Sedimentology: Calgary, Canadian Society of Petroleum Geologists Memoir 16, p. 161–178.

ANNUAL SEDIMENTATION CYCLES IN RHYTHMITES OF CARBONIFEROUS TIDAL CHANNELS

STEPHEN F. GREB
Kentucky Geological Survey, University of Kentucky, Lexington, Kentucky 40506-0107 USA
AND
ALLEN W. ARCHER
Department of Geology, Kansas State University, Manhattan, Kansas 66506-3201 USA

ABSTRACT: Three occurrences of rhythmites in Carboniferous tidal channels from the Appalachian and Illinois Basins were compared in order to assess the controls on the rapid accumulation of tidal sedimentation in paleoenvironments in which channelized flow and daily reworking normally preclude such preservation. In all three channels, rhythmites consist of stacked composite bedsets. Daily tidal sedimentation in these rhythmites is marked by submillimeter- to millimeter-scale laminae couplets. Bundles of 12 or fewer couplets represent fortnightly neap-spring deposition, alternating thick and thin neap-spring bundles represent perigeen and apogeen cycles, and the association of 24 or fewer neap-spring bundles in each bedset represents annual sedimentation cycles. The most complete annual bedsets are often divided into sand- and shale-dominated halves, which are inferred to represent seasonal differences in current energy during aggradation.

Interpreting the various orders of tidal cyclicity is complicated in many bedsets because of the thinness of daily laminae couplets and typical incompleteness of continuous daily sedimentation records in neap-spring bundles of the three channels studied. In many cases, bundles of laminae or ripples with foreset drapes are superficially very similar to laminae couplets. In such examples, fortnightly cycles can be misinterpreted as daily deposits. Furthermore, annual cycles can be misinterpreted as monthly deposits because 24 weekly cycles look like 24 daily deposits.

In all three channel-fill deposits, annual bedsets thin upward, through (1) a vertical decrease in the number of daily couplets per neap-spring bundle, (2) loss of neap bundles per monthly pair, and (3) loss of monthly bundle pairs in the shalier, low-energy, seasonal parts of each bedset. These vertical changes reflect a strong accommodation control, especially considering the lack of rhythmic sedimentation in lateral deposits. In each of the channels, the most complete tidal records were recorded toward the base and axes of the fills, and daily, neap-spring, monthly, seasonal, and annual cycles amalgamated upward and laterally in response to shallowing. This is significant because accommodation-controlled thinning and amalgamation causes substantial changes in the rhythmicity of the fill and results in subtle hiatal surfaces, which could lead to incorrect inferences of temporal duration spatially across the channel, even within the same bedset.

INTRODUCTION

Ever since thick-thin foreset pairs or bundles were first recognized as recording neap-spring tidal sedimentation (Visser, 1980), there has been interest in documenting different types of tidal rhythmites and interpreting the various orders of tidal periodicity that formed them. By far, the most complete records of tidal periodicities have been recorded in vertically accreted, heterolithic strata (Kvale et al., 1989). In most investigations of vertically accreted rhythmites, couplets of coarse and fine laminae are counted and then compared (often by spectral analysis) to known oscillations in tidal sedimentation (Archer, 1994). These oscillations occur because of the elliptical orbit of the moon around the earth, and the earth-moon system around the sun. Daily (diurnal), twice daily (semidiurnal), fortnightly (neap-spring), paired thick-and-thin fortnightly (lunar procession or monthly), 6-month (equinoctial), annual (12-month), and 18-year (lunar-nodal) cycles have all been recorded in the rock record (Williams, 1991). Analysis of cyclic rhythmites has invariably focused on the most complete and continuous rhythmites because they contain the most complete record of tidal periodicity. These cyclic tidal rhythmites show (1) repetitive vertical thickening and thinning of alternating sandstone-shale laminae couplets, (2) a consistent number of laminae between sandstone thickness maxima, and (3) a sinusoidal thickness distribution related to the lunar orbital cycle (Greb and Archer, 1995). However, even where cyclic rhythmites are documented, noncyclic rhythmites and other forms of tidally produced lamination are usually more common. Indeed, in the context of tidal stratification, cyclic rhythmites are relatively rare and appear to be restricted to specific depositional environments and facies.

Cyclic rhythmites have been noted on modern and ancient tidal flats of inner estuaries (Kuecher et al., 1990; Dalrymple et al., 1991; Kvale and Archer, 1991), on subtidal delta slopes (Smith et al., 1990; Williams, 1991; Jaeger and Nittrouer, 1995), above presumably compacting peats in Carboniferous coal measures (Kvale and Archer, 1 990), and within tidal channels (Van den Berg, 1981; Tessier and Gigot, 1989; Brown et al., 1990; Roep, 1991; Greb and Archer, 1995). Tidal channels are usually thought of as containing coarser, cross-bedded sands deposited in high-velocity channelized flows rather than as laminated rhythmites deposited at least in part from suspension. However, several modern abandoned tidal channels (Goldring et al., 1978; Van den Berg, 1981; Roep, 1991) and active channels in which dense mud is deposited by tractive flow (Dalrymple et al., 1996) have been documented, and they contain both noncyclic and cyclic tidal rhythmites. Because channel fills have sharply defined limits, the occurrence of various orders of tidal rhythmites within any single channel can be used to determine the controls on rhythmite preservation within a single, well-constrained depositional facies. Three Carboniferous tidal channels are discussed here (Fig. 1) in order to illustrate the variability in rhythmites in similar channel facies. They will also aid in interpreting controls on the preservation of near-complete records of ancient tidal sedimentation in tidal environments in which daily reworking, seasonal storms, and other factors normally preclude such preservation.

TIDAL-CHANNEL RHYTHMITES

Location 1

The channel-fill deposit at location 1 occurs midway through the Hazel Patch sandstone (an informal unit) of the Breathitt Group (Langsettian), central Appalachian Basin, Eastern Kentucky Coal Field. The channel fill is more than 100-m wide and 3.2-m thick (Fig. 2A) and contains rhythmites, which illustrate the difference between cyclic and noncyclic deposits in mixed tide and wave settings (Greb and Archer, 1995). Cyclic rhythmites exhibit vertical thickening and thinning trends in apparent

Tidalites: Processes and Products, SEPM Special Publication No. 61

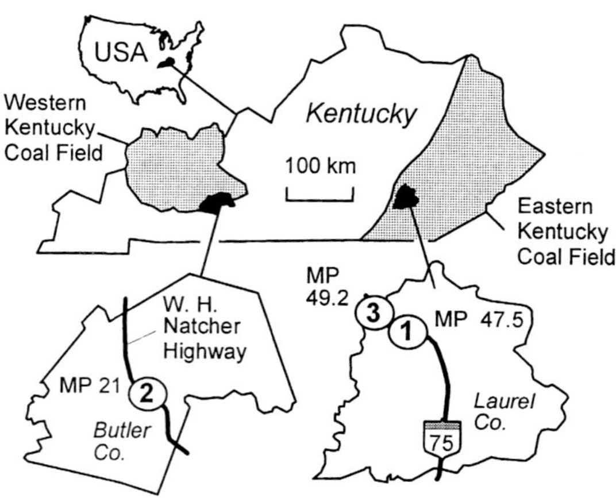

FIG. 1.—Location of study tidal channels (1 to 3) in the Kentucky portion of the Illinois and Appalachian Basins.

laminae (Figs. 2B-C). However, examination of a fresh surface shows that the cyclic rhythmites are composed of thinning and thickening bundles of laminae, each composed of thinner (0.04 to 2 mm) laminae couplets (Figs. 2D-E). Computer-based image analysis incorporates digital scanning of outcrop photographs and statistical analysis of the gray-scale values recorded in the scan, which can be termed optical density (Archer, 1994). When applied to photographs of rhythmic bedsets, two scales of bedding can be statistically extracted. A low-resolution scan of the bedset shown in Fig. 2C approximates the appearance of bedding on slightly more weathered surfaces and shows 24 bundles between points of minimum sandstone laminae thickness (Fig. 3A). A previous spectral analysis of a sample bedset found a frequency distribution of 25.71, which was interpreted as the frequency of laminae couplets per bedset (Greb and Archer, 1995). However, high-resolution scans show that individual bundles in the sandy and shaly parts of the bedset (Figs. 3B-C) actually consist of five to eight thinner laminae couplets (sometimes as mudstone-draped foresets on ripples). Analysis of numerous samples determined a maximum of 12 couplets and a minimum of one erosively based ripple lamina per bundle. In the shaly parts of rhythmic bedsets, there is commonly a weak alternation in bundle thickness between slightly sandier and shalier bundles (Fig. 3C).

In the channel-fill deposit, bedsets are thickest toward the base (14 cm) and thin upwards to less than 5 cm (Fig. 4A). Thinning is accompanied by a general decrease in shale content (Fig. 4A), a decrease in the number of bundles per bedset from 24 near the base to fewer than 7 near the top (Fig. 4B), and a general increase in ripple stratification per bedset (Fig. 4C).

Interpretation.—

The rhythmites were deposited in a broad, shallow channel that cut across a mixed tide-and wave-influenced sand flat. Previous interpretation of the laminae bundles as laminae couplets led to the misinterpretation that bedsets represent neap-spring tidal deposits (Greb and Archer, 1995). Because the layers are actually composed of bundles of laminae, they record more than twice-daily reversals and do not represent a single day's sedimentation. Therefore, the association of 24 bundles per bedset (close to the spectral analysis of 25.71) does not represent 24 days of sedimentation, but instead represents a higher order, repeatable cycle with a frequency of 24, such as the number of weekly neap-spring cycles in a year. The maximum number of laminae couplets in a bundle is 12, which is close to the expected 14 days of sedimentation in a fortnightly cycle in a diurnal tidal regime. Bundles of five to eight laminae couplets exhibit sinusoidal thickening and thinning trends typical of neap-spring cycles, even if they contain only half of the expected daily events. Many of these ostensibly incomplete bundles contain ripples with multiple foreset drapes, which could easily account for the apparent missing days of sedimentation.

The fine-scale sinusoidal distribution of laminae couplets in some bundles, especially the alternation of thick, sandy bundles

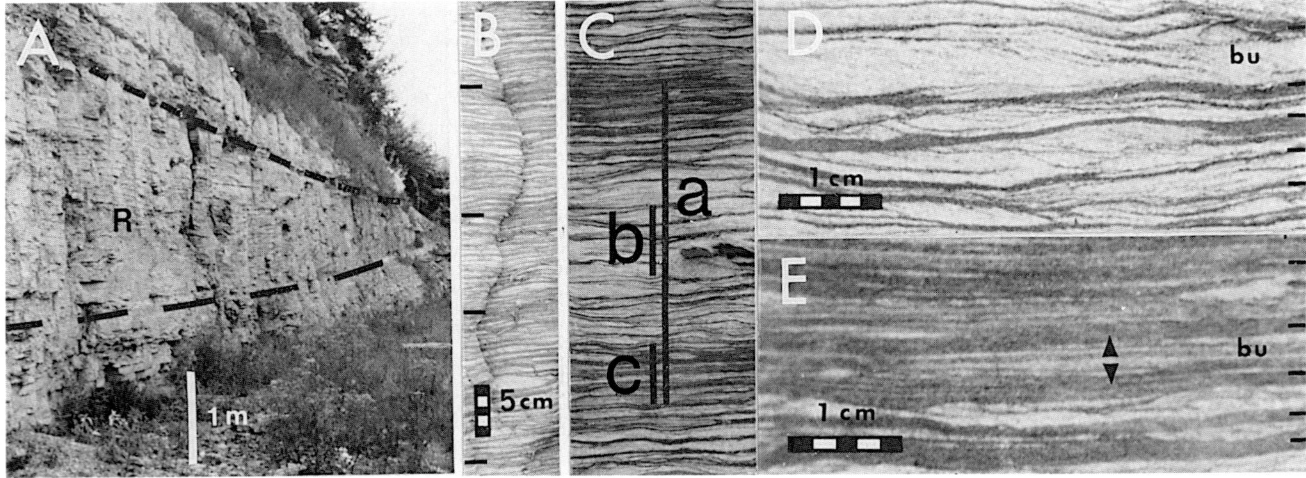

FIG. 2.—Photographs from channel 1. (A) View of channel fill, showing basal scour (long dashes), rhythmites (R), and contact with overlying sandstone (short dashes). (B) Rhythmites in fill, showing sinusoidal thickening and thinning of cycles. (C) Close-up of a cyclic rhythmite bedset in the lower part of the channel fill. Lines and letters (a, b, c) indicate areas of analysis shown in Fig. 3. (D) Detail of apparent laminae couplets in the sandy portion of the bedset. Each layer contains numerous foreset drapes and is actually a bundle (bu) of laminae (each delineated by dark line on margin). (E) Detail of laminae bundles (bu, each separated by dark lines) from the upper part of the composite bedset. Bundles consist of thickening and thinning laminae couplets (arrows); many less than 1 mm in thickness.

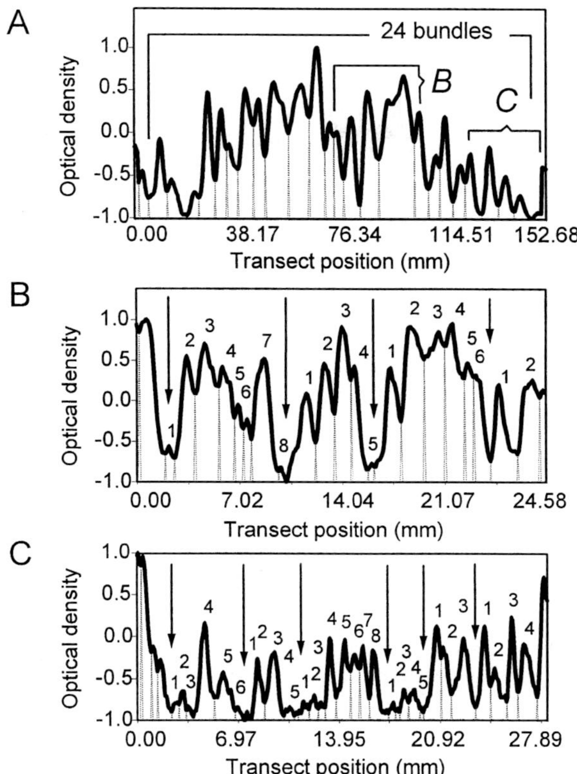

FIG. 3.—Computer image analysis from scans of Fig. 2B. Greater optical density values indicate lighter hue and define sandstone laminae. Lower values indicate claystones and mudstones. (A) Low-resolution scan of bedset and resultant sinusoidal distribution consisting of 24 mudstone-draped bundles, which could be misinterpreted as laminae couplets. (B) High-resolution scan from sandy part of bedset showing that each of the units in Fig. 3A are composed of five to eight laminae couplets arranged in sinusoidal bundles, separated by dark mudstone bands (indicated by arrows). (C) High-resolution scan from shaly part of the bedset showing similar bundled laminae; each is less than 1 mm thick.

and thin, shaly bundles (Fig. 3C), may document fortnightly inequality (Archer et al., 1991; Archer, 1995). At perigee the moon is closer to the earth, resulting in a high spring tide and thicker, sandier cycles. At apogee the moon is farther away, resulting in a lower spring tide and thinner, shalier cycles. Fortnight pairs of apogeen and perigeen bundles represent lunar monthly cycles. Dark and persistent claystone drapes between bundles (indicated by arrows in Fig. 3C) may be similar to dark clay bands noted in other tidal deposits, and apparently accumulated during neap-stage emersion, either from bacterial activity in suspended diatomaceous films (Tessier, 1993) or from organic matter settling with clays during lower energy neap tides (Brown et al., 1990).

The arrangement of 10–14 bundles in the sandy part of each bedset represents five- to seven-month, high-energy depositional cycles in which tractive bed load deposition dominated. The shaly part of each bedset is inferred to represent a five- to six-month cycle of dominantly suspension deposition. Similar cycles in tidal channels of The Netherlands document seasonality, in which higher energy conditions accompany winter storms and suspension deposition dominates during the rest of the year (Van den Berg, 1981; Roep, 1991).

The vertical decrease in the shale content of each bedset (Fig. 4A) indicates a temporal loss of the lower energy seasonal cycle with time. The concurrent decrease in the number of bundles per rhythmic bedset (Fig. 4B) and increase in ripple stratification (Fig. 4C) reflects (1) an increase in erosion of daily sedimentation by the highest spring tides in fortnightly cycles, (2) erosion of perigeen cycles by apogeen cycles during monthly cycles, and (3) erosion of low-energy seasonal cycles by spring tides during the highest energy seasonal cycles, perhaps related to six-month equinoctial cycles.

In addition to the vertical amalgamation of cycles, at least two surfaces were reworked (dashed lines in Figs. 4A-C) and the underlying annual bedset was partially removed. Surfaces were mostly form-concordant and were indicated by a sudden asymmetry in the vertical thickening and thinning of laminae bundles. These subtle hiatal surfaces suggest that although 24 annual cycles can be counted in the channel, it took more than 24 years to fill (24 bedsets were preserved).

Location 2

The second channel-fill deposit occurs in the lower Tradewater Formation of the Raccoon Creek Group (Duckmantlan), Illinois Basin, Western Kentucky Coal Field. The deposit is more than 230-m wide and 6-m thick (Fig. 5A). The base of the channel is a sharp scour surface with a sideritic sandstone lag. A thin coal drapes the scour toward the north, but it is truncated toward the center of the channel. Although the fill is rhythmic (Fig. 5B-C), individual bedsets vary considerably, and both cyclic (Fig. 5B) and noncyclic (Fig. 5C) rhythmites are noted. The channel fill contains at least three low-relief discontinuities, which divide the fill into lower, middle, and upper units (Fig. 5A).

Cyclic rhythmites 36 cm above the base of the channel exhibit two thinning-and-thickening cycles consisting of six to

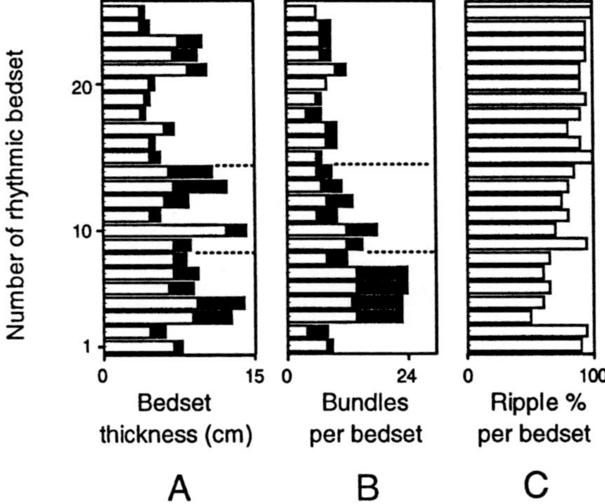

FIG. 4.—Bar charts of rhythmic bedding trends in channel 1. (A) Composite vertical bedset thicknesses. White and black bars indicate estimated sandstone and shale content, respectively. (B) Vertical changes in the number of laminae bundles per composite bedset. White and black bars indicate the number of bundles in the sand- and shale-dominated halves of each composite bedset, respectively. (C) Vertical changes in the percentage of ripple stratification per bedset. Dashed horizontal lines mark the positions of two subtle discontinuities that partially erode into an underlying bedset.

FIG. 5.—Photographs from channel 2. (A) View of the channel, showing approximate base of channel (dashed line) hidden in shadow, and three fill intervals (L = lower, M = middle, U = upper). The white arrow marks the low-angle discontinuity in the middle unit. The channel fill is approximately 6-m thick. (B) Close-up of successive cyclic rhythmites in the lower part of the channel fill. Scale in photo shows inches and centimeters. (C) Stacked rhythmites in the middle interval, showing noncyclic bedsets (bs) consisting of a lower sandy part and an upper shaly part. (D) Details of the sandy part of a composite bedset in the middle interval showing bundles (bu) of laminae (separated by dark lines) with persistent claystone drapes. (E) Detail of bedding in the shaly part of a bedset in the middle interval, where evenly laminated couplets are abundant. Bundles of laminae (bu) are less clearly defined but consist of thickening-and-thinning (arrows) laminae couplets. Ticks on scale are in millimeters.

eight laminae bundles (Fig. 5B). Each bundle is separated by a dark mudstone drape. The cycles are separated from each other by a relatively thick (3 cm) shale layer, which contains very thin (0.05 mm) sandy streaks (Fig. 5B). These bundles of laminae could be mistaken for individual laminae couplets, but careful examination shows that individual ripple laminae contain numerous claystone-draped foresets (as many as five laminae couplets per bundle).

The base of the middle unit is a sharp, flat discordance between underlying shaly rhythmites (Fig. 5B) and the sandstone-dominated rhythmites (Fig. 5C) that characterize the remainder of the fill. Rather than being flat-lying or form concordant, rhythmic bedsets in the middle unit dip at a low angle from the southern channel margin into the channel axis (Fig. 5A). Individual bedsets in the middle unit thin updip by as much as 50%. At least one subtle discordance occurs in the middle unit (indicated by arrows in Fig. 5A). This discordance is difficult to delineate in the center of the channel, but truncates at least four composite bedsets laterally toward the channel margin. Composite bedsets in the lower part of the middle unit pinch out against this discordance, then drape the surface toward the channel margin.

Noncyclic composite bedsets in the middle unit are 14.9 ± 4.7 cm thick (Fig. 6A). Most bedsets consist of a lower, thicker, sandstone-dominated part and an upper, thin, shale-dominated part (Fig. 5C), a composition similar to that of the composite bedsets in channel 1, but with less gradation between halves. Rather than having a sinusoidal distribution, bedding in individual bedsets is often asymmetric, and sandstone thickness increases sharply at the base (sometimes eroding into the underlying bedset), gradually and variably thinning upward within each bedset (Fig. 5C).

As in the first channel, layering in each composite bedset consists of bundles of thinner laminae couplets (Figs. 5D-E).

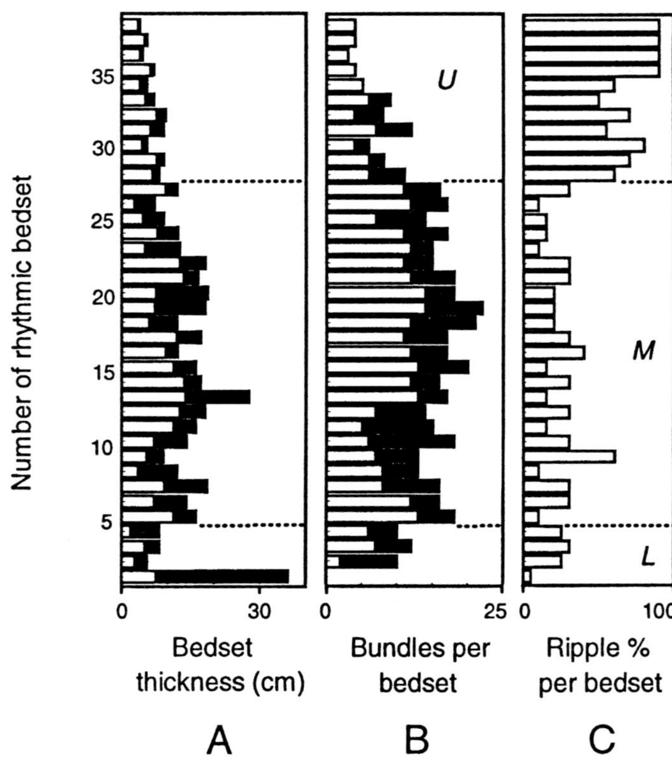

FIG. 6.—Bar charts of rhythmic bedding trends in channel 2. (A) Vertical composite bedset thicknesses. White and black bars indicate estimated sandstone and shale content, respectively. Dashed horizontal lines mark the position of the three intervals described in the text. (B) Vertical changes in the number of laminae bundles per composite bedset. White and black bars indicate estimated number of beds in the lower sand-dominated and upper shale-dominated parts of each composite bedset, respectively. (C) Vertical changes in the percentage of ripple stratification per composite bedset.

The sandy part of each bedset consists of as many as 14 laminae bundles and a mean of 10.6 ± 2.6 laminae bundles per bedset (Fig. 6B). Statistical analysis of a computer-based image scan (see Archer, 1994) of a sandy bedset was plotted as a function of power spectral density and frequency (Fig. 7). On such plots, significant peaks indicate the presence of well-developed cyclicities at specific periods. The larger number, 7.31, appears to be the number of bundles in the bedset, whereas the lower number, 2.25, may represent the sandstone-shale couplet nature of the bundles. Sandy laminae bundles vary from ripple-dominated to relatively horizontally laminated; ripple-dominated bundles have fewer couplets than horizontally-laminated bundles (Fig. 5D). Bundles in the shalier part of the bedsets are mostly evenly laminated, with each bundle composed of five to ten siltstone-shale laminae couplets. However, the fine-grained nature of the bundles makes delination of a single, persistent claystone drape between bundles problematic (Fig. 5E). Many laminae couplets are less than 1-mm thick (Figs. 5D-E).

The maximum number of bundles in a bedset is 23, and occurs in the middle unit (Fig. 6B). Although variable, bedset thickness and shale percentage generally decrease upward in the middle unit, through the loss and thinning of the shaly part of each composite bedset (Fig. 6A) and the number of bundles per bedset (Fig. 6B).

The upward-thinning trend continues into the upper unit (u in Fig. 5A). Rather than being scour-based, the base of the upper unit is marked by a sharp increase in the percentage of ripple stratification per composite bedset (Fig. 6C). Increasing ripple stratification is associated with a decrease in shale percentage and number of laminae bundles preserved per bedset. Composite bedsets thin vertically to the point where they are no longer rhythmic and become part of a ripple-bedded sandstone that extends beyond the channel fill (light-colored layer in Fig. 5A).

Interpretation.—

The coal that drapes the scour on the north margin of the channel indicates abandonment and at least partial filling by peat prior to inundation. The peat was then drowned in tidally influenced waters, a succession similar to the type seen in transgressive channel fills (Greb and Chesnut, 1992). Rhythmites were deposited in the preexisting channel. At first glance, the shaly, cyclic rhythmites at the base of the channel fill look like neap-spring cycles with six to eight laminae couplets per cycle. However, ripple cross-stratification contains numerous reactivation surfaces and thin claystone drapes that would make forming in a single or twice-daily tidal reversal difficult. Likewise, the shale laminae separating the two cycles are much thicker than typical clay drapes that characterize slack-water deposition between tidal reversals (Fig. 5B). Close examination shows these shale layers contain bundles of laminae as in the first channel. Therefore, these bedsets more probably record annual sedimentation cycles.

Deposition of annual cycles was interrupted early in the history of channel filling by lateral accretion of the middle unit across the lower fill. Lateral accretion of the middle unit resulted in thicker (and more complete) rhythmites toward the center of the channel, and thinner (more incomplete) rhythmites toward the margin. Similar trends have been noted in an abandoned tidal channel in The Netherlands (Van den Berg, 1981). The subtle discontinuity in the middle unit marks a change in deposition from lateral to vertical accretion; some reworking occurs along the contact. The sharp change between bedsets in the middle and upper units marks the point at which the channel was mostly filled and bedsets became increasingly more reworked.

As in the first channel, the bedsets in the middle part of the fill are inferred to represent annual cycles composed of 23 or fewer neap-spring laminae bundles. The arrangement of 10–14 neap-spring cycles in the sandy part of each composite bedset likewise represents five- to seven-month high-energy depositional cycles. The relatively sharp base of the sandy seasonal cycles and upward transition to shalier seasonal cycles in each bedset is similar to annual cycles documented in tidal channels from North Sea estuaries, where wave energy during winter storms can scour underlying summer deposits (Van den Berg, 1981), but it might also be a function of lateral accretion.

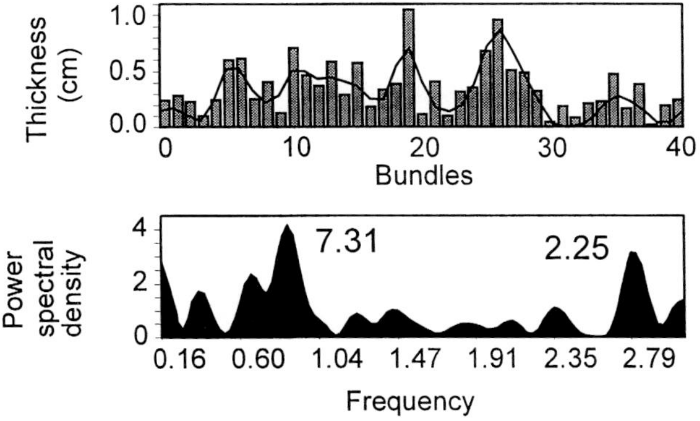

FIG. 7.—Computer-based image analysis of the sandy part of a composite bedset from the middle of channel 2 (as in Fig. 5D) showing distributions of 7.31 laminae bundles. The 2.25 density value may be the result of the alternating sandstone-dominant and shale-dominant nature of the bundles.

Location 3

The third channel-fill deposit occurs in the same unit as the first channel-fill deposit (Fig. 1), but at the top rather than the middle of the sandstone. The deposit is 5-m thick and more than 75-m wide. The north margin of the channel is covered. On the south margin of the channel, beds are slumped (Fig. 8A) and contain *Planolites* trace fossils. Composite bedsets in the fill are similar in scale to the bedsets in the second channel (except at the base) and can similarly be divided into lower sand-dominant and upper shale-dominant parts (Fig. 8B). However, in this deposit the sandy part of each bedset is dominated by cross-cutting ripple stratification (more than 80%); only rare laminae are grouped into thickening and thinning bundles between ripples (Fig. 8C). Slabs cut through the shalier parts of bedsets show thinning and thickening trends of five to eight, millimeter- to submillimeter-scale laminae couplets (Fig. 8D-E), similar to the fine-scale couplets in the first two channels. Bioturbation and soft-sediment deformation appear to rework some couplets (Fig. 8E). As in the first two channels, bedsets thin upwards (Fig. 9A) and shale percentage decreases upwards (Fig. 9B). The number of bundles per bedset and laminae couplets per bundle were not counted vertically throughout the fill because the high percentage of ripple stratification led to considerable lateral variability in individual bedsets.

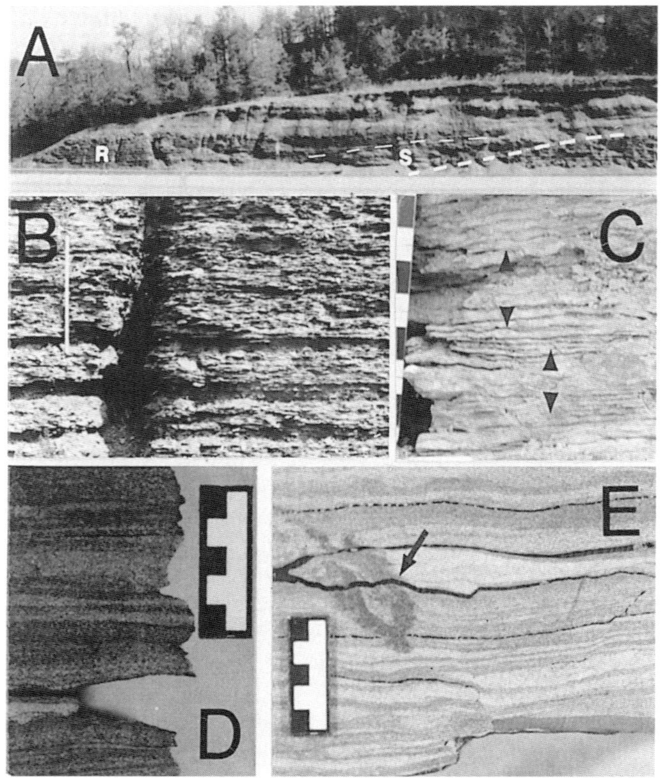

FIG. 8.—Photographs from channel 3. (A) View of the channel, showing marginal slump (S) and rhythmites (R). The outcrop is approximately 6 m tall. (B) Stacked bedsets in the fill consist of a lower sand-dominant interval and an upper shale-dominant interval (as in channel 2) but channel 3 is more ripple bedded. Yardstick for scale (1 yd = 0.9 m). (C) Thickening and thinning laminae couplets between ripple cross-stratification in outcrop (black bars = 1 cm). (D) Detail of prepared and slabbed sample from shaly part of a bedset, showing two laminae bundles with submillimeter-scale laminae couplets (black bars = 2 mm). (E) Another prepared and slabbed sample, showing bundles and laminae couplets disrupted by bioturbation (arrow) (black bars = 2 mm).

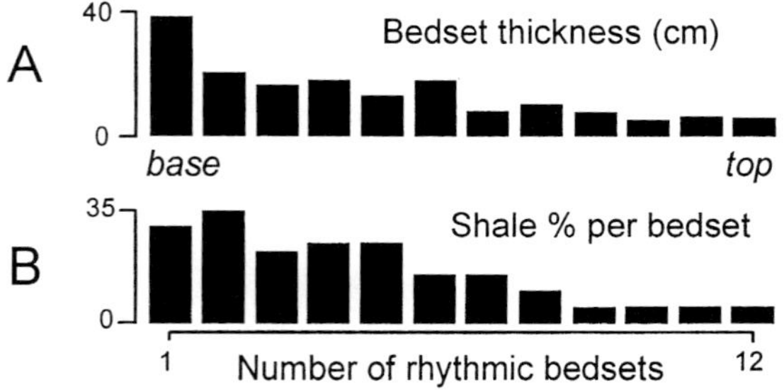

FIG. 9.—Bar charts of (A) bedding thickness and (B) shale percentage per bedset in channel 3.

Interpretation.—

The third channel fill occurs in the same sandstone as the first channel fill and contains vertically thinning bedsets of sandstone and shale laminae, a trend similar to that of bedsets in the second channel-fill deposit. These bedsets may also represent annual deposits composed of winter-summer seasonal cycles, but interpretations are less certain here than for the first two channels. Numerous thickening-upward laminae bundles in the sandy parts of bedsets might represent partial neap-spring cycles, but erosion by subsequent current ripples precludes a definitive measure of each bedset's periodicity. Rare thickening and thinning bundles of laminae couplets in the shaly parts of each bedset were also strongly suggestive of neap-spring cycles, but they were often partially reworked. In the context of the channels studied, these bedsets represent an end member because they are noncyclic and lack well-defined neap-spring cycles, but they exhibit partly eroded remnants of fine-scale laminae couplets and laminae bundles similar to the more rhythmic fills in the other channels. Because this channel occurs in the same unit as the first channel, we can suggest that the fine-scale laminae couplets record daily tidal lamination, rare thickening and thinning laminae bundles record fortnightly cycles, and the larger bedsets represent a repetitive alternation of higher energy (sandy) and lower energy (shaly) conditions, comparable to the seasonal fluctuations seen in the annual rhythmites of the other two channels. However, without a nearby comparison, this interpretation would be difficult to substantiate.

The tidal signature in this channel fill may have been degraded or superimposed by a strong seasonal energy or sediment flux. In modern tidal deposits of the North Sea (Van den Berg, 1981) and Bay of Fundy (Dalrymple et al., 1991), incomplete or amalgamated tidal cycles are often associated with increases in biogenic reworking, changes in shell test sizes, and mortality layers that can be directly related to changes in water temperature that accompany the changing seasons. Yet, in ancient tidal deposits where water temperatures during sedimentation cannot be reconstructed, or in equatorial latitudes (as are assumed for all three channels) where water temperature may not change significantly during the year, lithologic differences caused by current and wave energy may be the only evidence for seasonality. Seasonal changes in sediment flux, either from the seaward or landward side of the tidal channel, could lead to an abundance of ripple stratification and only minor evidence of neap-spring cyclicity preserved in the shalier parts of bedsets, and reworking in the remainder of the bedset, as in channel 3.

DISCUSSION

Although heterolithic tidal rhythmites are often perceived as deposits of tidal flats and subtidal deltas, they also can occur in channel facies. Several examples of interlayered sand and mud bedsets similar to those in this study have been documented in modern tidal channels (Terwindt, 1971; Goldring et al., 1978), and are inferred to be neap-spring and annual tidal cycles (Van den Berg, 1981; Tessier and Gigot, 1989; Roep, 1991). In channel fills described by Terwindt (1971), the interlayered bedsets were most common in the deepest parts of the channels, similar to the study channels. Likewise, the thickest bedsets in a channel studied by Van den Berg (1981) were associated with the passage of the channel thalweg. Whereas tidal channels have a high preservation potential in modern tidal environments (Oomkens, 1974), they are ideal for studies of rhythmic successions. Also, because modern abandoned channels that contain rhythmite deposits can connect to sandier tidal channels (Roep, 1991), the occurrence of isolated tidal rhythmites in the subsurface may indicate larger tidal channels nearby, which are hydrocarbon reservoirs in many basins.

Research in modern tidal environments has indicated that cyclic rhythmites are best developed in the inner reaches of macrotidal estuaries, where suspended-sediment concentrations are high and both bioturbation and wave reworking are minimal (Tessier and Gigot, 1989; Dalrymple et al., 1991, 1996). All three of the study channel fills contain abundant mudstone drapes, are relatively unbioturbated, and contain little or no evidence of primary sedimentary structures formed by wave action. Facies above and beneath the channel fills show considerably more evidence of current and wave reworking and do not exhibit cyclic rhythmites. Presumably, daily, neap-spring, and other tidal periods affected sedimentation in contemporaneous lateral facies in these three study intervals, but only the channels were conducive to cyclic rhythmite preservation. This uniqueness suggests that accommodation space may play a significant role in rhythmite deposition, and possibly in the periodicity of preserved tidal sedimentation.

In all three channel fills, composite bedsets thin upward through a decrease in shale content, the number of laminae couplets per laminae bundle, and the number of laminae bundles per bedset, as well as an increase in ripple stratification (Figs. 4, 6, 9). The occurrence of the thickest bedsets toward the base of the channels and the vertical transition from cyclic rhythmites into increasingly more incomplete, noncyclic rhythmites in the first two channels is inferred to have been caused by decreasing accommodation space as the channels filled. In the first two channels, the upward change from cyclic rhythmites with relatively complete fortnightly sedimentation into noncyclic rhythmites with very incomplete sedimentation records marks the point at which lack of accommodation space began to inhibit preservation of fortnightly sedimentation. In the first channel, the change is relatively gradual. In the second channel, sharp changes occur between the lower and middle unit because of lateral accretion, and then at the top of the fill when the space in front of the laterally accreting margin was filled.

In well-exposed outcrops, channel forms can be readily delineated and the temporal decrease in rhythmite preservation can be interpreted as a response to decreasing accommodation. Without the exposure of a scour, heterolithic tidal rhythmites are more likely to be assigned to a tidal flat or subtidal delta facies than a channel. This is significant, because upward thinning and coarsening rhythmite trends in nonchannel environments can be used to infer changes in the rate of progradation or paleoposition within an estuary. Increasing thickness and completeness of neap-spring bedsets in modern estuaries often reflect a shift toward more inner estuarine positions, where funneling heightens tidal range (Tessier and Gigot, 1989; Dalrymple et al., 1991). Likewise, in inner estuarine positions, a lateral decrease in the number of daily tidal laminae has been noted from the lower intertidal to upper intertidal zone, caused by reworking and nondeposition of weaker daily tides in the upper intertidal zone (Tessier, 1993). Within channels, such differences might simply reflect a loss of accommodation space without a change in the relative position of the rhythmites in the estuary.

Knowledge about the type of accommodation space that leads to the preservation of a given rhythmite is also critical when inferring ancient sediment accumulation rates from tidal rhythmites. Daily, monthly, or annual sedimentation rates of channel fill, calculated from rhythmites, simply reflect the amount of space available to be filled during deposition. Sedimentation rates calculated from the thickest tidal cycles toward the base of the channel are 15 cm/yr for the first channel, 10 cm/yr for the second channel, and as much as 80 cm/yr for the third channel. In contrast, sedimentation rates calculated from the top of the channel are less than 6 cm/yr for the first channel, less than 5 cm/yr for the second channel, and less than 4 cm/yr for the third channel. Likewise, the lateral change in bedset thickness from the axis to the margin of the channels would lead to a decrease in the inferred sedimentation rates for the same cycles. These results suggest that the rate of cyclic rhythmite deposition should be looked at as a local accumulation rate rather than an average accumulation or sedimentation rate related to sediment flux into the larger depositional system. Accumulation rates were certainly smaller outside of the channels. This is important because it shows that a vertical decrease in bedset thickness and number of laminae per bedset in a tidal environment need not reflect a shift in depositional environment or relative sea level, but may simply reflect a decrease in the rate at which any sediment could accumulate because of limited space.

Two other points of discussion bear mention concerning interpretation of tidal cyclicities. First, previous interpretation of neap-spring periodicity for the annual cycles in the first channel (Greb and Archer, 1995) illustrates the need for descriptive (e.g., bedset, cyclic versus noncyclic rhythmite) rather than interpretive (e.g., neap-spring or presumed neap-spring cycles) terminology when describing tidal rhythmites. New appreciation of increasingly more complex tidal cycles may change how units are interpreted with time, and nongenetic descriptions would aid in comparisons with future data.

Second, the common incomplete nature of the preserved tidal record may inherently bias interpretations of ancient tidal periodicities. In all three channels, subtle discontinuities highlight the incompleteness of the sedimentary record. In some cases, discontinuities occur at such low angles that they can only be observed by stepping back from the outcrop. These surfaces could be formed during a single change in current energy. For instance, in channel 2, the discontinuity in the middle unit may simply mark a shift or halt to lateral accretion and slight reworking of the surface through time, although the amount of the hiatus across the surface increases toward the channel margin. More flat-lying discontinuities, as in channel 1, could be created by storms and storm-enhanced tides with little interruption in sedimentation. However, the same type of surface could also mark much longer hiatuses, with little suggestion or evidence of the amount of missing sedimentary record. For example, the slump along the margin of the third channel records sedimentation on the channel margin that predated the overlying fill, but it is only preserved in the slump and not toward the center of the channel, suggesting a significant missing section in the center of the channel.

At a smaller scale, erosion or nondeposition of daily or half-daily couplets can cause complications in identifying diurnal and semidiurnal regimes (DeBoer et al., 1989; Tessier, 1993; Archer, et al., 1995; Archer, 1996). Partial preservation of twice-daily laminae couplets in a semidiurnal regime can mimic the deposits produced by once-daily laminae couplets in a diurnal regime, which can lead to misinterpretation of either cycle

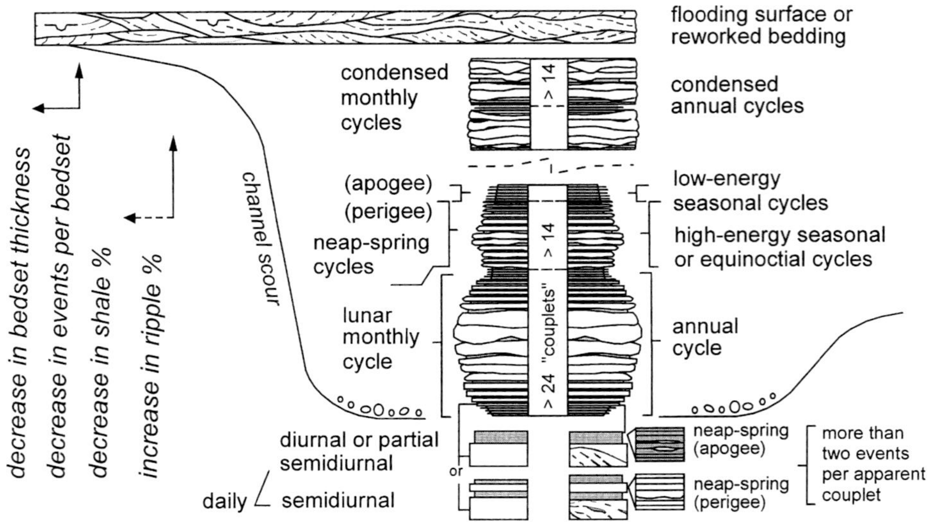

FIG. 10.—Trends noted in the channel fills with respect to the channel margins and axes, focusing on the similarity of the criteria that could be used to infer fortnightly versus seasonal and monthly versus annual tidal cycles. The key to interpretation is identifying more than two events per bundle, or apparent couplet, at the smallest (sometimes less than 1 mm) scale of bedding in the most complete bedsets. Amalgamation of cycles vertically and laterally in the channel fills makes interpretation more difficult.

order. When daily laminae couplets are only millimeters thick, interpretations can be even more complicated. An annual deposit optimally contains 12 monthly cycles or 24 neap-spring cycles. When neap-spring cycles are themselves only millimeters thick, as in the first two channels, bundles of laminae or ripples representative of a laminae bundle can be misinterpreted as daily cycles, regardless of the original tidal regime (Fig. 10). When this happens, composite bedsets are likely to be interpreted as monthly deposits, because 24 is less than, but close to, the optimal number of laminae expected for a lunar month in a diurnal system. Likewise, if these bedsets are misinterpreted as monthly deposits, the sandy and shaly parts of each bedset are inferred to represent apogeen and perigeen cycles caused by fortnightly inequality, rather than seasonal or equinoctial differences in annual cycles (Fig. 10).

The possibility of misinterpretation between daily and neap-spring cyclicity increases vertically within rhythmic channel fills as single ripple laminae replace submillimeter-scale laminae bundles. Also, as the shaly seasonal part of each composite bedset is lost, successive sandy bedsets may be superimposed on each other. The amalgamated bedsets might appear to be stacked, ripple-dominated neap-spring cycles, rather than stacked annual cycles, in which five to six months of sedimentation were not preserved. These problems are further complicated if the bedding is bioturbated. Bioturbation can rework the millimeter-scale laminae in bundles, such that they look like a single bioturbated laminae and are misinterpreted as a daily rather than fortnightly deposit.

All of these factors illustrate the need for careful analysis of the finest scale of bedding before inferring temporal significance to tidal cycles. In channels or other environments where accommodation space exerts strong controls on sedimentation, spatial changes in rhythmite completeness and appearance can be dramatic across short distances and may be unrelated to paleoposition, tidal regime, temporal duration, or other factors that can affect the completeness of the tidal sedimentation record.

ACKNOWLEDGMENTS

The authors wish to thank P. Myrow and an anonymous reviewer, as well as D. R. Chesnut and C. F. Eble, for their thoughtful comments and suggestions. We also acknowledge Meg Smath for timely editorial review and Shirley Dawson for typing. Funding for this project was provided by the Kentucky Geological Survey.

REFERENCES

ARCHER, A. W., 1994, Extraction of sedimentological information via computer-based image analyses of gray shales in Carboniferous coal-bearing sections of Indiana and Kansas, U.S.A: Mathematical Geology, v. 26, p. 47–65.

ARCHER, A. W., 1995, Modeling of cyclic tidal rhythmites based on a range of diurnal to semidiurnal tidal-station data: Marine Geology, v. 123, p. 1–10.

ARCHER, A. W., 1996, Reliability of lunar orbital periods extracted from ancient cyclic tidal rhythmites: Earth and Planetary Science Letters, v. 141, p. 1–10.

ARCHER, A. W., KVALE, E. P., AND JOHNSON, H. R., 1991, Analysis of modern equatorial tidal periodicities as a test of information encoded in tidal rhythmites, *in* Smith, D. G., Reinson. G. E., Zaitlin, B. A., and Rahmani, R. A., eds., Clastic Tidal Sedimentology: Calgary, Canadian Society of Petroleum Geologists Memoir 16, p. 189–196.

ARCHER, A. W., KUECHER, G. J., AND KVALE, E. P., 1995, The role of tidal-velocity asymmetries in the deposition of silty tidal rhyihmites (Carboniferous, Eastern Interior Coal Basin, U.S.A.): Journal of Sedimentary Research, v. B65, p. 408–416.

BROWN, M. A., ARCHER, A. W., AND KVALE, E. P., 1990, Neap-spring tidal cyclicity in laminated carbonate channel-fill deposits and its implications: Salem Limestone (Mississippian), south-central Indiana, U.S.A.: Journal of Sedimentary Petrology, v. 60, p.152–159.

DALRYMPLE, R. W., MAKINO, Y., AND ZAITLIN, B. A., 1991, Temporal and spatial patterns of rhythmite deposition on mud flats in the macrotidal Cobequid Bay-Salmon River estuary, Bay of Fundy, Canada, *in* Smith, D. G., Reinson, G. E., Zaitlin, B. A., and Rahmani, R. A., Clastic Tidal Sedimentology: Calgary, Canadian Society of Petroleum Geologists Memoir 16, p. 137–160.

DALRYMPLE, R. W., BAKER, E. K., HUGHES, M., AND HARRIS, P. T., 1996, Sedimentation in the Fly River Delta (Papua New Guinea), a muddy, tide-dominated delta in a foreland-basin setting (abs.): Tidalites 1996, International Conference on Tidal Sedimentology, p. 19–20.

DEBOER, P. L., OOST, A. P., AND VISSER, M. J., 1989, The diurnal inequality of the tide as a parameter for recognizing tidal influences: Journal of Sedimentary Petrology, v. 59, p. 912–921.

GOLDRING, R., BOSENCE, D. W. J., AND BLAKE, T., 1978, Estuarine sedimentation in the Eocene of southern England: Sedimentology, v. 25, p. 861–876.

GREB, S. F., AND ARCHER, A. W., 1995, Rhythmic sedimentation in a mixed tide and wave deposit, Hazel Patch sandstone (Lower Pennsylvanian), Eastern Kentucky Coal Field: Journal of Sedimentary Research, v. 65, p. 96–106.

Greb, S. F., and Chesnut, D. R., Jr., 1992, Transgressive channel filling in the Breathitt Formation (Upper Carboniferous), Eastern Kentucky Coal Field, U.S.A.: Sedimentary Geology, v. 75, p. 209–221.

Jaeger, J. M., and Nittrouer, C. A., 1995, Tidal controls on the formation of fine-scale sedimentary strata near the Amazon River mouth: Marine Geology, v. 125, p. 259–281.

Kuecher, G. J., Woodland, B. G., and Broadhurst, F. M., 1990, Evidence of deposition from individual tides and of tidal cycles from the Francis Creek Shale (host rock to the Mazon Creek biota), Westphalian D (Pennsylvanian), northeastern Illinois: Sedimentary Geology, v. 68, p. 211–221.

Kvale, E. P., Archer, A. W., and Johnson, H. R., 1989, Daily, monthly, and yearly tidal cycles in laminated siltstones of the Mansfield Formation (Pennsylvanian) of Indiana: Geology, v. 17, p. 365–368.

Kvale, E. P., and Archer, A. W., 1990, Tidal deposits associated with low-sulfur coals, Brazil Formation (Lower Pennsylvanian), Indiana: Journal of Sedimentary Petrology, v. 60, p. 563–574.

Kvale, E. P., and Archer, A. W., 1991, Characteristics of two Pennsylvanian-age, semidiurnal tidal deposits in the Illinois Basin, U.S.A, *in* Smith, D. G., Reinson, G. E., Zaitlin, B. A., and Rahmani, R. A., Clastic Tidal Sedimentology: Calgary, Canadian Society of Petroleum Geologists Memoir 16, p. 179–188.

Oomkens, E., 1974, Lithofacies in the late Quaternary Niger Delta complex: Sedimentology, v. 21, p. 195–222.

Roep, T. B., 1991, Neap-spring cycles in a subrecent tidal channel fill (3665 BP) at Schoorldam, northwest Netherlands: Sedimentary Geology, v. 71, p. 213–230.

Smith, N. D., Phillips, A. C., and Powell, R. D., 1990, Tidal drawdown: A mechanism for producing cyclic sediment laminations in glaciomarine deltas: Geology, v. 18, p. 10–13.

Terwindt, J. H. J., 1971, Lithofacies of inshore estuarine and tidal-inlet deposits: Geologie Mijnbouw, v. 50, p. 515–526.

Tessier, B., 1993, Upper intertidal rhythmites in the Mont-Saint-Michel Bay (northwest France)—Perspectives for paleoreconstruction: Marine Geology, v. 110, p. 355–367.

Tessier, B., and Gigot, P., 1989, A vertical record of different tidal cyclicities—An example from the Miocene marine molasse of Digne: Sedimentology, v. 36, p. 767–776.

Van den Berg, J. H., 1981, Rhythmic seasonal layering in a mesotidal channel fill sequence, Oosterschelde Mouth, the Netherlands: Oxford, International Association of Sedimentologists Special Publications, v. 5, p. 147–159.

Visser, M. J., 1980, Neap-spring cycles reflected in Holocene subtidal large-scale bed form deposits—A preliminary note: Geology, v. 8, p. 543–546.

Williams, G. E., 1991, Upper Proterozoic tidal rhythmites, South Australia—Sedimentary features, deposition, and implications for the earth's paleorotation, *in* Smith, D. G., Reinson, G. E., Zaitlin, B. A., and Rahmani, R. A., Clastic Tidal Sedimentology: Calgary, Canadian Society of Petroleum Geologists Memoir 16, p. 161–177.

RHYTHMIC SEDIMENTATION IN A MID-PENNSYLVANIAN DELTA-FRONT SUCCESSION, MAGOFFIN MEMBER (FOUR CORNERS FORMATION, BREATHITT GROUP), EASTERN KENTUCKY: A NEAR-COMPLETE RECORD OF DAILY, SEMI-MONTHLY, AND MONTHLY TIDAL PERIODICITIES

RHONDA M. ADKINS AND KENNETH A. ERIKSSON

Department of Geological Sciences, 4044 Derring Hall, Virginia Tech, Blacksburg, Virginia 24061

ABSTRACT: The Magoffin Member of the Four Corners Formation (Breathitt Group) outcrops in eastern Kentucky as a predominantly coarsening-upward succession of rhythmically interstratified sandstone, siltstone, and mudstone with minor occurrences of limestone at its base. Primary sedimentary structures, trace fossils, vertical successions of facies, and sediment body geometries suggest that these rhythmically bedded sediments were deposited in a delta-front/distributary-mouth-bar setting. The overall thickness of the Magoffin Member is highly variable. Within the study area, it ranges in thickness from less than 25 m to greater than 40 m. Evidence indicates that in areas where the Magoffin Member is thickest, it tends to be sandier and contain rhythmite intervals that are thicker and more complete than in areas where it is thin. The member displays several orders of centimeter- to decimeter-scale cycles that are consistent with semi-diurnal, diurnal, semi-monthly, and monthly tidal periodicities. Half-synodic (semi-monthly neap-spring-neap) and anomalistic (monthly) lunar periodicities are manifested by the systematic thickening and thinning of the shorter duration cycles. The rhythmite interval ranges in thickness from less than 5 m up to 20 m and records up to four months of deposition. Accumulation rates for the rhythmites typically ranged from 20–100 cm per neap-spring-neap event, but reached rates of over 30 cm per day in areas where the Magoffin Member is thickest and the most proximal deltaic facies are preserved. The nature of tidal bundling suggests that these rhythmites accumulated in a mixed, dominantly semi-diurnal tidal system where both lunar phases and lunar declination influenced tidal cyclicity.

INTRODUCTION

Siliciclastic tidal rhythmites have been identified from a range of geologic time periods, including from the Proterozoic of Australia (Williams, 1989, 1990) and the western United States (Chan et al., 1994); the Carboniferous of southeastern England (Allen, 1981) and the eastern United States (Kvale et al., 1989; Kvale and Archer, 1991; Martino and Sanderson, 1993; Greb and Archer, 1995); the Jurassic of the western United States (Kreisa and Moiola, 1986); and the Tertiary of France (Tessier and Gigot, 1989). Modern tidal rhythmite sequences have also been reported from the Bay of Fundy (Dalrymple et al., 1990, 1991); Schoorldam, northwest Netherlands (Roep, 1991); and Mont-Saint-Michel Bay (Tessier, 1993). Well-preserved tidal rhythmites provide high-resolution depositional records that can be used to gain information on paleotidal periodicities and on the evolution of the earth-moon system (Williams, 1991; Sonnet et al., 1996). Rhythmites have proven to be useful paleoenvironmental indicators, providing evidence of marine processes in the form of tidal rhythms in fine-grained nonfossiliferous facies. Absolute rates of local sedimentation can also be deduced from the information recorded within these successions (Williams, 1989). Recent work on rhythmites within Carboniferous strata of the mid-continental and eastern United States indicates that these successions are often characterized by rapid sedimentation rates and provide relatively complete records of continuous deposition, ranging from several months to tens of years (Kvale and Archer, 1990; Miller and Eriksson, 1997).

The mid-Pennsylvanian Breathitt Group of eastern Kentucky has traditionally been interpreted as a fluvio-deltaic deposit (e.g., Cobb et al., 1981; Tankard, 1986). The Breathitt Group contains several intervals of rhythmically interstratified sandstones, siltstones, and mudstones up to 20 m thick. One of these rhythmic intervals, within the upper portion of the Magoffin Member (Four Corners Formation, Breathitt Group), is the focus of this study. Within this interval, a record of tidal cyclicity is manifested by a systematic thickening and thinning of mudstone-draped sandstones. This interval provides a relatively short (up to four months) but complete record of mid-Pennsylvanian, ebb-tidal activity and displays laminae bundling patterns that are consistent with semi-diurnal, diurnal, semi-monthly, and monthly tidal periodicities. This paper discusses the nature of these depositional cycles and demonstrates that the thickest and most complete depositional cycles are developed where the Magoffin Member is the thickest.

METHOD OF STUDY

This study is primarily based upon field investigations of the Magoffin Member where it is exposed in a series of roadcuts along Kentucky Highway 80 and other secondary roads in Knott County, Kentucky, U.S.A. (Fig. 1). These outcrops were evaluated in terms of lithology, fossils, trace fossils, sedimentary structures including laminae and bed thickness variations, vertical stacking of facies, and sediment body geometries. Stratigraphic sections at locations A through G were measured from the base of the Magoffin Member (where identifiable) to the overlying fluvial-channel deposits to determine lateral thickness variations of the Magoffin Member within the study area. Subsurface (strip-log) data were also gathered from 22 surrounding 7-1/2 minute USGS quadrangles to aid in determining areal thickness variations of the Magoffin Member. Exposures of the Magoffin Member exhibit well-defined decimeter- to meter-scale cyclicity. Analyses of the cycles focused on determining the number of laminae/beds per neap-spring-neap bundle and the variations in laminae and bundle thicknesses. The rhythmite intervals were logged at a millimeter scale at five locations (A, C, D, E, and G; Fig. 1) along Kentucky Highway 80 (distance between outcrops ranges from 0.25 km to 3.0 km) to determine lateral changes in rhythmite thicknesses and bundling patterns. No detailed rhythmite analyses were conducted at locations B and F. The measurements were plotted on bar graphs to aid in the identification of cyclic patterns. Fast-Fourier Transform spectral analyses were run to detect tidal periodicities.

DEPOSITIONAL SETTING

The Breathitt Group outcrops along the western edge of the Pocahontas Basin (central part of the Appalachian Basin) in western West Virginia and eastern Kentucky. The Pocahontas

Tidalites: Processes and Products, SEPM Special Publication No. 61
 ISBN 1-56576-059-X

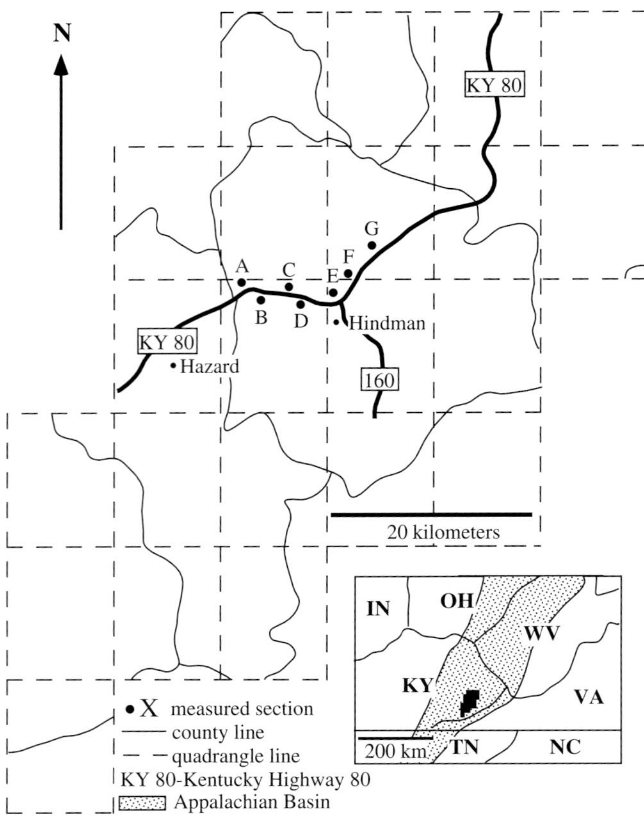

FIG. 1.—Location map of measured stratigraphic sections of Magoffin rhythmites in Knott County, Kentucky, U.S.A.

Basin, bounded by the Appalachian fold-and-thrust belt to the southeast and the Cincinnati-Waverly arch to the northwest (Tankard, 1986), was produced with the onset of Alleghenian (early Pennsylvanian) overthrusting as the eastern continental margin of North America collided with west Africa (Quinton and Beaumont, 1984). By Middle Pennsylvanian time, Alleghenian overthrusting had resulted in flexural downwarping of the basin and the deposition of immature sandstones, mudstones, and coals of the Breathitt Group. Marine oscillations resulted in these sediments being deposited in a basin that fluctuated between underfill (restricted marine) and overfill (nonmarine) conditions (Tankard, 1986). Sediment transport was west to northwest during this time (Englund and Thomas, 1989).

The Breathitt Group reaches a thickness of up to 950 m (Wanless, 1975) in eastern Kentucky, where it is divided into seven formations (Fig. 2). Thick orthoquartzose sandstone bodies divide the lower Breathitt Group, whereas widespread marine units divide the upper portion. The Magoffin Member, located at the base of the Four Corners Formation (Fig. 2), is the most laterally extensive of these marine members. It differentiates the Four Corners Formation from the underlying Hyden Formation (Chesnut, 1992) and contains the rhythmites discussed in this paper.

Within the study area, the Magoffin Member is typically 25–40 m thick and consists of a generally coarsening-upward succession (Fig. 3). At the base, the Magoffin Member is defined by a thin (less than 25 cm), dark, highly fossiliferous limestone, containing marine bivalves, crinoids, brachiopods, and ammonoids that directly overlies the Copeland Coal Zone. In the absence of the limestone, the base of the Magoffin is taken to be the top of the Copeland Coal Zone (Dennis, 1975; Chesnut, 1992; Bennington, 1996). The limestone, where present, is gradationally overlain by a 1- to 2-m-thick layer of dark gray shale with abundant marine fossils (same general fauna as limestone) and disseminated carbonaceous plant fragments. This basal fossiliferous zone is gradationally overlain by a succession of coarsening-upward, interstratified mudstones and siltstones up to 30-m thick (Fig. 3). This interval contains abundant siderite concretions, disseminated and carbonaceous plant fragments, and some marine fossils. Thin (up to 2 cm), laterally continuous siderite beds, and thin (up to 2 cm), discontinuous, knobby siderite stringers occur every few centimeters throughout this interval. Large siderite and limestone concretions, up to 1 m × 3 m (Aitken and Flint, 1995), occur along distinct horizons. Bidirectional tool marks located on the bottom of several siltstone beds at location A indicate a northwest/southeast paleoflow direction (major trend at 315°). There is some evidence of rhythmic cyclicity within this interval, but the beds/laminae are thin and laterally discontinuous, making it difficult to obtain thickness measurements.

The succession of interstratified mudstones and siltstones grades upwards into an interval of coarsening-upward, rhythmically interstratified mudstones and sandstones up to 20 m thick (Fig. 3). The sandstones are comprised of predominantly subangular quartz grains with abundant mica and disseminated carbonaceous plant fragments. Individual beds range in thick-

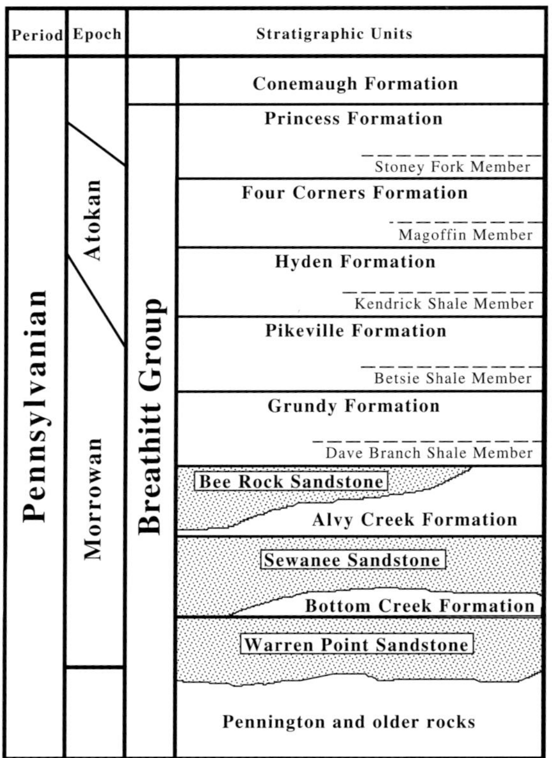

FIG. 2.—Stratigraphic framework of Pennsylvanian rocks in eastern Kentucky. Thick sandstone units subdivide the lower Breathitt Group and widespread marine units, such as the Magoffin Member, subdivide the Upper Breathitt Group into constituent formations. Adapted from Chesnut (1992).

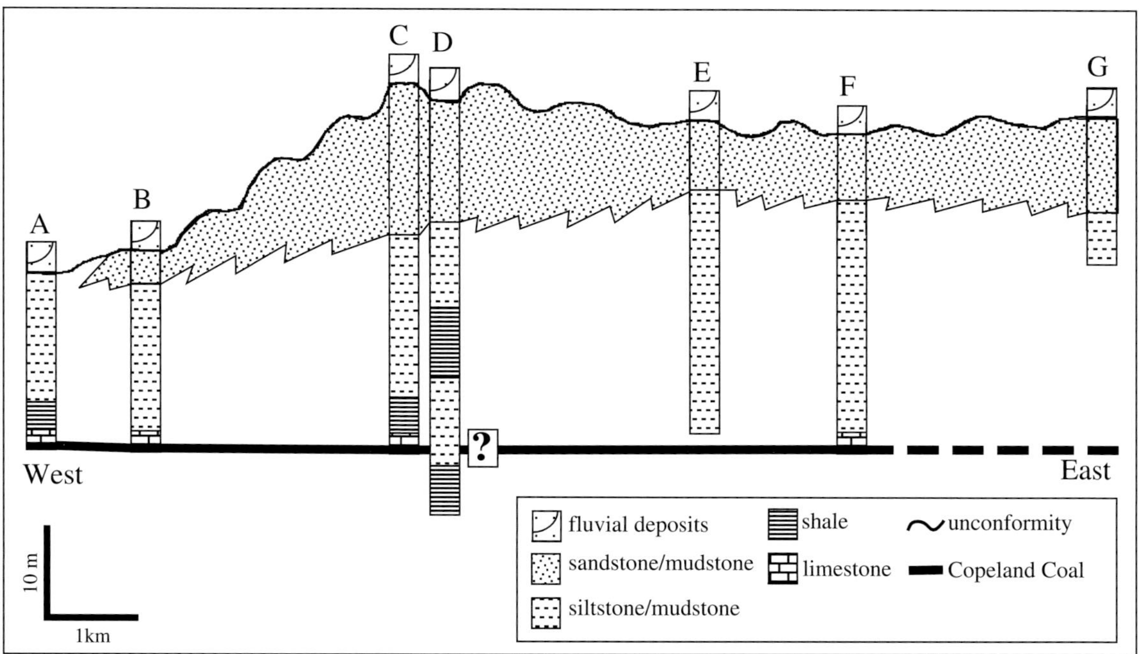

FIG. 3.—Schematic stratigraphic cross-section (down the east to west paleoslope) through measured Magoffin Member outcrops. This cross-section extends from the underlying Copeland Coal (datum) to the overlying incised-valley succession and shows east-to-west progradation of the deltaic facies. The thickness of the Magoffin Member is highly variable within the study area but, where the Magoffin Member is thickest, a thick succession of interlaminated mudstone and sandstone is developed at the top of the member. The rhythmite package is thickest, and also most complete, at these locations.

ness from less than 1 cm to greater than 20 cm. At the base, this interval contains some wavy and flaser bedding but towards the top all beds are tabular. Beds have a depositional dip of approximately 5° to the west to northwest (Aitken and Flint, 1995). Scattered siderite beds occur within this unit but are not as prolific as below. Large-scale siderite concretions, up to 4.5 m × 1.5 m (Aitken and Flint, 1995), also occur along distinct horizons. Small-scale dewatering structures are found throughout this interval. Rare climbing ripples, at the base of a few beds, indicate a generally westward paleoflow direction. Both vertical and horizontal trace fossils are rarely found within this interval. Vertical tubes (*Skolithos?*) generally occur in layers less than 5 cm thick. Horizontal feeding traces (*Psammichnites sp.*) are also found on the top surfaces of some beds. Multistory, cross-bedded sandstones erosionally truncate the Magoffin Member (Fig. 3). Each sandstone bed is 1–5 m thick. Erosional bounding surfaces separate each sandstone. Pockets of matrix-supported pebbles, coal spars, or heavy minerals occur at the base of many of these beds (Aitken and Flint, 1995).

The thickness of the Magoffin Member (from base of limestone or top of Copeland Coal Zone to incised sandstones) is highly variable (Fig. 4). Within the study area, three depocenters are recognizable. Four of the five measured rhythmite outcrops (C, D, E, and G) are located within the northern depocenter. The upper Magoffin Member is sandier where it is thick and muddier where it is thin (Figs. 3 and 5).

The Magoffin Member is a progradational succession that probably represents prodeltaic (basal fossiliferous zone) to delta-front/distributary-mouth-bar (upper rhythmic unit) deposition. The lower basal limestone and overlying dark shale represents a condensed section (Bennington, 1996) related to maximum flooding of the depositional interface. The section overlying the basal fossiliferous zone has been interpreted as a marine shelf deposit on the basis of its lithological relationships and fossil content (Aitken and Flint, 1995). The tool-marked beds are probably deltaic turbidite deposits. The large concretions are thought to have formed by early diagenetic precipitation of calcite and siderite. They can be used to distinguish marine-to-brackish from nonmarine sediments (Chesnut, 1981). The upper, non-fossiliferous section has been interpreted as a series of coalesced distributary mouth bars based on lateral facies geometries, the presence of vertical escape burrows, and the rhythmic nature of these deposits (Aitken and Flint, 1995)—which are herein interpreted as displaying cyclicity consistent with known tidal periodicities. The coalesced distributary-mouth-bar interpretation is consistent with Magoffin thickness variations depicted in Figs. 4 and 5, where each area of increased accumulation represents one depocenter. The inclined beds indicate the presence of large-scale clinoforms that prograded towards the northwest. The Magoffin Member, above the dark shale, probably comprises a Highstand Systems Tract. The erosional contact between the Magoffin Member and the overlying fluvial sandstones probably marks a lowstand period and represents a major fourth-order sequence boundary (Chesnut, 1989). The overlying thick sandstones are interpreted as stacked fluvial channel deposits that occupy an incised paleovalley (Aitken and Flint, 1995) and represent the base of a Transgressive Systems Tract. These overlying sandstones have also been interpreted as the deposits of distributary channels (Cobb et al., 1981) and shallow anastomosing streams (Rice et al., 1990).

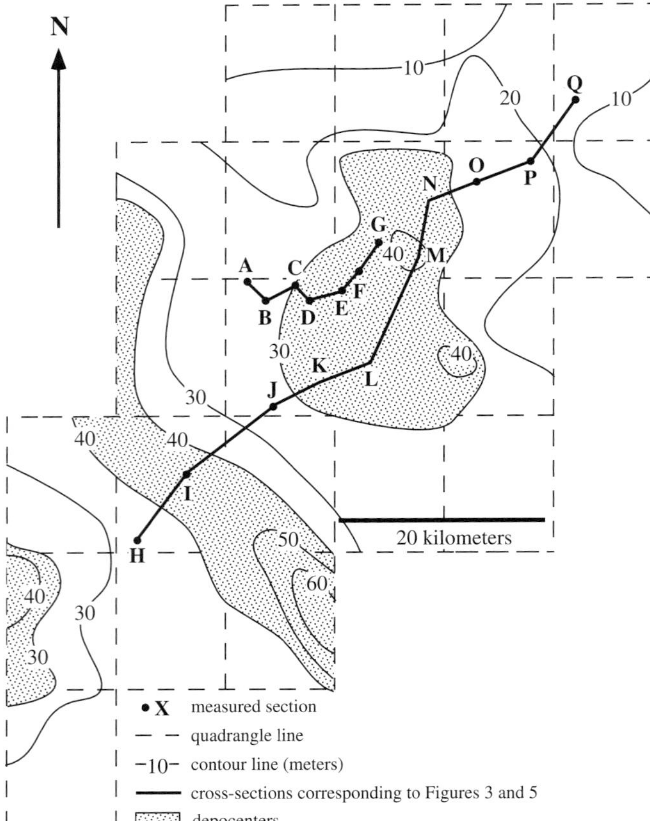

FIG. 4.—Contour map of study area. Three depocenters are located within this area. Most measured Magoffin sections are located within the northernmost depocenter (Locations C–G). Line A-G corresponds with Fig. 3. Line H-Q corresponds with Fig. 5.

TIDAL CYCLICITY

Four orders of cyclicity, representing semi-diurnal, diurnal, semi-monthly, and monthly tidal periodicities are recognized within the upper 20 m of the Magoffin Member (Fig. 6). Preservation of these cycles is particularly evident within the northern depocenter (locations C, D, E, and G), where sand-sized sediments and thick semi-diurnal rhythmite couplets (up to approximately 20 cm) are present. Outside the northern depocenter (location A), tidal cycles are less obvious. The presence of finer-grained sediments and thinner, more laterally discontinuous semi-diurnal couplets makes it difficult to identify semi-monthly and monthly tidal packages.

Semi-diurnal Cycle

Individual tidal events are represented by fining-upward sandstone/mudstone couplets (Fig. 7) that range in thickness from less than 0.5 cm to greater than 20 cm. The lower portion of each couplet is comprised of a thick layer of predominantly sub-angular, fine sand-sized quartz grains with minor amounts of coaly plant fragments and micas. The sandstone intervals are typically internally stratified, with coal fragments defining abundant thin (less than 0.5 mm), laterally discontinuous parallel laminae. Both thick (greater than 5 cm) and thin (less than 1 cm) beds/laminae display internal stratification. Rarely, climbing ripples are developed at the base of sandstone laminae/beds. The upper portion of each couplet is comprised of a thin, dark layer consisting of coaly plant fragments, clays, and micas. These couplets generally occur as laterally continuous layers with little to no evidence of wave reworking and only minor bioturbation.

The couplets are interpreted to have been deposited under the influence of tides. Sand-sized sediment was deposited during peak tidal flows, whereas the mud drapes accumulated through suspension fallout during slackwater periods (Kvale et

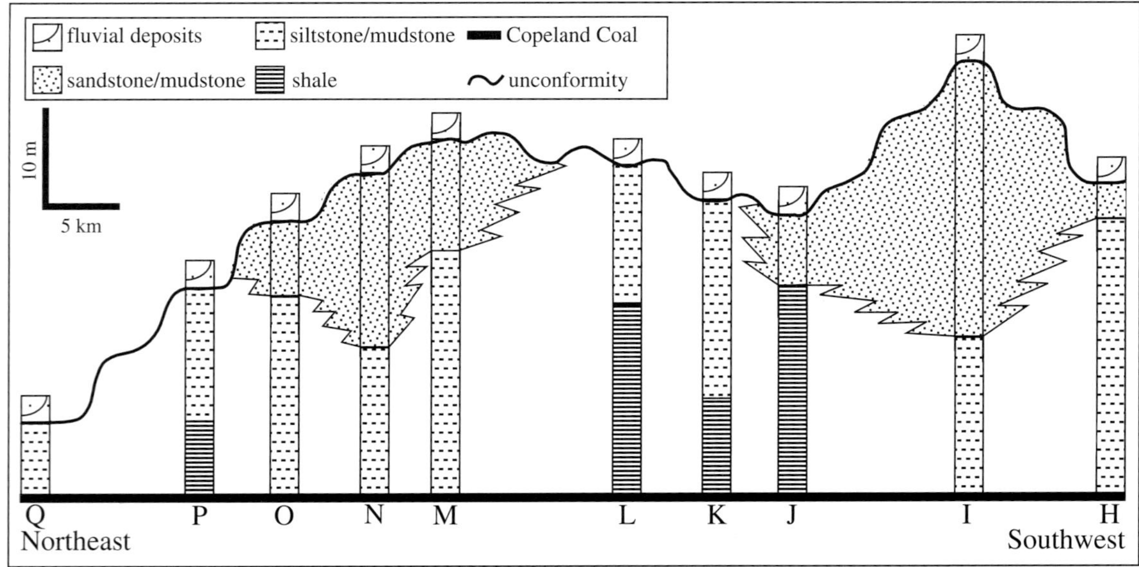

FIG. 5.—Schematic cross-section (parallel to inferred paleoshoreline) through Magoffin Member, based upon subsurface data. Sections extend from the underlying Copeland Coal (datum) to the overlying incised-valley succession. Where the Magoffin Member is thickest, a thick succession of interlaminated mudstone and sandstone is developed at the top of the member. No information concerning the rhythmite package is available for these sections.

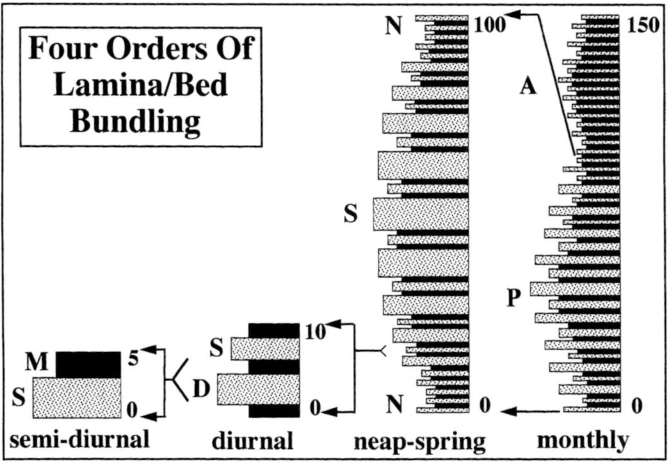

FIG. 6.—Idealized diagram of the hierarchy of depositional cycles found within the Magoffin Member. Sandstone (S)/mudstone (M) couplets record semi-diurnal cycles. Alternating dominant (D) and subordinate (S) couplets record diurnal cycles. Systematically thickening and thinning neap (N)/spring (S) cycles record semi-monthly cycles. Alternating apogean (A) and perigean (P) deposits record monthly cycles. Typical thicknesses are given in centimeters.

FIG. 7.—Photograph of typical semi-diurnal sandstone-mudstone couplets. Relatively thick sandstones (S) were deposited during tidal flows. Thin mudstones (M) were deposited during subsequent slackwater periods. Photograph was taken at location C. Scale units are in centimeters.

al., 1989). The origin of the internal stratification is discussed later in the paper.

Diurnal Cycle

Throughout most of the rhythmite interval, couplets are typically developed as alternating thick/thin pairs. This relationship can be seen in outcrop (Fig. 8) and on bar graphs of measured couplet thicknesses (Fig. 9).

The thick/thin pairing of individual couplets is demonstrative of the diurnal inequality of the tide found within mixed, predominantly semi-diurnal tidal settings (cf. Kvale et al., 1989). Within such tidal settings, lunar declination affects the strength of the semi-diurnal tides, resulting in one dominant and one subordinate daily tide (cf. DeBoer et al., 1989; Kvale et al., 1989). The thick/thin pairs of couplets are therefore interpreted to reflect deposition by both the dominant and subordinate daily tidal events.

Semi-monthly Cycle

Throughout the study area, semi-diurnal couplets are arranged in thickening and thinning cycles consisting of up to 28 couplets. Each thin-thick-thin cycle is typically 20–100 cm thick and rarely contains fewer than 20 sandstone/mudstone couplets. These bundles are apparent in outcrop (Fig. 10) and on bar graphs of measured rhythmite thicknesses (Fig. 9).

These packages are interpreted as semi-monthly (neap-spring-neap) tidal cycles. Neap-spring-neap cycles reflect the half-synodic lunar orbit around the earth (Allen, 1981). Thick couplets develop during periods of syzygy when the earth, moon, and sun are in alignment (Lanier et al., 1993; Kvale et al., 1995), whereas thin couplets develop when the earth, moon, and sun form a right angle (Kvale et al., 1995). In subtidal semi-diurnal settings, approximately 28 tidal inundations are expected per fortnight (Dalrymple, 1992). The presence of a maximum of 20 to 28 couplets per neap-spring-neap cycle indicates the preservation of a record of nearly every ebb or flood tidal event during the two weeks of deposition.

Monthly Cycle

Where several months of sediment accumulation are preserved, successive neap-spring-neap cycles commonly display a thick/thin relationship. The thicker semi-monthly package usually contains more abundant, coarser, and thicker couplets than the thinner semi-monthly package. This pairing of fortnightly deposits is apparent in outcrop (Fig. 11) and on bar graphs of measured couplet thicknesses (Fig. 9).

This thick/thin relationship reflects the anomalistic monthly periodicity of the moon, which is based upon its elongated elliptical orbit around the earth. Perigean (minimum lunar distance) periods are characterized by higher-than-average tides, resulting in the deposition of relatively thick neap-spring-neap bundles. Apogean (maximum lunar distance) periods, in con-

FIG. 8.—Photograph of thick/thin daily tidal deposits. This cyclicity records both the dominant (thick; D) and subordinate (thin; S) daily tides within a semi-diurnal tidal setting. Photograph taken at location D. Scale units are in centimeters.

trast, are characterized by lower-than-average tides, which result in the deposition of relatively thin neap-spring-neap bundles (Archer et al., 1991; Kvale et al., 1995).

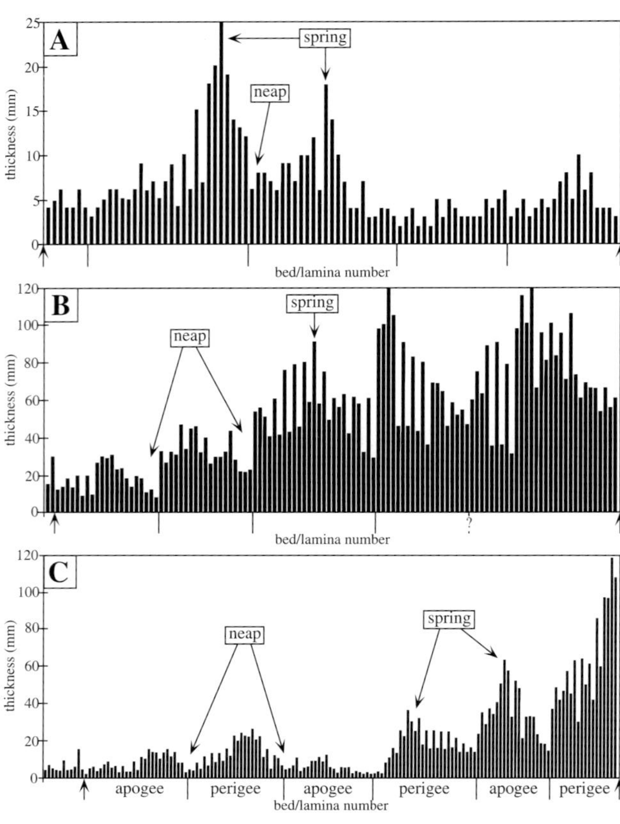

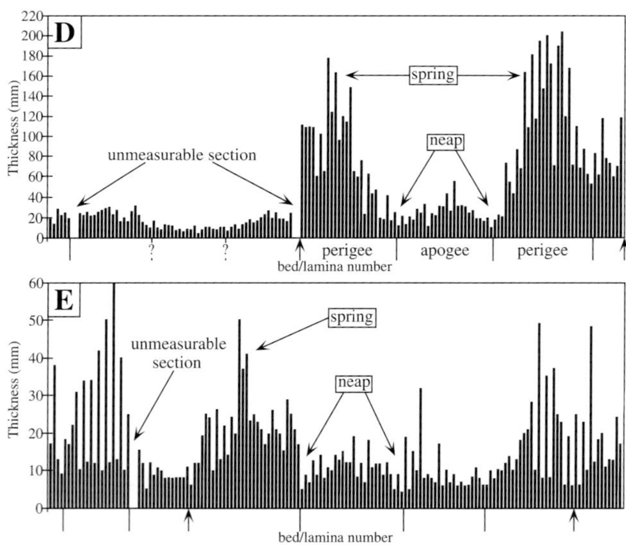

FIG. 9.—Bar graphs of measured rhythmite thicknesses (refer to Fig. 1 for locations). Bar graph A is from location A; bar graph B is from location C; bar graph C is from location D; bar graph D is from location E; bar graph E is from location G. Semi-monthly (neap-spring-neap) cycles are marked by tick marks along the x-axes. Monthly (perigee-apogee) cycles, where discernible, are marked along the x-axes. Also, note the daily thick/thin pairs within these intervals. Arrows along the x-axes indicate intervals used for harmonic analyses (Fig. 12).

HARMONIC ANALYSIS

Harmonic analyses using the Fast-Fourier Transform program described by Horn and Baliunas (1986) were performed on our data sets to test for tidal cyclicity (Fig. 12). This FFT

FIG. 10.—Photograph of semi-monthly neap-spring-neap cyclicity. Note the systematic thickening and thinning of strata. Neap (N) and spring (S) deposits are marked. Photograph was taken at location D. Scale units are in centimeters.

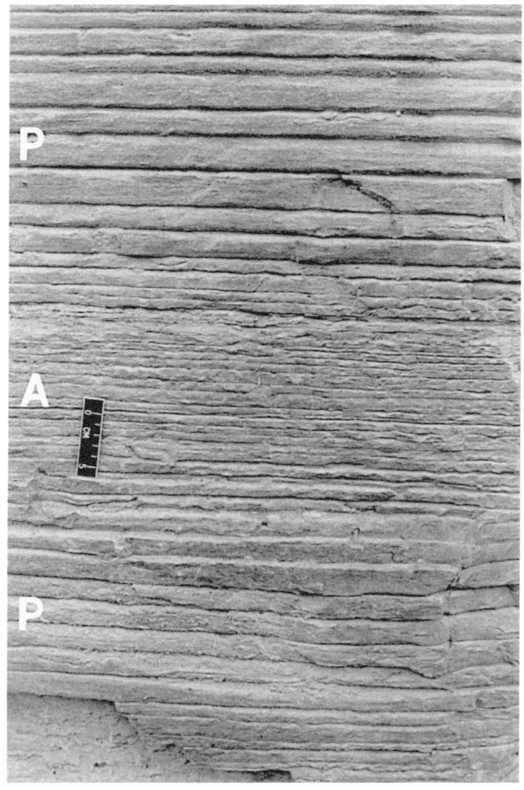

FIG. 11.—Photograph of monthly cyclicity. Thin apogean (A) deposits are shown between much thicker perigean (P) deposits. Photograph was taken at location D. Scale units are in centimeters.

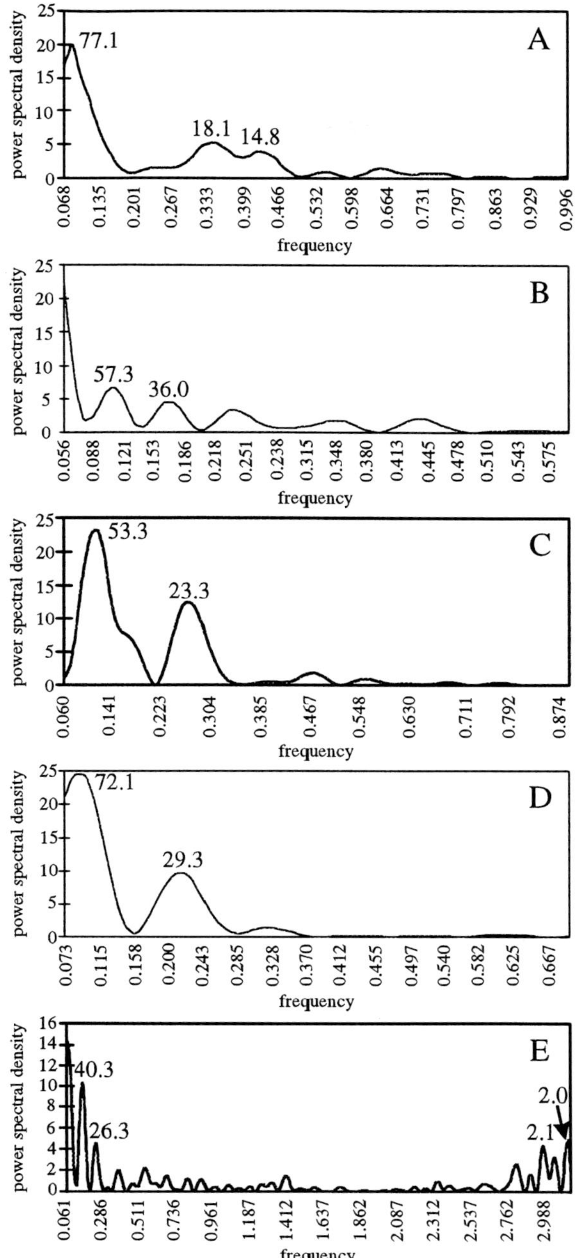

FIG. 12.—Power spectral plots of rhythmite intervals (refer to Fig. 1 for locations). Plot A is from location A; plot B is from location C; plot C is from location D; plot D is from location E; plot E is from location G. Plots B, C, and possibly A, D, and E show peaks suggestive of tidal processes related to monthly cyclicity. Plots B, C, D, E, and possibly A show peaks suggestive of tidal processes related to semi-monthly (neap-spring-neap) cyclicity. Plot E shows peaks suggestive of tidal processes related to semi-diurnal cyclicity.

program is able to differentiate and separate cycles with closely spaced periodicities as discussed by Archer (1994). These periodicities are expressed as beds or laminae per cycle. In a mixed, predominantly semi-diurnal tidal setting, up to 28 beds or laminae can be expected in each semi-monthly neap-spring-neap package (2 tides per day for 14 days). In a monthly cycle, up to 56 beds or laminae can be expected (2 tides per day for 28 days).

Location A shows significant peaks at 77.1 beds/laminae, 18.1 beds/laminae, and 14.8 beds/laminae (Fig. 12A). The 77.1 peak may suggest tidal processes that are related to a monthly tidal cycle. Either the 18.1 or the 14.8 peak may represent a semi-monthly cycle where only the strongest daily ebb tidal events are preserved. The other peak (18.1 or 14.8) may simply represent noise within the data set. Location C shows significant peaks at 57.3 beds/laminae and 36.0 beds/laminae (Fig. 12B). Location D shows significant peaks at 53.3 beds/laminae and 23.3 beds/laminae (Fig. 12C). The peaks for outcrops C and D are strongly suggestive of both monthly (57.3 for location C and 53.3 for location D) and semi-monthly (36.0 for location C and 23.3 for location D) tidal processes. Location E shows significant peaks at 72.1 beds/laminae and 29.3 beds/laminae (Fig. 12D). The 72.1 peak may represent monthly tidal cyclicity. The 29.3 peak is suggestive of semi-monthly neap-spring-neap tidal cyclicity. Location G shows significant peaks at 40.3 beds/laminae, 26.3 beds/laminae, 2.1 beds/laminae, and 2.0 beds/laminae (Fig. 12E). The 40.3 peak is suggestive of tidal processes related to a monthly tidal cycle. The 26.3 peak is suggestive of a semi-monthly tidal cycle, and the 2.1 and/or 2.0 peaks probably represent daily tidal cycles.

DISCUSSION

Flow Direction

Neap-spring-neap bundles contain a maximum of 28 couplets, indicating that one direction of tidal flow, either ebb or flood, is recorded by the Magoffin rhythmites. Based upon the paleoflow data from the tool marks and the climbing ripples, combined with rhythmite bundling patterns, the flow direction recorded by most of these rhythmite couplets can be assumed to be to the northwest. This direction is consistent with reported Middle Breathitt fluvial flow (Englund and Thomas, 1989). Because sediment transport for the rhythmite interval and fluvial flow was to the northwest during the time of deposition, the rhythmite interval can be assumed to record predominantly ebb tidal flow. Therefore, the Magoffin rhythmites is a series of sandstone beds deposited during ebb tidal flows separated by thin mudstone drapes that were deposited during subsequent slackwater periods (Fig. 13). No flood tidal deposits are identified in the Magoffin Member.

Internal Stratifications

The origin of the common parallel laminations within the lower portion of many of the sandstone beds/laminae (see semi-diurnal section above) is enigmatic. It is feasible that these laminated sandstones simply represent amalgamated rhythmite layers. However, based upon their widespread occurrence, ubiquitous nature (occurrence in rhythmites ranging in thickness from less than 1 cm to greater than 15 cm), and lateral continuity, combined with higher-order bundling patterns (semi-monthly and monthly), it is unlikely that more than a small percentage of these internal laminations represent amalgamated layers. It is more likely that these internal stratifications indicate that deposition occurred predominantly under upper-flow-regime tractional conditions (cf. Collinson and Thompson, 1989; Dalrymple, 1992), where unsteady flow is indicated by the presence of coal fragments along these lami-

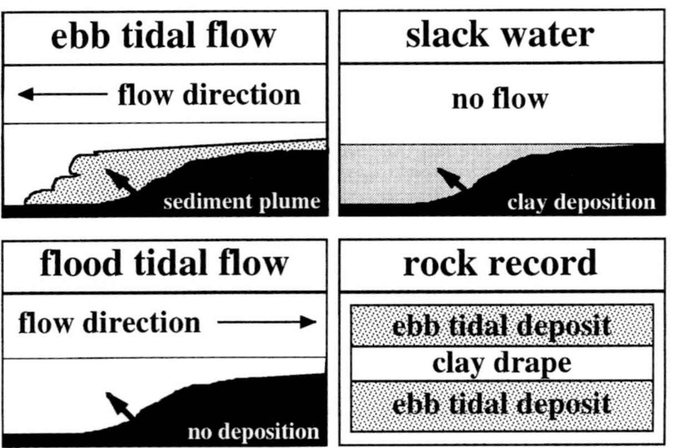

FIG. 13.—The Magoffin rhythmites record ebb tidal activity. During ebb tidal flows, sediment was carried across the study area in the form of large sediment plumes. During subsequent slack water periods, flow energy was reduced, allowing for the deposition of clay-sized particles. During flood tidal flows, no deposition occurred.

nae. The rare, climbing ripples indicate periodic lower flow regime conditions (cf. Dalrymple, 1992).

Sedimentation Rates

Sedimentation rates for the Magoffin rhythmites typically ranged from 20 cm to 100 cm per neap-spring-neap event. However, where the Magoffin Member is the thickest and the most proximal deltaic facies are preserved, sedimentation rates were probably at a maximum and reached over 30 cm per day (Fig. 14). Fig. 9D (location E) records individual rhythmite couplets that are up to 20 cm in thickness. Rhythmite bundling patterns for these thicker beds are comparable to what is observed in the thinner beds. Pairs of couplets display a thick/thin relationship, indicative of a diurnal tidal cycle. Neap-spring-neap bundles contain close to 28 couplets and there is evidence that a monthly tidal cyclicity is recorded. It is possible that some of these beds may represent amalgamated layers, but based upon rhythmite bundling patterns, it is unlikely that this is the case for more than a small percentage of the thicker beds. Other rhythmite sections (locations C, D, and G) do not record individual couplet thicknesses as great as those recorded at location E because of erosion at the sequence boundary and/or difficulty in measuring sections at the tops of benches.

Rhythmite Development

The thickest and most complete rhythmite cycles are developed where the Magoffin Member is thickest and where the interlaminated sandstone/mudstone interval is best preserved (Figs. 3 and 4). At locations C and D, where the Magoffin Member is greater than 40 m thick (Fig. 3), tidal cycles are easily recognizable (Figs. 9C and 9D). Harmonic analyses of the rhythmites at locations C and D are highly suggestive of semi-monthly and monthly tidal periodicities. The only measured outcrop that lies outside the northernmost depocenter is location A (Fig. 4), where the Magoffin Member is only 20 m thick. At this location, no interlaminated sandstone/mudstone interval is present and it is difficult to identify complete tidal packages on the bar graph of measured couplet thicknesses (Fig. 9A). Spectral analyses for the rhythmites at location A show peaks that are less consistent with known tidal periodicities (Fig. 12A) than do the spectral analyses for locations C (Fig. 12B) and D (Fig. 12C). The peaks on Fig. 12A are suggestive of tidal processes possibly related to monthly and semi-monthly tidal periodicities, but the short data set makes it difficult to draw any firm conclusions about the nature of the cycles preserved at this location.

Modern and Ancient Analogs

The most notable modern analog for the Magoffin Member is the Bay of Fundy, which contains a thick rhythmite package displaying diurnal and semi-monthly cycles (Dalrymple et al., 1991). The Bay of Fundy has estimated depositional rates of up to 7 m/yr on its upper tidal flats. Most reported ancient rhythmite successions contain individual couplets that range in thickness from less than 1 mm up to a few centimeters. These values are an order of magnitude lower than the values reported for the Magoffin Member. However, a few notable exceptions of thicker rhythmites are reported in the literature. Lanier et al. (1993) identified couplets that ranged in thickness from less than 1 mm up to 12.5 cm from the Pennsylvanian Tanganoxie Sandstone of the Douglas Group (Kansas). These rhythmites occur within a transgressive package and display well-developed, semi-monthly cyclicity. Kvale and Archer (1991) reported couplets from the Pennsylvanian Abbott Formation of

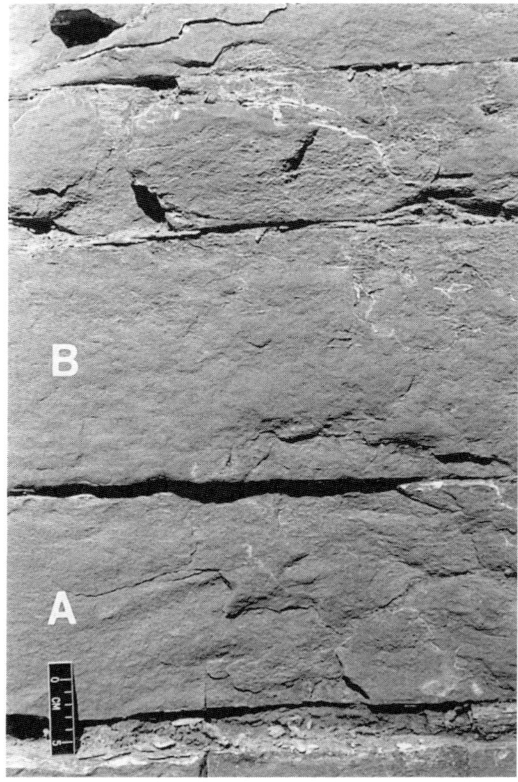

FIG. 14.—Photograph of thick rhythmite beds. Depositional rates for rhythmites typically ranged from 20 cm to 100 cm per neap-spring-neap cycle, but exceeded 30 cm per day in areas where the Magoffin Member is thickest and the most proximal deltaic facies are preserved. Beds A and B in this photograph, taken at location E, represent a dominant/subordinate diurnal deposit that is over 30 cm thick. Scale units are in centimeters.

Illinois that range in thickness from less than 2 cm up to 22 cm. These display well-developed diurnal and semi-monthly cyclicity.

The above-mentioned rhythmite packages occur within retrogradational successions. Rhythmites have also been identified from progradational settings such as the Pennsylvanian Brazil Formation of Indiana (Kvale and Archer, 1990) and the Mississippian Pride Shale of West Virginia (Miller and Eriksson, 1997). However, depositional rates for these progradational packages are much lower than depositional rates recorded by the Magoffin Member rhythmites. The Brazil Formation records depositional rates that range from 1 cm/yr to 1 m/yr and displays well-developed semi-monthly cyclicity. The Pride Shale records average depositional rates of 10 cm/yr but reached over 60 cm/yr. These rhythmites record diurnal, semi-monthly, monthly, annual, and 18.6-year nodal cycles.

CONCLUSIONS

The Magoffin Member is comprised of a coarsening-upward succession of sandstones, siltstones, and mudstones. Within the study area, it is interpreted to be a laterally extensive prodeltaic to delta-front/distributary-mouth-bar deposit that contains tidally influenced rhythmites within its upper 20 m. These rhythmites display several orders of centimeter- to decimeter-scale cycles that are consistent with known semi-diurnal, diurnal, semi-monthly, and monthly tidal periodicities. Based upon paleocurrent data and rhythmite bundling patterns, it is probable that deposition of sand occurred entirely in response to ebb tidal flow. The upper 20 m of the Magoffin Member appears to provide a near-complete record of short-term (up to four months) ebb tidal activity for the mid-Pennsylvanian, central Appalachian Basin. Tidal bundling patterns within these deposits are suggestive of a semi-diurnal tidal setting where semi-monthly and monthly cycles were forced by synodic and anomalistic lunar periodicities. Tidal cycles indicate that the rhythmite interval accumulated at rates of 20–100 cm per semi-monthly (neap-spring-neap) event. However, accumulation rates exceeded 30 cm per day for short periods of time (less than 1 week) in areas where the Magoffin Member is thickest and the most proximal deltaic facies are preserved. These rates compare favorably with other ancient and modern tidal rhythmite deposits. Within the study area, the most complete tidal cycles in the Magoffin Member appear to correlate with the overall thickness of this unit. In areas where the Magoffin Member is thickest, interlaminated sandstone/mudstone deposits comprise the upper portion of the interval. Tidal cycles are better developed therein, than in areas where the Magoffin Member is thinner and is mainly comprised of interlaminated siltstone/mudstone deposits.

ACKNOWLEDGMENTS

This research was supported by grants from the Geological Society of America, the Virginia Tech Graduate School, and the Department of Geological Sciences at Virginia Tech. The authors have benefitted greatly from discussions with Donald Chesnut (Kentucky Geological Survey), Stephen Greb (Kentucky Geological Survey), and Daniel Miller (Virginia Tech). The Kentucky Geological Survey has provided much appreciated data (subsurface strip-logs), assistance, and financial support. Thanks are also given to the two reviewers, Allen Archer and Stephen Greb, for their constructive comments.

REFERENCES

Aitken, J. F., and Flint, S. S., 1995, The application of high-resolution sequence stratigraphy to fluvial systems: a case study from the Upper Carboniferous Breathitt Group, eastern Kentucky, USA: Sedimentology, v. 42, p. 3–30.

Allen, J. R. L., 1981, Lower Cretaceous tides revealed by cross-bedding with mud drapes: Nature, v. 289, p. 579–581.

Archer, A. W., 1994, Extraction of sedimentological information via computer-based image analysis of gray shale in Carboniferous coal-bearing sections of Indiana and Kansas, USA: Mathematical Geology, v. 26, p. 47–65.

Archer, A. W., Kvale, E. P., and Johnson, H. R., 1991, Analysis of modern equatorial tidal periodicities as a test of information encoded in ancient tidal rhythmites, *in* Smith, D. G., Reinson, G. E., Zaitlin, B. A., and Rahmani, R. A., eds., Clastic Tidal Sedimentology: Calgary, Canadian Association of Petroleum Geologists Memoir 16, p. 189–196.

Bennington, J. B., 1996, Stratigraphic and biofacies patterns in the Middle Pennsylvanian Magoffin Marine Unit in the Appalachian Basin, U.S.A.: International Journal of Coal Geology, v. 31, p. 169–193.

Chan, M. A., Kvale, E. P., Archer, A. W., and Sonnet, C. P., 1994, Oldest direct evidence of lunar-solar tidal forcing encoded in sedimentary rhythmites, Proterozoic Big Cottonwood Formation, central Utah: Geology, v. 22, p. 791–794.

Chesnut, D. R., 1981, Marine zones of the Upper Carboniferous of eastern Kentucky, *in* Cobb, J. C., Chesnut, D. R., Hester, N. C., and Hower, J. C., eds., Coal and Coal-bearing Rocks of Eastern Kentucky (Annual Geological Society of America Coal Division Field Trip 1981): Lexington, Kentucky Geological Survey, p. 57–66.

Chesnut, D. R., 1989, Pennsylvanian rocks of the eastern Kentucky Coal Field, *in* Cecil, C. B., and Elbe, C., eds., Carboniferous Geology of the Eastern United States, 28th International Geological Congress, Field Trip T143: Coal and Hydrocarbon Resources of North America, v. 2, Washington, D.C., American Geophysical Union, p. 57–58.

Chesnut, D. R., 1992, Stratigraphic and Structural Framework of the Carboniferous Rocks of the Central Appalachian Basin in Kentucky, Bulletin 3, Series XI: Lexington, Kentucky Geological Survey, 42 p.

Cobb, J. C., Chesnut, D. R., Hester, N. C., and Hower, J. C., 1981, Field trip roadlog: Coal and Coal-bearing Rocks of Eastern Kentucky, Annual Geological Society of America Coal Division Field Trip: Lexington, Kentucky Geological Survey, p. 6–56.

Collinson, J. D., and Thompson, D. B., 1989, Sedimentary Structures: London, Chapman, and Hall, p. 100–102.

Dalrymple, R. W., 1992, Tidal depositional systems, *in* Walker, R. G., and James, N. P., eds., Facies Models: Response to Sea Level Change: Geological Association of Canada, p. 195–218.

Dalrymple, R. W., Knight, R. J., Zaitlin, B. A., and Middleton, G. V., 1990, Dynamics and facies of a macrotidal sand-bar complex, Cobequid Bay-Salmon River Estuary (Bay of Fundy): Sedimentology, v. 37, p. 577–612.

Dalrymple, R. W., Makino, Y., and Zaitlin, B. A., 1991, Temporal and spatial patterns of rhythmite deposition on mud flats in the macrotidal Cobequid Bay-Salmon River estuary, Bay of Fundy, Canada, *in* Smith, D. G., Reinson, G. E., Zaitlin, B. A., and Rahmani, R. A., eds., Clastic Tidal Sedimentology: Calgary, Canadian Association of Petroleum Geologists Memoir 16, p. 137–160.

DeBoer, P. L., Oost, A. P., and Visser, M. J., 1989, The diurnal inequality of the tide as a parameter for recognizing tidal influence: Journal of Sedimentary Petrology, v. 59, p. 912–921.

Dennis, A. M., 1975, Bay-fill paleoecology in the Carboniferous of eastern Kentucky: Unpublished M.S. Thesis, University of Kentucky, 55 p.

Englund, K. J., and Thomas, R. E., 1989, Late Palaeozoic depositional trends in the central Appalachian Basin: Bulletin of United States Geological Survey, v. 1839, p. F1–F19.

Greb, S. F., and Archer, A. W., 1995, Rhythmic sedimentation in a mixed tide and wave deposit, Hazel Patch Sandstone (Pennsylvanian), Eastern Kentucky Coal Field: Journal of Sedimentary Research, v. B65, p. 96–106.

Horne, J. H., and Baliunas, S. L., 1986, A prescription for period analysis of unevenly sampled time series: Astrophysics Journal, v. 302, p. 757–763.

Kreisa, R. D., and Moiola, R. J., 1986, Sigmoidal tidal bundles and other tide-generated sedimentary structures of the Curtis Formation, Utah: Geological Society of America Bulletin, v. 97, p. 381–387.

KVALE, E. P., AND ARCHER, A. W., 1990, Tidal deposits associated with low sulfur coals, Brazil Formation (Lower Pennsylvanian), Indiana: Journal of Sedimentary Petrology, v. 60, p. 563–574.

KVALE, E. P., AND ARCHER, A. W., 1991, Characteristics of two, Pennsylvanian-age, semidiurnal deposits in the Illinois Basin, U.S.A., *in* Smith, D. G., Reinson, G. E., Zaitlin, B. A., and Rahmani, R. A., eds., Clastic Tidal Sedimentology: Calgary, Canadian Association of Petroleum Geologists Memoir 16, p. 179–188.

KVALE, E. P., ARCHER, A. W., AND JOHNSON, H. R., 1989, Daily, monthly, and yearly tidal cycles within laminated siltstones of the Mansfield Formation (Pennsylvanian) of Indiana: Geology, v. 17, p. 365–368.

KVALE, E. P., CURTRIGHT, J., BILODEAU, D., ARCHER, A., JOHNSON, H. R., AND PICKET, B., 1995, Analysis of modern tides and implications for ancient tidalites: Continental Shelf Research, v. 15, p. 1921–1943.

LANIER, W. P., FELDMAN, H. R., AND ARCHER, A. W., 1993, Tidal sedimentation from a fluvial to estuarine transition, Douglas Group, Missourian-Virilian, Kansas: Journal of Sedimentary Petrology, v. 63, p. 860–873.

MARTINO, R. L., AND SANDERSON, D. D., 1993, Fourier and autocorrelation analysis of estuarine tidal rhythmites, Lower Breathitt Formation (Pennsylvanian), eastern Kentucky, USA: Journal of Sedimentary Petrology, v. 63, p. 105–119.

MILLER, D. J., AND ERIKSSON, K. A., 1997, Late Mississippian prodeltaic rhythmites in the Appalachian Basin: A hierarchical record of tidal and climatic periodicities: Journal of Sedimentary Research, v. 67, p. 653–660.

QUINLAN, G. M., AND BEAUMONT, C., 1984, Appalachian thrusting, lithospheric flexure, and the Paleozoic stratigraphy of the Eastern Interior of North America: Canadian Journal of Earth Science, v. 21, p. 973–996.

RICE, C. L., SABLE, E. G., DEVER, G. R., AND KEHN, T. M., 1980, The Mississippian and Pennsylvanian (Carboniferous) Systems in Kentucky, Series XI: Lexington, Kentucky Geological Survey, 32 p.

ROEP, T. B., 1991, Neap-spring cycles in a subrecent tidal channel fill (3665 BP) at Schoorldam, NW Netherlands: Sedimentary Geology, v. 71, p. 213–230.

SONNET, C. P., KVALE, E. P., ZAKHARIAN, A., CHAN, M. A., AND DEMKO, T. M., 1996, Late Proterozoic and Paleozoic tides, retreat of the moon, and rotation of the earth: Science, v. 273, p. 100–104.

TANKARD, A. J., 1986, Depositional response to foreland deformation in the Carboniferous of eastern Kentucky: The American Association of Petroleum Geologists Bulletin, v. 70, p. 853–868.

TESSIER, B., 1993, Upper intertidal rhythmites in the Mont-Saint-Michel Bay (NW France): Perspectives for paleoreconstruction: Marine Geology, v. 110, p. 355–367.

TESSIER, B., AND GIGOT, Z. P., 1989, A vertical record of different tidal cyclicities: An example from the Miocene Marine Molasse of Digne (Haute Provence, France): Sedimentology, v. 36, p. 767–776.

WANLESS, H. R., 1975, Paleotectonic investigations of the Pennsylvanian System in the United Sates; part I, Appalachian region: Washington, D.C., United States Geological Survey Professional Paper 853, p. 17–62.

WILLIAMS, G. E., 1989, Tidal rhythmites: Geochronometers for the ancient earth-moon system: Episodes, v. 12, p. 162–171.

WILLIAMS, G. E., 1990, Tidal rhythmites: Key to the history of the earth's rotation and the lunar orbit: Journal of Physics of the Earth, v. 38, p. 475–491.

WILLIAMS, G. E., 1991, Upper Protozoic tidal rhythmites, South Australia: Sedimentary features, deposition, and implications for the earth's paleorotation, *in* Smith, D. G., Reinson, G. E., Zaitlin, B. A., and Rahmani, R. A., eds., Clastic Tidal Sedimentology: Calgary, Canadian Association of Petroleum Geologists Memoir 16, p. 161–178.

EVIDENCE OF ANCIENT FRESHWATER TIDAL DEPOSITS

ERIK P. KVALE AND MARIA MASTALERZ
Indiana Geological Survey, Indiana University, Bloomington, Indiana 47405

ABSTRACT: Well-developed tidal rhythmites associated with tidal mudflats overlie some of the lowest-sulfur Pennsylvanian coals in the eastern part of the Illinois Basin, U.S.A. Such an observation is seemingly at odds with the traditionally held view that low-sulfur coals are not typically associated with marine-influenced roof rock in the Illinois Basin and elsewhere. This association can be explained, however, if the tidal mudflats formed in a freshwater or low-salinity marine setting. Such deposits are known from modern systems but have not been adequately documented from the rock record.

Geochemical, petrographic, and sedimentologic analyses of tidal deposits immediately above the low-sulfur (sulfur values <1%) Lower Block Coal Member of the Pennsylvanian Brazil Formation (Atokan), Daviess County, Indiana, confirm the potential for the preservation of extensive (>480 km^2 area) freshwater tidal flat deposits in the rock record. A strong tidal (marine) signal, manifested as tidal rhythmites preserving small-scale neap-spring cycles, is preserved within the laminated mudstone and interbedded sandstone and mudstone immediately above the Lower Block Coal. These cycles indicate deposition within a mixed, predominantly diurnal tidal system in which sedimentation rates were as high as 1 m/yr. Carbon to sulfur ratios, maceral types, and the dominance of terrestrial organic markers within the rhythmite facies reveal that the Lower Block peat (coal) was initially onlapped by a freshwater (low salinity) tidal flat. The presence of tidal rhythmites indicates that at least mesotidal conditions prevailed during transgression.

The areal distribution, sedimentology, stratigraphy, petrography, and geochemistry of the succession from the Lower Block Coal to the next younger coal (Upper Block Coal Member of the Brazil Formation) suggest that the tidal flat facies formed within an embayed coastal setting that experienced significant rainfall and runoff typical of an ever-wet climate. The embayments likely formed by the collapse and transgression of the coastal peat mires (Lower Block Coal) and ultimately filled with mudflat and mixed sandflat and mudflat facies. Very high rainfall and associated runoff resulted in the formation of coast-hugging freshwater plumes. These plumes pushed the salt water wedge offshore and prevented it from entering the embayments until the coastal peats had already been covered by at least a meter of tidal flat mud.

The results of this study may have implications for paleogeographic reconstructions in other Carboniferous basins. Similar tide-dominated facies may have gone unrecognized in some Carboniferous successions in Europe, and elsewhere, that are currently interpreted to be ancient nonmarine mud-dominated tropical systems. Such recognition would have important sequence stratigraphic and paleogeographic implications.

INTRODUCTION

Tides are known to propagate hundreds of kilometers inland with the potential to affect sedimentation patterns along the tidal pathway. The extent to which tidal currents can influence inland depositional processes is related to the degree of tidal amplification and the relative intensities of the marine and fluvial processes within the tidal basin. The St. Lawrence River separating Canada and the United States provides an example of the extreme inland propagation of tidal influence. Macrotidal conditions exist more than 500 km inland from the Gulf of St. Lawrence; the gulf itself experiences only microtidal conditions (LeBlond, 1979; Archer *et al.,* 1995). Conversely, riverine-sourced freshwater plumes are known to extend tens of kilometers offshore in tide-dominated settings. The mouth of the Amazon River is such an example (Lentz and Limeburner, 1995). In such freshwater, tide-dominated environments, water chemistries may be completely fresh with marine and brackish water bioturbators absent. Given the absence of marine or brackish water fossils, ancient freshwater tidal mudflats or sandflats might be confused for lacustrine or fluvial overbank deposits. In fact, the presence of *in situ* freshwater micro and macro fossils [see dicussion of the depositional setting of the Braidwood-type fauna of Richardson and Johnson (1971) in Archer and Kvale (1993)] may support these interpretations.

Recent work on modern and ancient fluvial-tidal transitions (e.g., Kvale and Archer, 1990; Shanley, et al., 1992; Dalrymple et al., 1992; Greb, 1993; Archer et al., 1994; Räsänen et al., 1995; Eberth, 1996; Marshall and Lundberg, 1996) have suggested the possibility of preserving freshwater tidal deposits in the rock record. However, direct evidence of ancient freshwater tidal settings has not been clearly established. We present herein unambiguous examples of ancient tidal deposits, yet, the deposits are characterized by geochemical, petrographical, and biological evidence indicating freshwater depositional chemistries thus providing direct evidence for the preservation of ancient freshwater tidal deposits. The acknowledgement that freshwater tidal deposits exist in the rock record—and the establishment of criteria for their recognition—has important implications for sequence stratigraphic, facies, and paleobiologic studies, as well as for paleogeographic reconstructions.

STUDY AREA AND GEOLOGIC SETTING

This study focused on the lower part of the Brazil Formation (Pennsylvanian) in Daviess County, Indiana, U.S.A. (Figs. 1 and 2). Daviess County is situated on the east flank of the Illinois Basin, which is located in central and southern Illinois, southwestern Indiana, and western Kentucky. The basin is spoon-shaped with a long dimension extending northwest-southeast for 600 km and with a maximum width of 320 km. During the Pennsylvanian, the basin was located about 300 km north of the margin of the North American craton (Collinson, et al., 1988) in a near-equatorial setting characterized by a tropical to temperate rainy climate (Scotese et al., 1979; Phillips and Peppers, 1984; Cecil, 1990; Crowley et al., 1996).

The Brazil Formation and upper part of the underlying Mansfield Formation were examined in detail over a 500 km^2 area utilizing core, driller's logs, geophysical logs, and coal strip-mine exposures as part of an Indiana Geological Survey coal evaluation program. The upper Mansfield and Brazil interval is characterized by a stacked succession of primarily fine-grained deposits of mudstones, shales, thinly interbedded mudstones and sandstones, thin fine-grained quartz arenites and sparse, thin, and discontinuous arenaceous packestones to wackestones that occur between the Ferdinand Bed (Mansfield Formation) and the top of the Buffaloville Coal Member (Brazil Formation) (Fig. 2). The section is punctuated by a series of paleosols. Most of the paleosols are correlatable across the study area and typically are overlain by coals (Fig. 3) (Kvale et al., 1996). A

Tidalites: Processes and Products, SEPM Special Publication No. 61

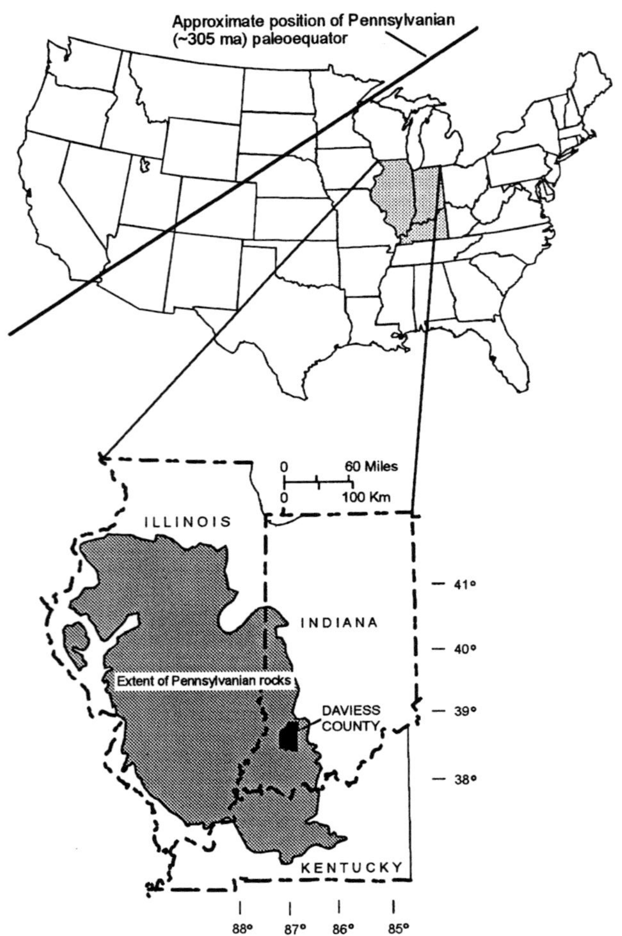

FIG. 1.—Map of the Illinois Basin, U.S.A., showing distribution of Pennsylvanian sediments; Daviess County, Indiana, shaded. Approximate position of Pennsylvanian paleoequator from Crowley and Baum (1991)

marine (in the sense of sedimentation processes) flooding surface caps each paleosol (or coal when present). These bounding surfaces package stratal units (parasequences?) (*sensu* Van Wagoner et al., 1990) that are structurally inclined on the average approximately 6–7 m/km to the southwest towards the center of the basin. The siliciclastic facies between the bounding surfaces are dominated by small-scale sedimentary structures such as ripple bedding in the sandstones, and tidal structures such as lenticular, wavy, and flaser bedding and tidal rhythmites.

Tidal rhythmites are herein defined as structures consisting of: (1) stacked sets of clay draped ripples that as cosets can be classified as lenticular, wavy, and flaser bedding; and (2) flat, laminated siltstone and argillaceous siltstone with thin intercalated claystone layers wherein the tidal origin of these rhythmites is indicated by the progressive thickening and thinning of individually accreted sets or bundles in response to changing current velocities associated with lunar cycles. Tidal rhythmites consistent with this definition have been well documented from modern and ancient tide-dominated distributary channels, tide-dominated delta fronts, tidal channels, and estuarine systems (see summaries in Feldman et al., 1993, and Kvale et al., 1997).

Studies of modern mud-rich, tide-dominated tropical coastlines reveal the difficulty of differentiating facies associated with active distributary channels from those of prograding coastal tidal flats (e.g., Coleman et al., 1970; Cecil et al., 1993; Staub and Esterle, 1993). The absence of obvious large-scale channels that could be interpreted as deltaic distributaries between flooding surfaces 11 and 15 (small segment of the regional study shown in Fig. 3) suggest that the overall setting may not be deltaic. Rather, the interval can be interpreted as deposits of a shifting mud-dominated coastal plain in which the bulk of the sediments were delivered via long-shore currents. Bioturbation is sparse or patchy between surfaces 11 and 15 with the exception of the units above surface 11 (Ferdinand Bed position), where the most marine facies are present (Fig. 3). The Ferdinand Bed limestone, a lenticular packestone containing echinoid and coral fragments, is laterally replaced by an erosive-based bioturbated sandstone dominated by wavy to flaser bedding and a lenticular-bedded bioturbated mudstone. Traces in the sandstone and mudstone include an assemblage of *Rosselia, Conostichus, Zoophycus,* and *Teichichnus,* which is typically considered to represent normal marine salinity conditions. This stratigraphic interval is not observable in outcrop, so paleocurrent data on the sandstone are not obtainable other than to say that some of the ripple forms in rock core appear to be bipolar. Bioturbation is much less intense and traces are less diverse in the other stratal packages. Traces include *Planolites* (above surfaces 12, common; 13, rare; and 14, common), *Teichichnus* (above surfaces 12 and 14, common), *Zoophycos* (above surface 14, rare); and *Locheia* (above surface 13, rare).

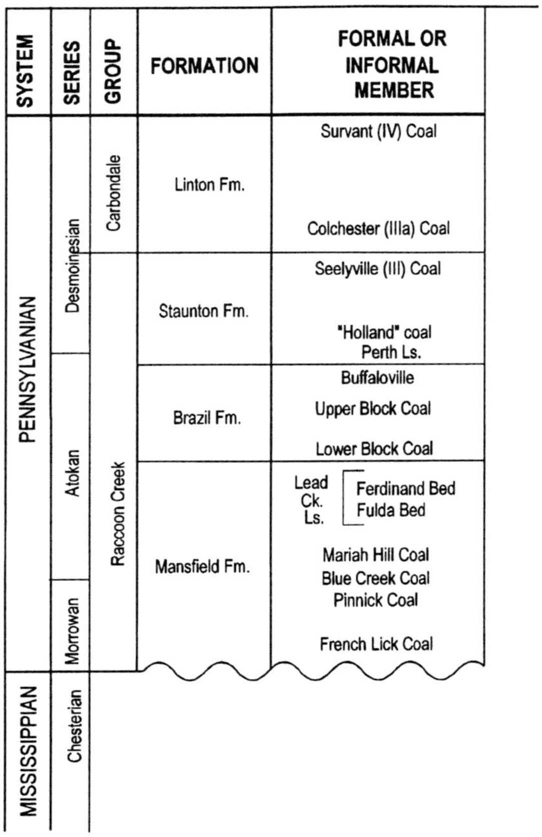

SYSTEM	SERIES	GROUP	FORMATION	FORMAL OR INFORMAL MEMBER
PENNSYLVANIAN	Desmoinesian	Carbondale	Linton Fm.	Survant (IV) Coal Colchester (IIIa) Coal
		Raccoon Creek	Staunton Fm.	Seelyville (III) Coal "Holland" coal Perth Ls.
	Atokan		Brazil Fm.	Buffaloville Upper Block Coal Lower Block Coal
	Atokan / Morrowan		Mansfield Fm.	Lead Ck. Ls. [Ferdinand Bed, Fulda Bed] Mariah Hill Coal Blue Creek Coal Pinnick Coal French Lick Coal
MISSISSIPPIAN	Chesterian			

FIG. 2.—Stratigraphic section listing the prominent mapable Pennsylvanian coal and limestone positions in southeastern Daviess County.

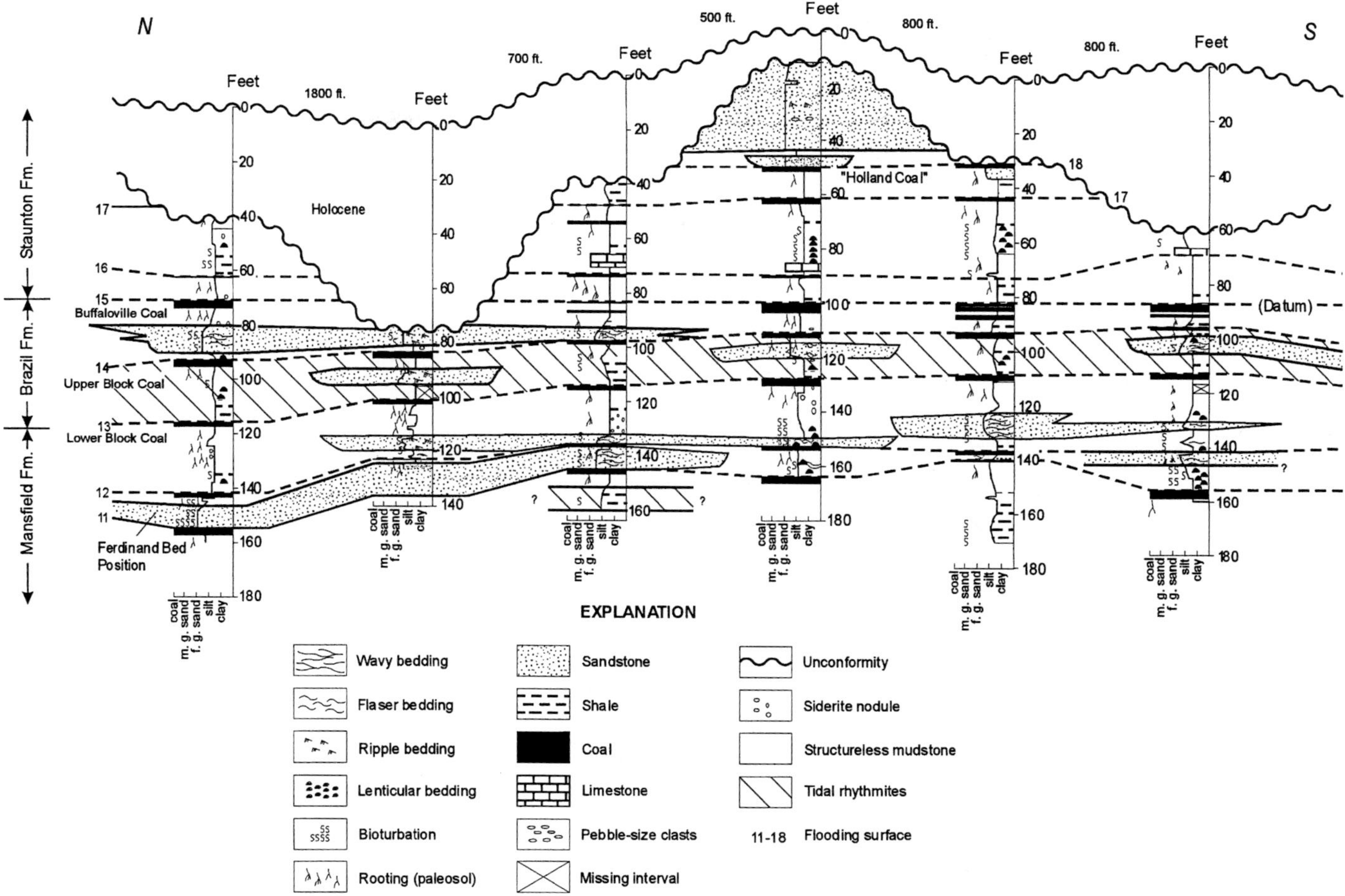

FIG. 3.—Detailed cross-section showing flooding surface boundaries and lithofacies distribution. Descriptions are from proprietary core data. Spacing between cores is noted. Total north-south length of cross-section is approximately 1 km. Thicknesses are in feet.

LOWER BLOCK COAL TIDAL FLAT

Sedimentology

The interval between the Upper and Lower Block Coals is clearly tide-dominated. A typical vertical sequence above the Lower Block Coal consists of the following: (1) a basal dark gray rhythmic-bedded mudstone overlain by (2) a lenticular-bedded shale to wavy and flaser-bedded sandstone interval with associated tidal channels, and, lastly (3) a rhythmic-bedded mudstone and rooted underclay interval that underlies the Upper Block Coal. This overall sequence ranges from 4.5–10 m thick and shows a consistent coarsening-upward followed by a fining-upward trend over most of the study area (Figs. 3 and 4). Well developed tidal rhythmites within the sandflat and mudflat facies can be found in mine highwalls and rock core over an area spanning 16 km in an east-west direction and 30 km in a north-south direction, indicating an extensive tidal flat complex with a minimum coverage of almost 500 km^2.

The best and most continuous records of neap-spring tidal cycles are found in the finely laminated (laminae thickness $<$ 1 cm) fine-grained facies immediately above the coal (Figs. 5 and 6). Sparse rootlets in this facies suggest an intertidal setting. Trace fossils and desiccation features (such as mudcracks) are absent. Little reworking of the mudstone-dominated rhythmites is apparent, and the system appears to have been totally aggradational immediately above the coal.

The coarser sandstone exhibits more familiar tidal bedding features, including wavy to flaser bedding, and within tidal channels, large-scale tidal bundles (Fig. 7). Limited paleocurrent measurements collected from ripple foresets exposed in surface mine high-walls along the east central edge of the study area indicate north to northeast (landward) paleoflows, thus suggesting flood dominance. Broad, shallow, tidal-channel deposits (a few meters to tens of meters wide) are dominated by small- and medium-scale trough cross-bedding. Paleocurrent measurements within the channel deposits indicate bimodal paleoflows and evidence of flow-segregation. However, ebb-oriented (south- to southwest-directed) paleoflows are more common. Trace fossils, consisting principally of *Lockeia* and *Planolites* are locally present within this facies, but bioturbation is rare. Mudstones of the upper facies, observable in coal mine high-walls (accessible only in core), overlying the wavy and flaser bedded sandstone and lenticular shale interval, are often laminated and show good preservation of neap-spring cyclicity. Sedimentation cycles are neither as complete nor as continuous as those from the lower, laminated mudstone facies. The upper mudstones are typically rooted with root density increasing upwards. The facies grades upward into a structureless, greenish-

FIG. 4.—Photograph of core 24AE-6 from a depth of 130 ft to 150 ft below surface. Bottom of core is lower left and top is upper right. Lower Block Coal (3rd slot from left) has been removed. Note that the core coarsens upwards (to the right) above the Lower Block Coal, and then fines upwards to the mudstone-rich paleosol (present in the last two slots to the right) present below the Upper Block Coal (not shown in these core boxes). Arrows point to some of the neap-tide events in the transgressive (three arrows in slot left of 10-cm bar scale) and regressive facies (three arrows in the third slot to the right of the 10-cm scale). Sandiest portion of the core (in the two slots to the right of the 10-cm bar scale) represents the facies deposited during maximum transgression.

gray, rooted paleosol. Flood or ebb dominance has not yet been established for the upper tidal rhythmite facies.

The vertical sequence described above is very similar to the transgressive-regressive lithofacies sequence recognized for the tidal flat deposits in the tide-dominated German Bight (Reineck, 1972; Davis and Clifton, 1987). The transgressive component is represented by the depositional model shown in Fig. 8. In this model, an initial relative sea-level rise causes drowning of the coal-forming peat mire, followed by landward migration of the mudflat, mixed flat, and sandflat. An increase in sedimentation rate relative to subsidence results in seaward progradation, and a reversal in the ordering of the facies. The consequence is a coarsening-upward followed by a fining-upwards trend.

Tidal rhythmites above the Lower Block Coal are present in every core and surface coal mine high-wall exposure along the central and eastern portions of the study area in a north-south direction. Unfortunately, core or mine exposures are not yet available for the westernmost portion of the study area with the exception of a single core in the western extreme of the study area. In this particular core (Indiana Geological Survey drill hole SDH-377), the transgressive component is much sandier overall with the Lower Block Coal capped by a thin (30 cm) dark grey disrupted (dewatered) to laminated mudstone that is overlain by a dominantly wavy to flaser-bedded sandstone with less well developed neap-spring cycles than those observed further east. Sulfur content within the coal is relatively higher than the Lower Block Coal to the east, suggesting the presence of more saline water during deposition of the sandier transgressive facies in the western part of the study area. Petroleum geophysical logs indicate that the interval between the Upper and Lower Block Coals in the western area may be coarser overall.

Geophysical logs and drill cuttings taken from southwestern Indiana approximately 40 km to the southwest of the study area indicate the presence of a marine carbonate facies (wackestone to grainstone) spanning the interval from the Lead Creek Limestone to the Perth Limestone Member of the Staunton Formation (Droste and Horowitz, 1996). Although breaks in the geophysical log response suggest the possibility of low-stand erosion and subaerial exposure, there is no conclusive evidence to support this. Thus, it appears possible that this area in southwestern Indiana could have remained subtidal during much of the deposition of the upper part of the Mansfield and Brazil formations. This suggests a minimum seaward shift of approximately 40 km for the intertidal-subtidal zone during transgression.

Tidal Cycles and Depositional Rates

Tidal rhythmites above the Lower Block Coal are present in every core that penetrated the Lower Block Coal interval within the study area (approximately 40 cores) but were analyzed in varying detail in ten cores. Specific locations of nonproprietary cores are summarized in Kvale et al. (1996). In every case, the cores taken above the Lower Block Coal record deposition in a mixed, predominantly diurnal setting, preserving semidaily, daily, semimonthly, and semiannual cycles (Fig. 6). The presence of the tidal cycles is very important to the investigation because it clearly establishes a link to marine processes. The

FIG. 5.—Photograph of core showing tidal rhythmites within a meter above the Lower Block Coal. Small arrows indicate positions of neap-tide deposits. Large arrow points to a possible storm deposit of slightly coarser-grained siltstone that is thicker than its position in the neap-spring cycle suggests it should be. Top of the core is to the left. Total length of the core shown is approximately 20 cm.

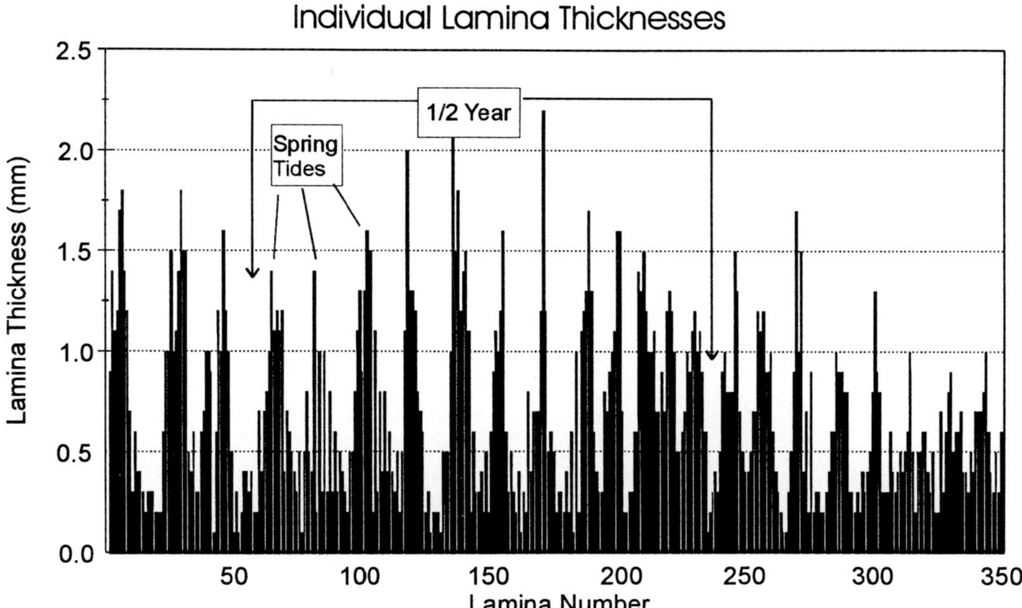

FIG. 6.—Bar chart of individual laminae from core represented by Fig. 5. Semiannual cycle and spring tidal events are labeled.

interpretation of a predominantly diurnal setting is based on the number of individual events preserved in a neap-spring cycle but, more importantly, on the overall pattern of individual events within the neap-spring cycle. The neap-spring cycles exhibited above the Upper Block Coal were controlled by the declinational changes of the moon (relative to the earth's equator) associated with its orbit about the earth (tropical cycle) rather than by the more familiar synodic cycle of changes in lunar phase (see discussions of the origin of neap-spring cycles in Archer et al., 1991; Kvale et al., 1995; and Kvale et al., 1997). The evidence for this is the preservation of so-called crossover events in the neap tide portions of the tidal cycles (Fig. 9). This is typical of dominantly diurnal tidal systems where tropical monthly tides are responsible for generating neap-spring cycles. In such cases, the dominant tidal force depends on the declination of the moon with the force being greatest when the moon is at zenith. In these systems, the equatorial passages of the moon (crossovers) occur at the same rate as the generation of neap tides. Similar crossovers developed in predominantly semidiurnal systems move through successive neap-spring cycles (Fig. 9). The distinction between a diurnal signal and a truncated semidiurnal signal can be tenuous (DeBoer et al., 1989; Archer, 1995; Kvale et al., 1995). However, the coincidence of the crossover events with the neap tides in the Brazil Formation substantiates a mixed, predominantly diurnal interpretation for this deposit.

Depositional rates for the complete sequence appear to be very high. Upright stumps of pteridosperms, calamites, and locally *Lepidodendron wortheni* more than 4 m tall were observed in growth position in some highball exposures. Anecdotal evidence from mine engineers indicates that upright trees as tall as 8 m have been seen. Measurements of sedimentation rates (and by inference minimum estimates of subsidence rates) based on neap-spring cycle thicknesses indicate rates of 1–6 cm/lunar month for the rhythmic bedded mudstone units, and 6–50 cm/lunar month for the wavy to flaser-bedded units. Complete semiannual cycles have been measured in the mudflat-to-mixed-flat facies, with annual thicknesses (thickness of two semiannual cycles) approximating 1 m. All of these sedimentation rates are for lithified sediments.

FIG. 7.—High wall exposure above the Upper Block Coal Member, Montgomery Quadrangle, Daviess County, Indiana, showing channel sandstone that exhibits large-scale trough cross-stratification with clay-draped foresets (large arrow just above person's right hand). Base of the channel is at the level of the person's shoulder. Coal is approximately 1 m below the rubble at bottom of photograph.

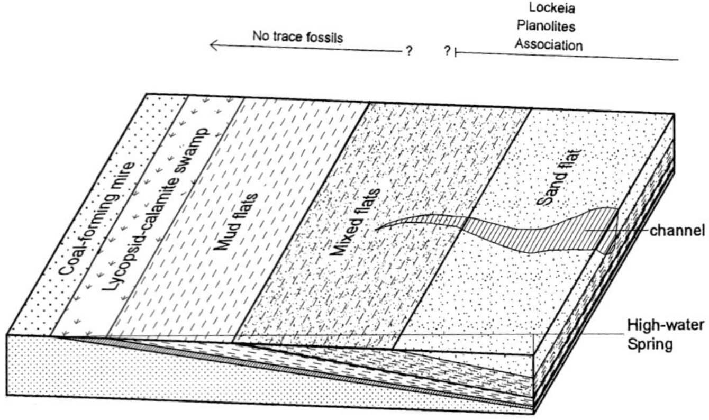

FIG. 8.—Depositional model for interval between Upper and Lower Block Coals. Vertical facies profile between coals results from transgression and landward migration of tidal flats. Regression (the consequence of a seaward progradation of the facies shown in this figure) would reverse the vertical succession, resulting in a fining upwards.

Generation of Accomodation Space

One of the problems associated with the development of the tidal facies in the Brazil Formation is to explain how the accommodation space was generated so rapidly. The Daviess County study area is located along the margin of the Illinois Basin and there is little evidence here of rapid tectonic subsidence that would generate sufficient accommodation space (such as along major fault blocks). Generally speaking, the Pennsylvanian section is thin, and although faults are mapped in the immediate vicinity, they are local and apparently do not cut the Pennsylvanian section. The most likely source of accommodation space was the rapid dewatering and compaction of the coal, forming peat mires (Kvale and Archer, 1990).

A recent study describing the rapid generation of so-called marsh-loss hotspots in the Mississippi River Deltaic Plain via plant mortality supports this interpretation (DeLaune et al., 1994). When the vertical accretion of plant material (which locally averaged 0.98 cm/yr over a 27-year period) in the Mississippi Delta Plain cannot keep pace with the submergence of the coastal plain, a critical point is reached in which plant mortality and peat collapse occur by autocompaction. Increased rates of plant mortality can be driven by ecological stresses induced from saltwater intrusion, flooding, or inadequate mineral sediment supply. This causes marsh surface elevations to decrease, resulting in the formation of depressions that are likely to flood excessively with rain or brackish to marine water. Submergence rates that exceed 1 cm/yr have been documented in areas along the Mississippi River, the result of a combination of peat collapse, the consolidation of underlying sediments, and global sea-level rise. In the DeLaune et al. study (1994), peat collapse originated from an increase in the decomposition rates of root tissue and/or a loss of root turgor (air passages). Elevation losses were greatest in areas where plant mortality and soil organic matter contents were high (marshes) and less in areas where vegetation persisted and soil organic contents were low (such as along natural levees). Although not addressed specifically, their data (DeLaune et al., 1994, Fig. 1, p. 1023) suggest areas effected by peat collapse can be tens of square kilometers. The potential of differential compaction of fine-grained, organic-rich deposits versus coarser grained sediments effecting coastline morphology is further illustrated by the work of Oertel et al. (1992). This study proposes that much of the barrier and back-barrier systems of the east coast of the U.S.A. are the result of the differential compaction of coarser grained coastal foreshore (now barrier islands) and organic-rich, fine-grained backshore deposits (now back-barrier lagoons).

The subsidence rates noted by DeLaune et al. (1994) are two orders of magnitude less than that interpreted for the transgressive facies above the Lower Block Coal. This can be explained by the presence of much thicker peat deposits (Lower Block Coal) and their potential for a much greater and perhaps more rapid collapse than that reported for the relatively thin and more inorganic-rich coastal peats of the Mississippi delta. Peat compaction has been suggested as the major factor in generating accommodation space for other rapidly deposited sediments in other Carboniferous basins (e.g., Gradzinski and Doktor, 1995).

Geochemistry and Petrography

The Lower Block Coal is a high quality coal of great interest to the local mining companies. Its sulfur and ash contents are

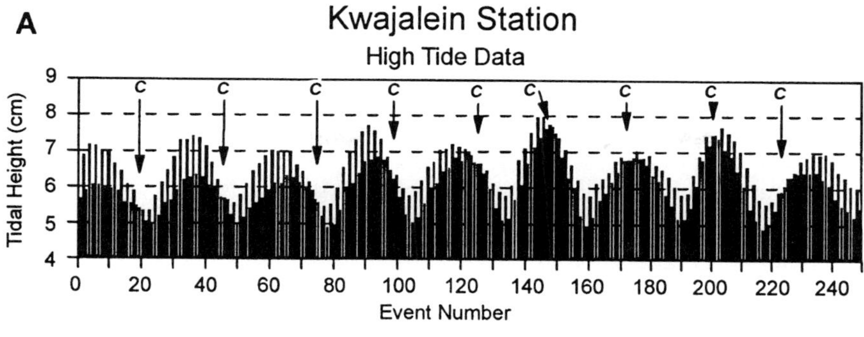

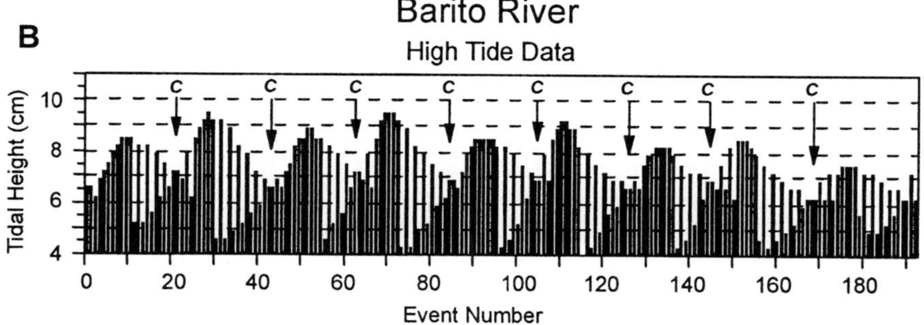

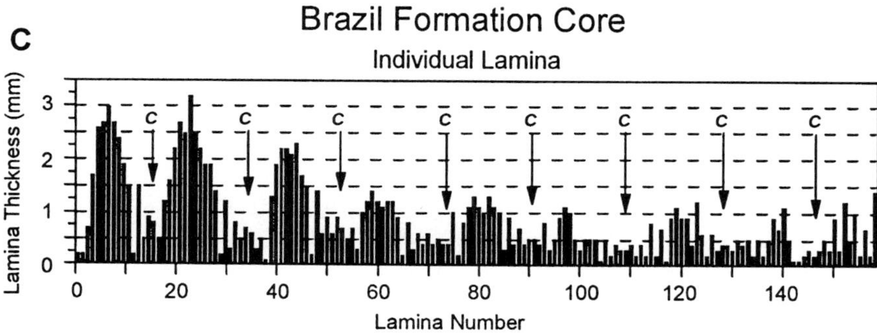

FIG. 9.—Bar charts illustrating the relative positions of crossover events (c) from modern mixed, predominantly semidiurnal (A) and mixed, predominantly diurnal tidal systems (B). The position of subsequent crossover events moves through the neap-spring tidal cycle in the predominantly semidiurnal system (A) and remains fixed to the neap-tide events in the predominantly diurnal system (B). Data in A are from predicted high tide tables from Kwajalein Atoll, Pacific Ocean. Data in B are from predicted high tide tables from Barito River, Borneo. Note the crossover positions from the Brazil Formation tidal rhythmites (C) are fixed to the neap tides, indicating that the periods of crossover events and neap tides are the same. Thus, neap-spring cycles for the Brazil tidal rhythmites are generated by tropical tidal periods rather than synodic tidal periods, confirming the interpretation of a true mixed, predominantly diurnal system as being responsible for the generation of the Brazil tidal rhythmites.

typically below 1% and 6%, respectively. Individual lithotypes typically include 63–78% vitrinite, and an average of 15% inertinite and up to 7% liptinite. The plant community was dominated by lycopod trees and tree ferns, with a higher lycopod to tree fern ratio in the upper half of the seam. Localized doming and ombrogenous conditions have been suggested as a depositional model for this coal (Kvale and Archer, 1990; Kvale et al., 1996; and Padgett et al., 1996).

Other than sedimentologic evidence, little else indicates a marine influence within the succession. Kvale and Archer (1990) first proposed that the interval between the Lower and Upper Block Coal Members of the Brazil Formation records a transgression of a fresh to brackish water tidal flat over a coastal peat mire. Their interpretation of a low salinity marine deposit was based primarily on the low-sulfur nature of the underlying Lower Block Coal and the absence of marine body and trace fossils within the bedrock immediately overlying it. Later work revealed the presence of microscopic agglutinated forams in the laminated mudstone (Winton Wightman, personal communication) and sparse burrows in the coarser facies in the middle of the succession, suggesting a possible brackish water influence. However, factors other than salinity constraints (such as sedimentation rates) can influence the faunal content of a sedimentary succession. Thus, other than the low-sulfur nature of the underlying Lower Block Coal, the low salinity interpretation for the overlying rock was equivocal.

In order to understand the absence of brackish to marine salinity indicators, two or three core samples from several locations were collected from above the Lower Block Coal and analyzed geochemically and petrographically. For those cores from which the samples were collected, the samples closest to the coal represent the transgressive mudflat facies, whereas the samples collected furthest from the coal represent the regressive, mixed-flat facies. In some cases, a third sample was collected from the transgressive-regressive transitional mixed flat to sandflat facies. Types of organic matter were identified in 12 selected samples with a special emphasis on determining the proportions between terrestrial and marine inputs. Standard techniques of sample preparation for reflected and fluorescent light microscopy were used (Mackowsky, 1982).

Bulk geochemical characteristics, such as total organic carbon (TOC) and total sulfur (S), were determined on raw and HCl-treated samples using a LECO instrument. Using these parameters, a weight ratio of carbon to sulfur (C/S) was calculated. This ratio can be used to differentiate between marine and nonmarine sediments in cases where syngenetic pyritic sulfur is the dominant sulfur species (Berner and Raiswell, 1983; Berner, 1984). In addition, pyrolysis-GC/MS was applied to all 12 samples after an initial extraction with CH_2C_{12}. The pyrolysis-GC/MS was used to reveal chemical differences between unextractable organic matter (predominantly as a result of different organic matter types) and also to detect organic sulfur compounds. The detection of organic sulfur compounds is of special significance because it gives an indication of whether or not the C/S ratio can be used as a salinity indicator.

The quantity of TOC in dark, mud-rich tidal rhythmite laminae ranges from 0.1% to 6.9% (Fig. 10). Sulfur content is usually low and only in a few samples does it exceed 0.1% (Fig. 10). Small and angular fragments of vitrinite comprise the dominant organic matter type in the majority of dark laminae and they account for 50–90% of the total organic matter. Inertinite macerals are represented mainly by semifusinite that usually does not exceed 5% of total organic matter and in many samples occurs in trace proportion. Fusinite is very rare. In the laminae with an increased proportion of inertinite macerals, vitrinite appears to be more oxidized that in those samples from which inertinite macerals are absent. Liptinite macerals occur as sporinite (pollen and spores), alginite and liptodetrinite. Sporinite content varies from 0% to 25% of total organic matter. Alginite is represented mainly by *Botryococcus* bodies. *Botryococcus* occurs in selected laminae only, but if present, numerous bodies are found in close proximity. Liptodetrinite usually is the most abundant liptinite maceral, but, because of very small size, its origin is difficult to define. In summary, woody-like terrestrial material is the dominant organic matter in the tidal sediments above the Lower Block Coal, although there is a very general tendency of increasing amount of alginite towards the regressive facies of the sequence.

Pyrograms further substantiate that terrestrial material dominates the organic component of the tidal flat facies. Fig. 11 illustrates pyrograms of selected samples from the rhythmite sequences. Pyrograms A and B were selected from 12 pyrograms obtained for this study. All the pyrograms show striking similarities. They show the dominance of phenols, (alkyl)benzenes, and (alkyl)naphthalenes, suggestive of lignin-like material, which corroborates the petrographic observations.

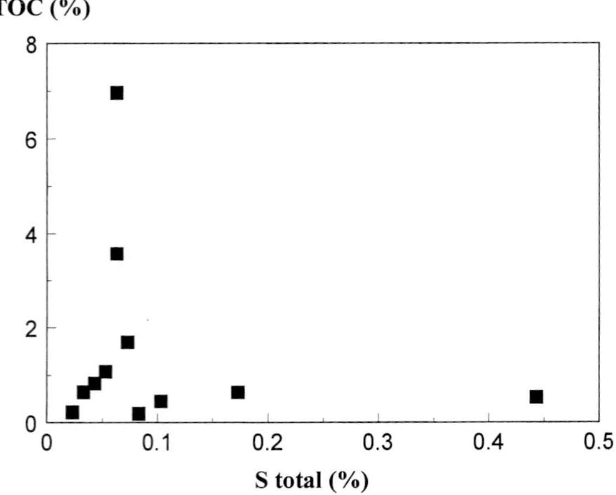

FIG. 10.—Total Organic Carbon (TOC) as a function of total sulfur (S total) content in rhythmites above the Lower Block Coal.

n-Alkanes are subordinate and the dominance of mid-chain n-alkene/n-alkane over n-C_{25}-n-C_{33} suggests the input from sporinite and alginite as their source. The pyrolyzates of all samples show close similarity to those of the vitrinite concentrates of the Desmoinesian Herrin Coal Member (Carbondale Formation) in Illinois (Stankiewicz et al., 1994) and to those of the Atokan Blue Creek Coal Member (Mansfield Formation) in Indiana (Mastalerz et al., 1997).

A weight ratio of carbon to sulfur (C/S) for all the samples has a wide range, depending upon whether it represents transgressive or regressive facies. Fig. 12 presents C/S ratios for three locations where for each location, two or three samples were taken from the rhythmite sequence. In two of the three cases, the C/S ratio is much higher in samples closer to the Lower Block Coal (transgressive mudflat facies), with values of 138 to 30, and strongly suggests a freshwater environment. Further away from the Lower Block Coal (regressive mixed flat facies), the ratio drops to as low as 2. Values of 2 and 3.7 (Fig. 12) are considered to reflect marine to brackish water environments. In order for the C/S ratio to be considered a salinity indicator, however, it needs to be demonstrated that syngenetic pyrite was the major contributor to the total sulfur and that organic sulfur was minor (Berner and Raiswell, 1983; Berner, 1984). In the sediments studied, pyrite occurs in all samples as dispersed, very-fine framboids of syngenetic origin. Considering the low total sulfur content in the dark, mud-rich rhythmite laminae, it seems very likely that this syngenetic pyrite is the major contributor to the total sulfur. This was confirmed by pyrolysis-GC MS, which detected neither elemental nor thiophenic sulfur (Fig. 11). Consequently, the C/S ratio appears to be a good indicator of salinity for the interval studied, and the variations in C/S observed likely resulted from salinity changes. Thus, the transgressive mudflat facies of the rhythmite succession were deposited under freshwater conditions, whereas the regressive facies were laid down under brackish to marine conditions.

DISCUSSION

Depositional Models

Traditionally in the Illinois Basin, the sulfur values of coals are considered to reflect the depositional environments of the deposits that immediately overlie them (e.g., Eggert and Phillips, 1982). Low-sulfur coals are interpreted to have been buried by nonmarine sediments, whereas high-sulfur coals are considered to have been buried by marine sediments. The problem with this model is that water salinity and depositional processes are not separated. Although there is undoubtedly a correlation between the depositional chemistry of the facies that transgressed a peat and the sulfur content of the resulting coal, there is not necessarily a direct correlation between depositional processes of roof facies and coal quality.

The extent of the Lower Block tidal flat, with its well developed tidal rhythmites, is intriguing. Unlike many tidal rhythmites in the Carboniferous of the eastern and midwestern parts of the U.S. and the Cretaceous of the western U.S. and Canada, the Lower Block Coal and its overlying tidal rhythmite facies are not part of a larger incised valley fill, thus excluding an estuarine (*sensu* Dalrymple et al., 1992) interpretation for the succession. Rather, the lateral continuity of the tidal flat facies

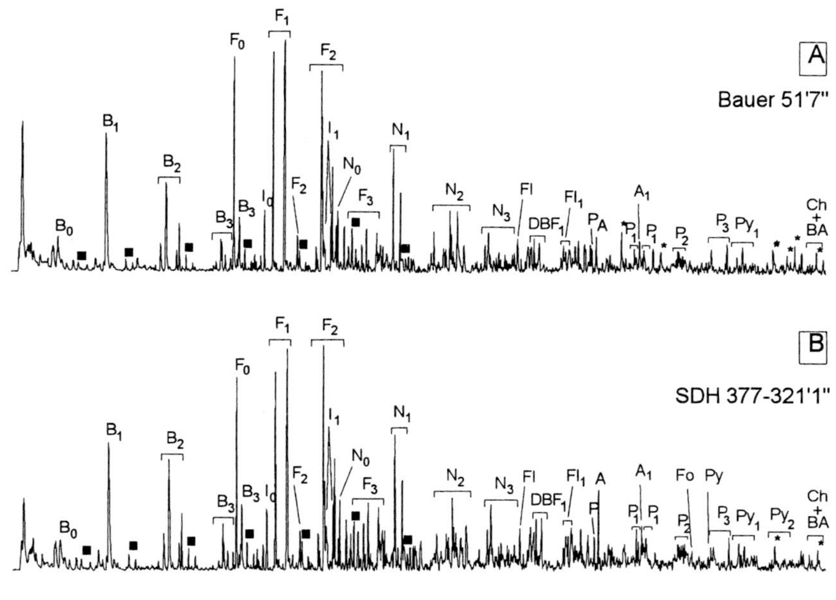

FIG. 11.—Reconstructed ion chromatograms of the pyrolysates (20 s at 700°C) of kerogen samples from Bauer 47 (2.44 m above coal) (A), and SDH 377 (12 cm above coal) (B). Key for the peaks: B_n—(alkyl)benzenes, F_n—(alkyl)phenols, I_n—(alkyl)indenes, N_n—(alkyl)naphthalenes, Fl_n—(alkyl)fluorenes, DBF_n—(alkyl)dibenzofurans, P_n—(alkyl)phenantrenes, A_n—(alkyl)anthracenes, Py_n—(alkyl)pyrenes, Fo—fluoranhene, BA—benzanthracene, Ch—chrysene, -*n*-alk-1-ene/*n*-alkane pairs, where n indicates extent of alkyl substitution (0: none, 1: methyl, 2: dimethyl or ethyl, etc.). Numbers below pyrograms indicate carbon numbers in alkene/alkane pairs.

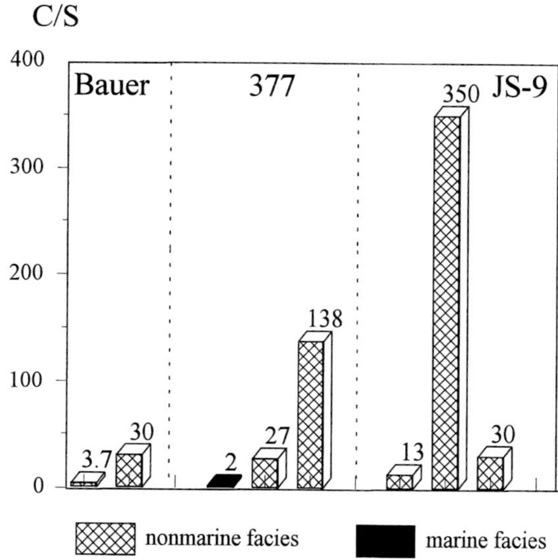

FIG. 12.—The C/S weight ratios from three rhythmite locations above the Lower Block Coal: Bauer 47 (2.44 m and 3.87 m above the coal), SDH 377 (12 cm, 2.68 m, and 3.93 m above the coal), and JS-9 (0.92 m, 3.69 m, and 5.09 m above the coal). In each location, the sample on the right is the closest to the underlying Lower Block Coal (transgressive facies) with the distance from the Lower Block Coal increasing towards the left (far left sample of each location represents the regressive facies). Note very high C/S in the samples closest to the coal and decreasing values of this ratio when distance from the coal increases, with the exception of location JS-9.

above the Lower Block Coal suggests a broad, uniform coastal tidal setting that preserves a thick and laterally continuous tidal rhythmite succession. The geochemistry of the roof facies indicates that the initial transgression of this peat mire was by a freshwater tidal flat deposit. Not until near maximum transgression (sandier subtidal facies) do the geochemistry indicators suggest brackish to normal marine salinity within the study area.

Two possibilities could explain the deposition and preservation of the Daviess County coastal freshwater tidal flats: (1) the Illinois Basin seaway during transgression of the Lower Block Coal in Indiana was not characterized by a normal marine salinity; or (2) extremely high rainfall and runoff in an ever-wet climate resulted in the development of a coast-hugging freshwater plume during the initial transgressive stages.

The presence of the lower Brazil Formation marine carbonate equivalents in southwestern Indiana does not support the first possibility. A direct modern analog is not known for the second model. However, coastal regions in the vicinity of significantly large rivers, such as the Amazon, are known for the propagation of freshwater or low-salinity surface plumes some hundreds of kilometers offshore, even along tide-dominated shorelines (see Lentz and Limeburner, 1995; Jaeger and Nittrouer, 1995). The width of the Amazon plume can exceed 200 km given the right wind conditions. Although the paleogeography during early Brazil Formation time is not well constrained, there is no evidence in outcrop or the subsurface to the north or south of the study area indicating the penecontemporaneous presence of a large, Amazon-type river. Our model proposes that a freshwater plume formed along the coast from extensive runoff from a poorly integrated drainage network feeding from the coastal peat mires and swamps. High runoff is consistent with the ever-wet climatic models and estimates of very high annual rainfall (on the order of 400–500 cm/yr) for the middle Pennsylvanian (Phillips and Peppers, 1984; Cecil, 1990; Crowley et al., 1996). Climatic models further indicate the possible existence of west to east blowing winds (southwesterly to northeasterly relative to modern geography) for the Pennsylvanian Illinois Basin (Crowley et al., 1996). Easterly directed winds could have been an important factor for maintaining a freshwater plume along the Pennsylvanian Indiana (southeasterly) coastline of the Illinois Basin and the development of estuarine (*sensu* Pritchard, 1967) water chemistry and circulation in that region.

Possibly, the Brazil Formation coastline was barred during at least a portion of the Lower Block peat transgression. However, large (macrotidal) tidal ranges would favor a more open coastal setting. Paleotidal ranges are difficult to assess at present. Modern tidal rhythmites are known only from areas where meso- to macrotidal ranges exist, suggesting that the Brazil Formation coastline was not barred (Kvale and Archer, 1996).

A very low paleotopographic slope can be inferred to have existed during the Pennsylvanian of the Illinois Basin (perhaps as low as the modern northern Australian coastal plain crossed by the south Alligator estuary, which is less than 0.005% according to Eberth, 1996). This is based on the basin-wide extent of certain late Atokan and Desmoinesian coals thought to be essentially contemporaneous from basin center to basin margin (e.g., Nelson et al., 1991; Wanless et al., 1970). A low paleotopographic slope coupled with a 300-km-wide continental shelf (Collinson et al., 1988) supports the interpretation of a relatively large tidal range (see Cram, 1979) during the time of deposition and further suggests that the Brazil Formation coastline was not barred.

The transgressive phase of the Lower Block Coal peat was initiated by the rapid collapse of the mire (consistent with the preservation of upright trees above the coal and high sedimentation rates of the overlying facies as calculated from the rhythmite tidal cycles) and the formation of embayed areas. Petrographic and palynologic evidence of the Lower Block Coal indicates that a high water table (indicated by the dominance of Lycopod trees) and nutrient-poor conditions prevailed prior to transgression. The vertical accretion of the Lower Block peat conceivably kept pace with an overall rise in sea level over many hundreds or even thousands of years, yet the collapse of the mire, the ensuing inundation, and subsequent maximum transgression occurred over a very short period—perhaps a few years.

A paleogeographic reconstruction of the phases of transgression over the Lower Block Coal is summarized in Fig. 13. Lower Block peat accumulated in the interfluve areas that separated inland tidal channels and creeks (Fig. 13A). None of these channels were encountered in the study area, but historical mining activity indicates that the coal becomes thinner and less laterally continuous several kilometers south of Daviess County, suggesting areas of more active channels. Transgression was initiated by an overall increase in the rate of plant mortality, the result of a lack of nutrients or freshwater flooding perhaps caused by unusually high spring tides, and the subsequent deflation and collapse of the peats and marsh deposits (Fig. 13B). Sediments were supplied by tidal currents and carried inland along the main tidal channels and creeks. As subsidence continued the tidal prism increased, resulting in the broadening and deepening of the channels (Fig. 13C) during the deposition of the lower intertidal and subtidal facies within Daviess County. Coincidental with increased subsidence was a more complete estuarine mixing and the arrival of the saltwater wedge into the study area. Once subsidence rates slowed, the embayed areas filled and eventually were colonized by fringing marsh deposits, which eventually evolved into coal-forming mires of the Upper Block Coal (Fig. 13D).

Sequence Stratigraphic Implications

From a sequence stratigraphic sense, the coarsening-upwards (and interpreted deepening-upwards) transgressive facies above the Lower Block Coal do not readily conform to standard sequence stratigraphic models. In most sequence stratigraphic models, coarsening-upwards siliciclastic parasequences are interpreted to be shoaling-upwards above a marine flooding surface. In such a scenario, there is typically evidence of an abrupt increase in water depth immediately above the flooding surface. However, the transgressive facies immediately above the Lower Block Coal records a gradual upwards increase in water depth from intertidal muds to subtidal sands rather than shoaling or shallowing upwards. The regressive (shoaling) facies above the Lower Block Coal is typically one half to one third the thickness of underlying transgressive facies. This somewhat unusual sequence of preservation of the flooding event can be explained by the fast rate of development of accommodation space resulting from the rapid collapse of the underlying peat (Lower Block Coal). Without this rapid development of accommodation space the thick transgressive rhythmite package would not have developed or been preserved.

CONCLUSIONS

The recognition of freshwater tidal flats in the rock record can be problematic. Modern examples are known that exist hundreds of kilometers inland, far beyond the saltwater wedge. More coastal freshwater tidal flats are not as well known. In such settings, saltwater wedges are pushed offshore by freshwater plumes. Both inland and coastal examples are not generally recognized from the rock record, yet their recognition has important implications for paleogeographic and depositional reconstructions, thus having an impact on sequence stratigraphic interpretations.

Tidal rhythmites preserving semidaily and neap-spring cycles can be an important component in recognizing tidal influences in the rock record, but their presence may be restricted to those environments that experienced meso- to macrotidal conditions and ironically may be best developed in settings least expected to produce them—within essentially freshwater environments and above relatively low-sulfur coals in coal-producing areas. Such settings are advantageous to the formation of rhythmites because they are beyond the saltwater wedge and the effects of burrowing marine organisms; accommodation space can be provided by either inland tidal or estuarine channels or by the rapid subsidence of autocompacting coastal or channel-fringing marshes and peat mires.

Many of the lithofacies (including ichnofacies) described for the Pennsylvanian of the central and eastern U.S. are similar to those described from the Carboniferous in western Europe and easternmost Canada. Although tidal influence has become increasingly recognized in the Carboniferous of the eastern U.S., it has generally not been considered as significant for the coal basins of northwestern Europe. Notable exceptions include works by Read (1992) on the Kincardine Basin, Scotland, who stated that "tidal deposits are probably much more widespread in the British Carboniferous than has hitherto been recognized" (Reed, 1992, p. 141), and Aitkenhead and Riley (1996) who have recognized well-developed tidal rhythmites in the Carboniferous of the Central Pennine Basin (England). Although it is quite possible that tidal influence during the Carboniferous was more significant in those basins west of the Appalachians (North American) than those formed east of the Appalachian

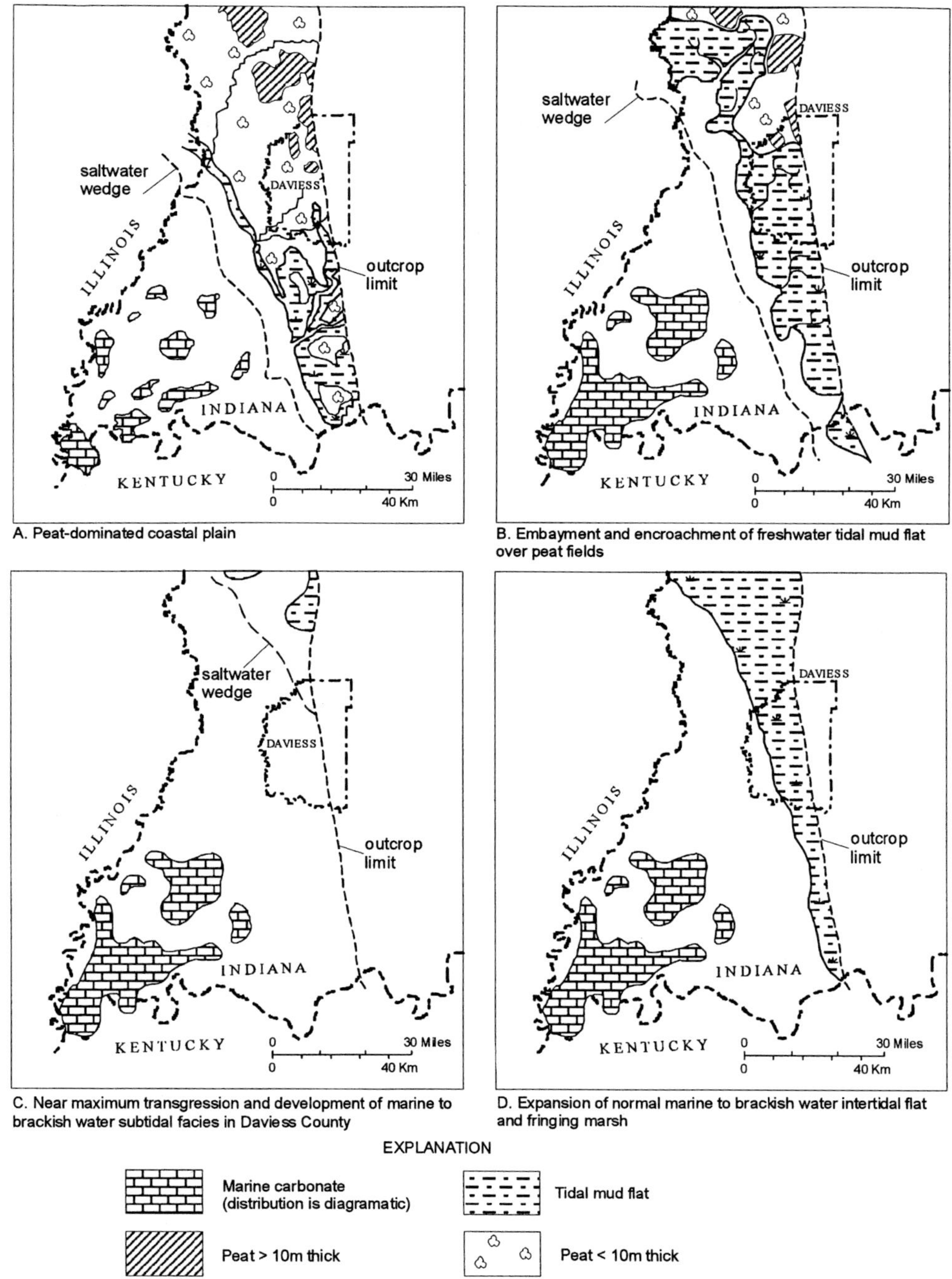

FIG. 13.—Generalized paleogeographic model of coastal inundation and fill above the Lower Block Coal. Marine carbonate distribution based on Droste and Horowitz (1996 and personal communication). Distribution of carbonate facies is very generalized.

orogenic belt (Nova Scotian and European), it is also possible that the subtleties for recognizing tidal influence in tropical mud-dominated systems have not been appreciated, particularly in freshwater settings. As such, tidal deposits may have been misinterpreted as lacustrine or flood plain sediments in the rock record.

ACKNOWLEDGMENTS

We thank A. Stankiewicz, H. Feldman, L. Furer, J. Devera, J. Droste, A. Archer, and M. Atkinson for technical discussions regarding this research; G. Klein, J. Rupp, and an anonymous reader for reviewing the paper; C. Mallard, and C. Rankin for collection of some of the field data; A. Zawistoski for some of the tidal rhythmite measurements; and R. Kersey, B. Hill, and J. Day for the final preparation of most of the figures for the paper. We also thank Solar Sources, Inc., for the use of proprietary core and for permission to publish our findings. Funding for this project was provided by a 1994 Indiana Department of Commerce Coal Research Grant to Indiana University and a grant from the National Aeronautics and Space

Administration (NASA) Innovative Research Program to E. K. through a contract with the University of Arizona.

This paper is dedicated to the memory of Rowena S. Paton, Shell, Wyoming.

REFERENCES

AITKENHEAD, N., AND RILEY, N. J., 1996, Kinderscoutian and Marsdenian successions in the Bradup and Hag Farm boreholes, near Ilkley, west Yorkshire: Proceedings of the Yorkshire Geological Society, v. 51, part 2, p. 115–125.

ARCHER, A. W., 1995, Modeling of cyclic tidal rhythmites based on a range of diurnal to semidiurnal tidal-station data: Marine Geology, v. 123, p. 1–10.

ARCHER, A. W., FELDMAN, H. R., KVALE, E. P., AND LANIER, W. P., 1994, Comparison of drier–to wetter-interval estuarine roof facies in the Eastern and Western Interior coal basins, USA: Palaeogeography, Palaeoclimatology, Palaeoecology, v. 106, p. 171–185.

ARCHER, A. W., KUECHER, G. J., AND KVALE, E. P., 1995, The role of tidal-velocity asymmetries in the deposition of silty tidal rhythmites (Carboniferous, Eastern Interior Coal Basin): Journal of Sedimentary Research, v. A65, p. 408–416.

ARCHER, A. W., AND KVALE, E. P., 1993, Origin of gray-shale lithofacies ("clastic wedges") in U.S. midcontinental coal measures (Pennsylvanian): An alternative explanation, *in* Cobb, J. C., and Cecil, C. B., eds., Modern and Ancient Coal-Forming Environments: Boulder, Geological Society of America Special Paper 286, p. 181–192.

ARCHER, A. W., KVALE, E. P., AND JOHNSON, H. R., 1991, Analysis of modern equatorial tidal periodicities as a test for information encoded in tidal rhythmites, *in* Smith, D. G., Reinson, G. E., Zaitlin, B. A., and Rahmani, R. A., eds., Clastic Tidal Sedimentology: Calgary, Canadian Society Petroleum Geology Memoir 16, p. 189–196.

BERNER, R. A., 1984, Sedimentary pyrite formation: An update: Geochimica et Cosmochimica Acta, v. 48, p. 605–615.

BERNER, R. A., AND RAISWELL, R., 1983, Burial of organic carbon and pyrite sulfur in sediments over Phanerozoic time: A new theory: Geochimica et Cosmochimica Acta, v. 47, p. 855–862.

CECIL, C. B., 1990, Paleoclimatic controls on stratigraphic repetition of chemical and siliciclastic rocks: Geology, v. 18, p. 533–536.

CECIL, C. B., DULONG, F. T., COBB, J. C., AND SUPARDI, 1993, Allogenic and autogenic controls on sedimentation in the central Sumatra basin as an analogue to Pennsylvanian coal-bearing strata in the Appalachian basin, *in* Cobb, J. C., and Cecil, C. B., eds., Modern and Ancient Coal-Forming Environments: Boulder, Geological Society of America Special Paper 286, p. 3–22.

COLEMAN, J. M., GAGLIANO, S. M., AND SMITH, W. G., 1970, Sedimentation in a Malaysian high tide tropical delta, *in* Morgan, J. P., ed., Deltaic Sedimentation Modern and Ancient: Tulsa, Society of Economic Paleontologists and Mineralogists Special Publication 15, p. 185–197.

COLLINSON, C., SARGENT, M., AND JENNINGS, J. R., 1988, Illinois Basin region, *in* Sloss, L. L., ed., Sedimentary Cover-North America Craton: Geological Society of America, Geology of North America, v. D-2, p. 383–426.

CRAM, J. M., 1979, The influence of continental shelf width on tidal range: Paleoceanographic implications: Journal of Geology, v. 87, p. 441–447.

CROWLEY, T. J., AND BAUM, S. K., 1991, Estimating carboniferous sea-level fluctuations from Gondwana ice extent: Geology, v. 19, p. 975–977.

CROWLEY, T. J., YIP, K.-J. J., BAUM, S. K., AND MOORE, S. B., 1996, Modelling Carboniferous coal formation: Palaeoclimates, v. 2, p. 159–177.

DALRYMPLE, R. W., ZAITLIN, B. A., AND BOYD, R., 1992, Estuarine facies models: Conceptual basis and stratigraphic implications: Journal of Sedimentary Petrology, v. 62, p. 1130–1146.

DAVIS, R. A., JR., AND CLIFTON, H. E., 1987, Sea-level change and the preservation potential of wave-dominated and tide-dominated coastal sequences, *in* Nummedal, D., Pilkey, D. H., and Howard, J. D., eds., Sea-level Fluctuations and Coastal Evolution: Tulsa, Society of Economic Paleontologists and Mineralogists Special Publication 4, p. 167–178.

DE BOER, P. L., OOST, A. P., AND VISSER, M. J., 1989, The diurnal inequality of the tides as a parameter for recognizing tidal influences: Journal of Sedimentary Petrology, v. 59, p. 912–921.

DELANNE, R. D., NYMAN, J. A., AND PATRICK, W. H., JR., 1994, Peat collapse, ponding and wetland loss in a rapidly submerging coastal marsh: Journal of Coastal Research, v. 10, p. 1021–1030.

DROSTE, J. B., AND HOROWITZ, A. S., 1996, Distribution of limestone in the Brazil Formation (Pennsylvanian) in the subsurface of southwestern Indiana and western Kentucky (abs.): Indiana Academy of Science 112th Annual Meeting, Programs and Abstracts, p. 78.

EBERTH, D. A., 1996, Origin and significance of mud-filled incised valleys (Upper Cretaceous) in southern Alberta, Canada: Sedimentology, v. 43, p. 459–477.

EGGERT, D. L., AND PHILLIPS, T. L., 1982, Environments of deposition of coal balls, cuticular shale and gray shale floras in Fountain and Parke counties, Indiana: Indiana Department of Natural Resources, Geological Survey Special Report. 30, Bloomington, Indiana, 43 p.

FELDMAN, H. R., ARCHER, A. W., KVALE, E. P., CUNNINGHAM, C. R., MAPLES, C. G., AND WEST, R. W., 1993, A tidal model for Carboniferous Conservat Lagerstatten Formation: Palaois, v. 8, p. 485–498.

GRADZINSKI, R., AND DOKTOR, M., 1995, Upright stems and their burial conditions in the coal-bearing Mudstone Series (Upper Carboniferous), Upper Silesia Coal Basin, Poland: Studia Geologica Plonica, v. 108, p. 129–147.

GREB, S. F., 1993, Peat accumulation and sedimentation in the Livingston Paleovalley (Morrowan), eastern Kentucky, *in* Archer, A. W., Feldman, H. R., and Lanier, W. P., eds., Incised Paleovalleys of the Douglas Group in Northeastern Kansas: Field Guide and Related Papers, Kansas Geological Survey Open File Report 93-24, p. 15.1–15.6.

JAEGER, J. M., AND NITTROUER, C. A., 1995, Tidal controls on the formation of fine-scale sedimentary strata near the Amazon river mouth: Marine Geology, v. 125, p. 259–281.

KVALE, E. P., AND ARCHER, A. W., 1990, Tidal deposits associated with low-sulfur coals, Brazil Formation (Lower Pennsylvanian), Indiana: Journal of Sedimentary Petrology, v. 60, p. 563–574.

KVALE, E. P., ARCHER, A. W., MASTALERZ, M., FELDMAN, H. R., AND HESTER, N. C., 1997, Tidal rhythmites and their applications: Indiana University, Indiana Geological Survey, Open File Report 96-6, Bloomington, Indiana, 101 p.

KVALE, E. P., CUTRIGHT, J., BILODEAU, D., ARCHER, A. W., JOHNSON, H. R., AND PICKETT, B., 1995, Analysis of modern tides and implications for ancient tidalites: Continental Shelf Research, v. 15, p. 1921–1943.

KVALE, E. P., FURER, L. C., AND MASTALERZ, M., 1996, Exploration models for selected economic coals in southeastern Daviess and southwestern Martin counties: Indiana University, Indiana Geological Survey, Open File Report 96-2, Bloomington, Indiana, 28 p.

LEBLOND, P. H., 1979, Forced fortnightly tides in shallow rivers: Atmosphere-Ocean, v. 17, p. 253–264.

LENTZ, S. J., AND LIMEBURNER, R., 1995, The Amazon River plume during AMASSEDS: Spatial characteristics and salinity variability: Journal of Geophysical Research, v. 100, p. 2355–2375.

MACKOWSKY, M.-TH., 1982, Methods of examination of coal, *in* Stach, E., Mackowsky, M.-Th., Teichmüller, M., Taylor, G. H., Chandra, D., Teichmüller, R., eds., Stach's Textbook of Coal Petrology, 3rd ed.: Berlin, Gebrüder Borntraeger, p. 300–348.

MARSHAL, L., AND LUNDBERG, J. G., 1996, Miocene deposits in the Amazonian foreland basin, technical comment: Science, v. 273, p. 123–124.

MASTALERZ, M., STANKIEWICZ, A. B., SALMON, G., KVALE, E. P., AND MILLARD, C. L., 1997, Organic geochemical study of sequences overlying coal seams; example from the Mansfield Formation (Lower Pennsylvanian), Indiana: International Journal of Coal Geology, v. 33, p. 275–299.

NELSON, W. J., TRASK, C. B., JACOBSON, R. J., DAMBERGER, H. H., WILLIAMSON, A. D., AND WILLIAMS, D. A., 1991, Absaroka Sequence, *in* Leighton, M., Kolata, D. R., Oltz, D. F., and Eidel, J. J., Interior Cratonic Basins: Tulsa, American Association of Petroleum Geologists Memoir 51, p. 143–164.

OERTEL, G. F., KRAFT, J. C., KEARNEY, M. S., AND WOO, H. J., 1992, A rational theory for barrier-lagoon development, *in* Fletcher, C. H., Wehmiller, J. F., eds., Quaternary Coasts of the United States: Marine and Lacustrine Systems: Tulsa, Society of Economic Paleontologists and Mineralogists Special Publication 48, p. 77–87.

PADGETT, P. L., RIMMER, S. M., EBLE, C. F., FERM, J. C., HOWER, J. C., MASTALERZ, M., AND THOMPSON, M., 1996, Petrology of the Lower Block Coal in southwestern Indiana: Implications for the depositional environment (abs.): Thirteenth Annual Meeting of The Society for Organic Petrology (TSOP), Abstract and Program, Carbondale, Illinois, v. 1B, p. 19–21.

PHILLIPS, T. L., AND PEPPERS, R. A., 1984, Changing patterns of Pennsylvania coal-swamp vegetation and implications of climatic control on coal occurrence: International Journal of Coal Geology, v. 3, p. 205–255.

PRITCHARD, D. W., 1967, What is an estuary: Physical standpoint?, *in* Lauff, G. H., ed., Estuaries: American Association for Advancement of Science, Publication 83, Washington, D.C., p. 3–5.

RÄSÄNEN, M. E., LINNA, A. M., SANTOS, J. C. R., AND NEGRI, F. R., 1995, Late

Miocene tidal deposits in the Amazonian foreland basin: Science, v. 269, p. 386–390.

READ, W. A., 1992, Evidence of tidal influences in Arnsbergian rhythmites in the Kincardine Basin: Scottish Journal of Geology, v. 28, p. 135–142.

REINECK, H.-E., 1972, Tidal flats, *in* Rigby, J. K., and Hamblin, W. K., eds., Recognition of ancient sedimentary environments: Society of Economic Paleontologists and Mineralogists Special Publication 16, p. 146–159.

RICHARDSON, E. S., JR., AND JOHNSON, R. G., 1971, The Mazon Creek faunas, Proceedings of the North American Paleontologic Convention I, p. 1222–1235.

SCOTESE, C. R., BAMBACK, R. K., BARTON, C., VAN DER VOO, R., AND ZIEGLER, A. M., 1979, Paleozoic base maps: Journal of Geology, v. 87, p. 217–233.

SHANLEY, K. W., MCCABE, P. J., AND HETTINGER, R. D., 1992, Tidal influence in Cretaceous fluvial strata from Utah, USA, a key to sequence stratigraphic interpretation: Sedimentology, v. 39, p. 905–930.

STANKIEWICZ, B. A., KRUGE, M. A., CRELLING, J. C., AND SALMON, G. L., 1994, Density gradient centrifugation: Application to the separation of macerals of Type I, II, and III sedimentary organic matter: Energy Fuels, v. 8, p. 1513–1521.

STAUB, J. R., AND ESTERLE, J. S., 1993, Provenance and sediment dispersal in the Rajang River delta/coastal plain system, Sarawak, East Malaysia: Sedimentary Geology, v. 85, p. 191–201.

VAN WAGONER, J. C., MITCHUM, R. M., CAMPION, K. M., AND RAHMANIAN, V. D., 1990, Siliciclastic sequence stratigraphy in well logs, cores, and outcrops: Concepts for high-resolution correlation of time and facies: American Association of Petroleum Geologists Methods in Exploration Series, No. 7, 55 p.

WANLESS, H. R., BAROFFIO, J. R., GAMBLE, J. C., HORNE, J. C., ORLOPP, D. R., ROCHA-CAMPOS, A., SOUTER, J. E., TRESCOTT, P. C., VAIL, R. S., AND WRIGHT, C. R., 1970, Late Paleozoic deltas in the central and eastern United States, *in* Morgan, J. P., ed., Deltaic Sedimentation: Society of Economic Paleontologists and Mineralogists Special Publication 15, p. 215–245.

WELLS, J. T., AND COLEMAN, J. M., 1981, Physical processes and fine-grained sediment dynamics, coast of Surinam, South America: Journal of Sedimentary Petrology, v. 51, p. 1053–1068.

CLIMBING-RIPPLE BEDDING IN THE FLUVIO-ESTUARINE TRANSITION: A COMMON FEATURE ASSOCIATED WITH TIDAL DYNAMICS (MODERN AND ANCIENT ANALOGUES)

WILLIAM P. LANIER†
Department of Earth Sciences, Emporia State University, Emporia, Kansas 66801, USA
AND
BERNADETTE TESSIER‡
Université de Lille 1, Laboratoire de Sédimentologie et Géodynamique, URA 719 CNRS. 59 655 Villeneuve d'Ascq, France

ABSTRACT: Climbing ripples characterize a variety of sedimentary depositional settings in which suspension sedimentation exceeds the rate of traction transport, but are poorly documented from tidal environments. Research within a modern macrotidal estuary (Bay of Mont-Saint-Michel, France) in comparison with a Carboniferous example of climbing ripples from tidally influenced sedimentary rocks (Tonganoxie Sandstone, eastern Kansas, USA), demonstrates that this form of stratification is very common and also closely associated with tidal dynamics along the fluvio-tidal transition zone of macrotidal estuaries.

In the modern example, flood- and ebb-dominated climbing-ripple facies (CRF) have been distinguished. Successive climbing-ripple units are up to 10 cm in thickness. Flood dominated CRF are associated with tidal channel levees found in the inner/straight channel zone of the fluvio-estuarine transition. In thicker CRF units, sedimentary structures indicate very high suspended sediment loads and rapidly decelerating flow velocity. Ebb-dominated CRF are found in chute channels and chute bars associated with the meandering zone of the fluvial-estuarine transition. The role of tidal dynamics in the formation of these CRF, both flood- and ebb-dominated, is indicated by the vertical organization and thickness evolution of the successive climbing-ripple units. They are frequently arranged in packages of strata that thicken and thin progressively. These packages are tidal rhythmites and correspond to the sedimentary record of the neap-spring-neap cycle. The increasing energy from neap to spring tides is indicated by an overall decreasing angle of climb in the generalized bedding sequence (progradation is dominant), whereas the decreasing energy from spring to neap is evidenced by an increase of this angle (vertical accretion is dominant).

Facies within the Tonganoxie Sandstone (Carboniferous) have identical climbing-ripple successions and similar vertical progressive thickening and thinning of strata in outcrop and in core, indicating a strong tidal influence on sediment deposition. Single sedimentation units in the Tonganoxie are up to 16 cm in thickness, but show an identical vertical progression of sedimentary structures as those from the modern facies in Mont-Saint-Michel. The physical sedimentary structures of both the modern and ancient are strongly comparable on a hydrodynamic basis insofar as silt-sized sediments dominate both systems. Furthermore, a variety of additional physical and biogenic sedimentary structures that require periodic or episodic exposure have been described from both the modern and the ancient. Sedimentation patterns in both systems suggest relatively rapid aggradation within the existing accommodation space, with soils and rooted horizons capping the units once this accommodation space is filled.

INTRODUCTION

Climbing ripples or ripple-drift cross-stratification are strata that are produced when ripples advance upslope, showing a positive angle of climb (Harms et al., 1982). Their formation requires a high suspended sediment load and begins when deposition from suspension exceeds the rate of traction deposition. These sedimentary structures are generally associated with waning flow conditions and their aggradation is favored by rapid deceleration of flow velocity when the transport capacity of the system is greatly exceeded. Climbing-ripple facies are very common in turbidic, fluvial, and deltaic overbank/crevass splay, and glacial outwash deposits where these fundamental physical parameters for their formation are commonplace. However, climbing ripples are poorly documented in tidal environments, both modern and ancient (Wunderlich, 1969; Yokokawa et al., 1995). This paper describes occurrences of climbing ripples from sedimentary facies located along the fluvio-estuarine transition from the Bay of Mont-Saint-Michel (France) and compares these with a Carboniferous example of climbing ripples that appears to have been deposited in a similar sedimentary environment under comparable hydrodynamic conditions.

GEOLOGIC SETTING

Modern—The Mont-Saint-Michel Estuary (France)

The Bay of Mont-Saint Michel, located in the embayment formed by the Cotentin and Britanny peninsulas of northwestern France (Fig. 1), is a macrotidal environment with tidal range up to 15.3 m during high spring tides.

The eastern part of the Bay of Mont-Saint-Michel defines the estuarine portion of the system and is fed by two perennial river systems, the Sée and Sélune (Fig. 1A). The mean annual fluvial discharge is very low (10 m^3/s to 20 m^3/s), and the fluvial sediment input is similarly insignificant. The system is tide-dominated. Maximum flood tidal currents during spring tides in this estuary can be as high as 2.5 m/s in the channels and range from 0.5–1 m/s on the flats. Flood velocities exceed those of the ebb flow (see Larsonneur, 1989, 1994, for further description of the Mont-Saint-Michel Bay).

According to Dalrymple et al. (1992), tide-dominated estuaries can be subdivided into three main zones from the mouth to the head on the basis of energy distribution (Fig. 1B): (1) an outer zone where elongated sand bars are developed; (2) an intermediate zone characterized by a braided system constituted by multiple channels and shoals; and (3) an inner zone represented by a single channel. This last, or inner portion of the estuary, is called the fluvio-tidal transitional zone, which evolves from a relatively straight to a meandering channel, becoming straight again towards the opening of the outer estuary. The outer/straight part of the fluvio-transitional zone is tidally dominated, the inner/straight part is fluvially dominated, and the meandering portion is a mixed zone that coincides with the lowest energy in the system (minimum tidal and fluvial energies).

This morphosedimentary model, which has been established for the macrotidal Cobequid Bay-Salmon River Estuary (Canada), can also be applied to the Bay of Mont-Saint-Michel Estuary (Fig. 1A compared with Fig. 1B). The major differences between the two macrotidal systems is with regard to the lack of elongated sand-bar systems within the outer portions (mouth)

†Deceased ‡Corresponding author
Tidalites: Processes and Products, SEPM Special Publication No. 61

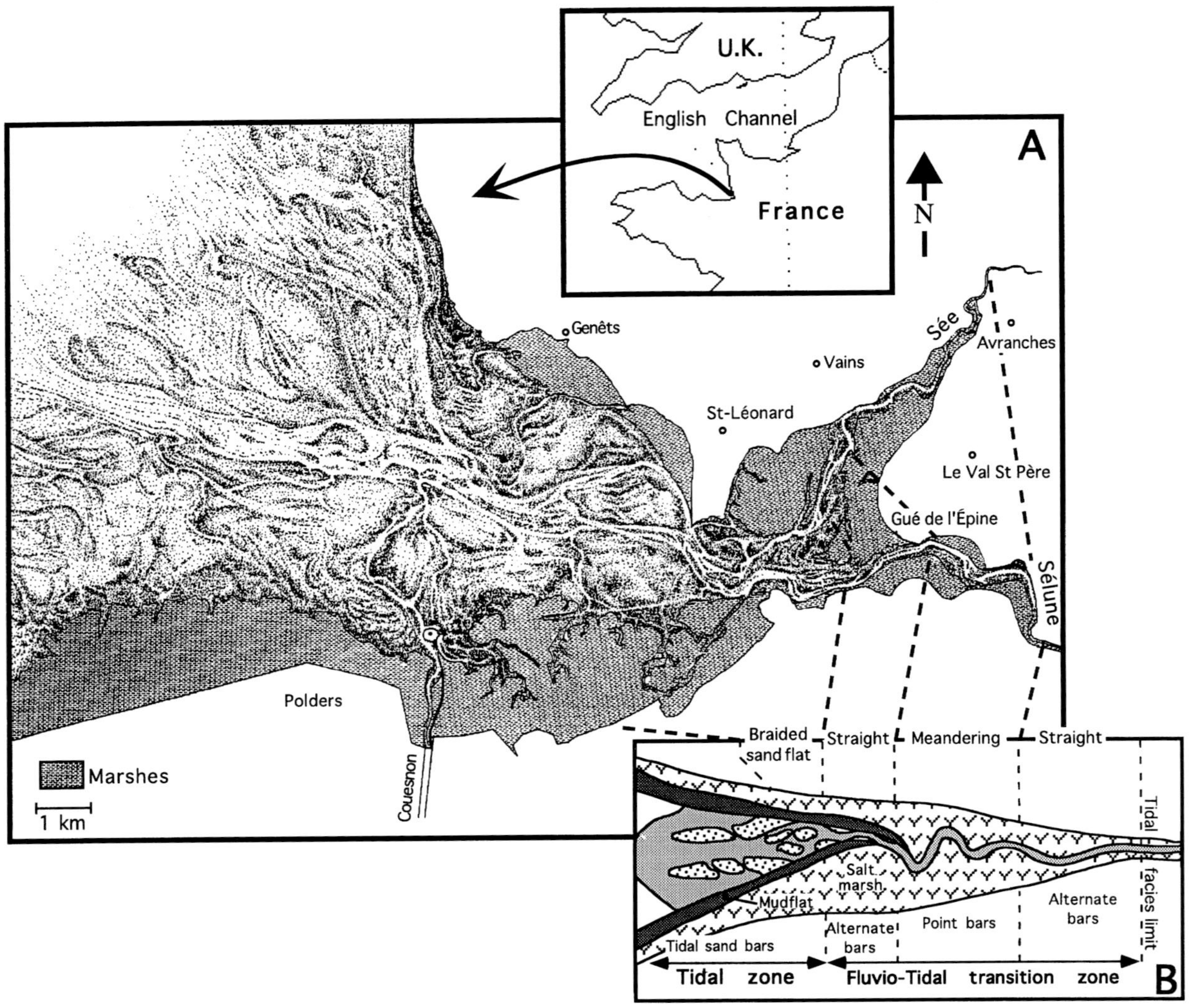

FIG. 1.—(A) Locality and morphosedimentary map for the Mont-Saint-Michel Estuary in northwest France (after Larsonneur, 1989); (B) the generalized facies distribution for tidal zonations related to the Mont-Saint-Michel system as per the model of Dalrymple et al. (1992).

of the Mont-Saint-Michel Estuary. Unlike Cobequid Bay in Nova Scotia, Mont-Saint-Michel Bay is located along an open coast, and, as such, wave dynamics from the English Channel preclude the formation of elongate sand-bar systems. Moreover, sediments in this area of the estuary are medium- to fine-grained sands, which do not allow for the formation of large dunes. Thus, the mouth of the estuary consists essentially of a wide braided-bar system.

Climbing-ripple stratification is very common upstream from the braided facies where the two river channels form the fluvial to estuarine transition. These two channels display a well developed straight-meandering-straight morphology (Fig. 1A), which is similar to the model of Dalrymple et al. (1992).

Along the fluvio-tidal zone, more or less extended intertidal and marsh areas are developed. They are characterized by silt-dominated sediments (very-fine-grained sand to very-fine-grained silt) that can contain up to 60% bioclastic marine carbonate (shell debris) (Bourcart and Charlier, 1959; Larsonneur, 1994). This sediment is highly thixotrophic and easily reworked by currents. As a consequence, current stages during a typical tidal cycle are characterized by a very high suspended sediment load that can approach several thousand mg/l (Giraud, 1996).

Ancient—The Tonganoxie Sandstone, Douglas Group, Stephanian (Buildex Quarry, Eastern Kansas, USA)

The Douglas Group (Stephanian) of eastern Kansas (Fig. 2) contains several paleovalleys that were eroded during sea-level lowstands and filled during the subsequent transgressions (Archer et al., 1994; Feldman et al., 1995). The Stranger Formation (lower Douglas Group, Fig. 2A) is characterized by a tripartite sedimentological facies distribution pattern that records the fluvial-estuarine-marine transition (Archer et al., 1994). These facies are exposed within the Douglas Group rocks in a generally northeast to southwest trending outcrop belt (Fig. 2B). The fluvial component of the lower Douglas Group system is best developed in northeastern Kansas (Feldman et al., 1995). The fluvio-estuarine transition of the Stranger Formation, which contains silt-dominated tidal rhythmites (Lanier et al., 1993) is particularly well exposed at the Buildex Quarry in Franklin County, Kansas (Fig. 2B).

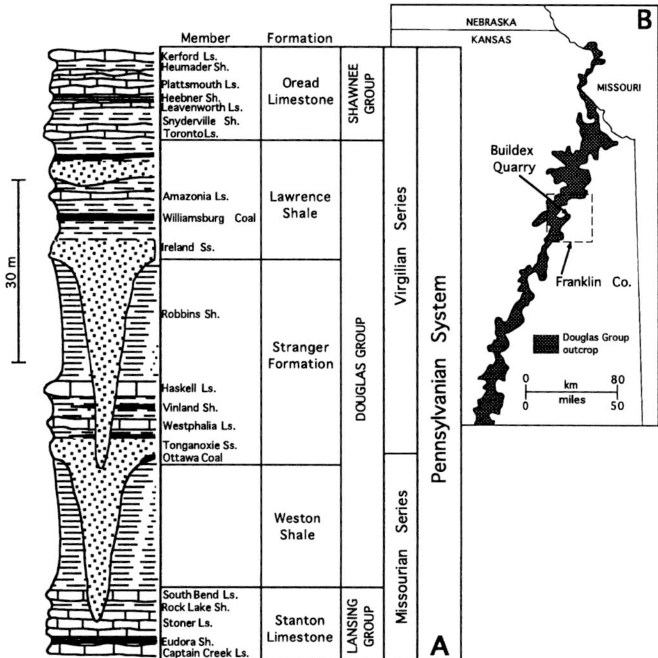

FIG. 2.—(A) Stratigraphy of the Carboniferous system of Kansas, USA (modified from Lanier et al., 1993). (B) outcrop map of the Douglas Group in eastern Kansas, and locality of the Buildex Quarry in Franklin County, USA.

The outcrop in the Buildex Quarry exposes the Weston Shale, which is disconformably overlain by the Ottawa Coal and the Tonganoxie Sandstone that belong to the lower part of the Stranger Formation (Fig. 2A). The Ottawa Coal is thin, laterally discontinuous, and likely accumulated relatively high along the valley wall within floodplain of the Tonganoxie fluvial system (Lanier et al., 1993). About 9 m of the Tonganoxie Sandstone is exposed in the quarry. The sequence consists of a well-defined package of gently inclined (about 2–3° SW) planar beds to flat, planar laminae of poorly indurated, carbonate-cemented siltstone and mudstone. The sequence is bounded above and below by thin coals and paleosols with upright plant remains. A striking feature of the exposure is the lateral continuity of bedding throughout the approximately 400-m-long exposure in the quarry, and the apparently systematic fashion in which the beds thicken and thin.

The lower 5 m of the exposed Tonganoxie Sandstone is considered in the present description (Fig. 3). It is composed entirely of siltstones with a maximum of about 10% very-fine-grained slightly micaceous quartz sand. It can be subdivided from base to top into two main facies on the basis of bed/laminae thickness variations and lateral continuity of stratification, primary physical and biogenic sedimentary structures, and the presence of upright plants and rooted horizons (Lanier et al., 1993). Well-developed climbing ripples are found throughout this 5-m-thick succession.

The lower portion of the succession is the planar bedded/laminated facies (PBL, Fig. 3). It is 2.8 m thick and consists of vertically accreted planar, locally convoluted, beds to flat planar laminae of lithologically homogeneous gray (buff weathering) siltstones. Bed/laminae thicknesses range from 0.05 mm to 16 cm. Each bedform is normally graded and thicker units are laterally continuous through the extent of the quarry exposure. On the basis of systematic changes in stratum thickness and sedimentary structures, three units (A1–A3, Fig. 3) have been distinguished in the PLB facies (Lanier et al., 1993).

The overlying 1.8 m of succession defines the channel/levee facies (CL, Fig. 3). The transition from the PBL facies appears to be gradational, conformable, and characterized by increasing siltstone bed thicknesses. When traced laterally, however, many beds in the CL facies thin and pinch out, defining a channel-form geometry with a width of approximately 10–12 m and a thickness of 1 m (B1, Fig. 3). The axis of the channel is towards the southwest (215°) and is roughly parallel to the axis of the Tonganoxie paleovalley. Where lateral contacts have been exposed in excavations, these channel facies siltstones onlap and may erosionally truncate a channel levee sequence (B2, Fig. 3) of vertically accreted, planar-bedded and laminated siltstones that contain upright plants with attached leaves (largely pte-

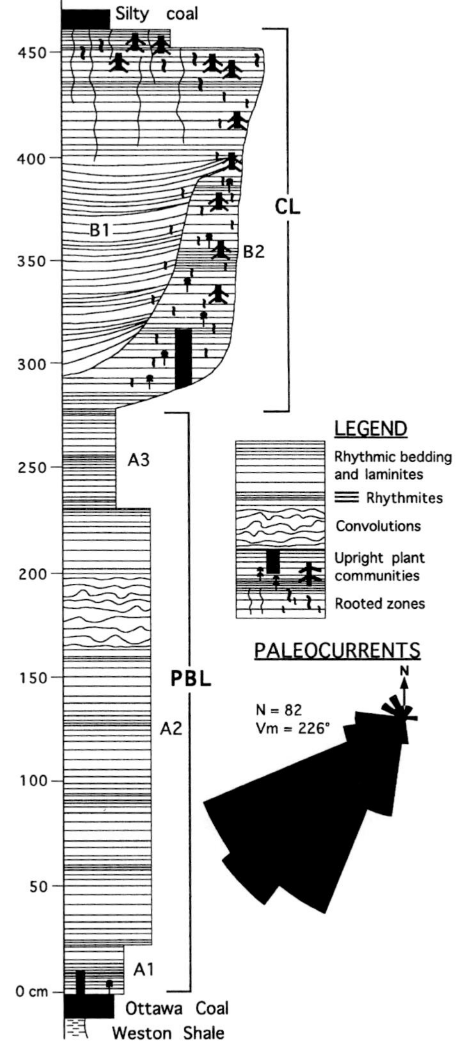

FIG. 3.—Generalized sedimentologic log and paleocurrent rose diagram of the Tonganoxie sandstones exposed on the west wall of the Buildex Quarry (after Lanier et al., 1993). PBL—Planar-bedded-and-laminated facies; A1, A2, A3—the different units in PLB; CL—Channel-and-levee facies; B1, B2—the different units in CL. See text for description.

ridosperms and small lycopods). Rooted horizons are very common. The top of the CL facies is a thin (5–8 cm thick), laterally discontinuous, silty coal. The siltstone beds beneath the coal contain upright 2–6-cm diameter lycopods at a minimum density of 28 plants per square meter. These strata are pervasively rooted and solitary roots can extend down as deep as 50–75 cm into the CL facies (Fig. 3).

CLIMBING-RIPPLE FACIES: DESCRIPTION AND OCCURRENCE

Modern

Facies characterizing the upper intertidal domain along the fluvio-tidal channels of the Mont-Saint-Michel Estuary are typically composed of heterolithic beddings such as planar-, lenticular-, wavy-, and flaser-bedding. These heterolithic beddings are commonly associated with tidal rhythmites (TR) that correspond to the sedimentary record of neap-spring-neap cycles and can be considered as the most characteristic sedimentary facies along the fluvio-tidal domain in the Mont-Saint-Michel Estuary (Tessier, 1993, Tessier et al., 1995; Tessier, this volume). Climbing-ripple stratification is also very common within the fluvio-tidal domain of the Mont-Saint-Michel Estuary, and is frequently associated as well with tidal rhythmites. Both flood- and ebb-dominated climbing-ripple facies (CRF) have been distinguished.

Flood-dominated CRF.—

Flood-dominated CRF are associated with channel levees found in the inner/straight channel zone of the fluvial to estuarine transition (Fig. 1A). Individual sedimentation units from the flood-dominated CRF can reach a maximum of 10 cm in thickness. The thickest units show a vertical sedimentary structure sequence from base to top consisting of structureless (massive) stratification to planar laminae, through a transition to climbing laminae, and a mud drape (Figs. 4A and B). Climbing laminae display an upward evolution from a low to a progressively higher angle of climb. This evolution corresponds to a passage from erosional-stoss side (Type A) to depositional stoss side (Type B), climbing-ripple bedding (Jopling and Walker, 1968). The transition from climbing laminae to mud drape may be gradual, incorporating mud in the upper portions of the climbing-ripple sedimentation unit (Fig. 4B), but the contact is frequently sharp (Fig. 4C). The contact between successive units is sharp and generally nonerosive. The upper mud drape may be as thick as 1 cm and may create a smooth, planar upper surface that masks the underlying ripple morphology (Fig. 4C). In these cases, the morphology of the underlying rippled bedform has no control on the internal structuring of the subsequent sedimentation unit.

Successive units associated with flood-dominated CRF frequently thicken and thin progressively upwards (Figs. 4C and D). This pattern indicates the association of flood CRF with tidal rhythmites. The thickening is related to increasing energy from neap to spring and thinning related to decreasing energy from the spring to the neap. Thinner units in these neap-spring-neap records are normally graded with a thin mud drape.

Ebb-dominated CRF.—

Ebb-dominated CRF at Mont-Saint-Michel is associated with point bars in the meandering domain of fluvial-estuarine transition (Fig. 1A). This facies more specifically characterizes the upstream areas of point bars and the chute channel/bar systems. Type B (Jopling and Walker, 1968) climbing-ripple stratification (Figs. 4E and F) is dominant in this facies and is formed by the superimposition of successive linguoid to sinuous crested ripple trains. Successive ripple crests are advanced to the lee side of the underlying ripple crest and the climb angle of the ripples is strongly controlled by the amount of sediment deposited from suspension and the substrate morphology. Mud drapes are generally absent or very thin in this facies because of its intertidal position. However, when deposition takes place in a topographic depression where standing water is retained during low tide, thin mud drapes may be formed and preserved.

Ebb-dominated, climbing-ripple facies are also associated with tidal rhythmites. This association is more obvious in the upper intertidal facies deposited at the upstream zones of tidal point bars. Neap-spring-neap cycles are recorded by progressive thickening and thinning of successive strata that vary, in this vertical succession, from thin parallel/wavy bedding to Type B climbing ripples through thin wavy/parallel bedding (Fig. 4F). The increasing and decreasing energy during the neap-spring-neap cycle may also be materialized by progressive decreasing and increasing angle of climb in the Type B ripple succession. Thick mud drapes are lacking. Ebb-dominated CRF within chute channel/bar facies of the point bars consist exclusively of Type B climbing ripples and are not as frequently associated with TR.

Ancient

Climbing ripples are the dominant physical sedimentary structure in the thin Carboniferous sedimentary sequence that is found at the Buildex Quarry in eastern Kansas.

Silt-dominated strata within the Al and basal A2 units of the PBL facies (Fig. 3) are organized into packages of 6-14 sedimentation units with thicknesses of 0.5 mm–1.4 cm, varying in thickness upwards in a cyclic and systematic fashion (Fig. 6 of Lanier et al., 1993). A variety of Type B climbing ripples characterizes thicker sedimentation units within this facies. Both stoss and lee sides are preserved; each unit is normally graded, and the angle of climb is measurable as a preferred concentration of fines and organic matter at the ripple crests (Fig. 5A). Mud drapes are absent. Consequently, each successive ripple crest advances upslope from the crest of the previously deposited ripple. There is a vertical, gradational transition within this portion of the section from relatively thin (Fig. 5A) through well-developed climbing silt waves (Fig. 5B). Contacts between strata are sharp and generally nonerosive. Climb angles, as manifested by preferential accumulation of organic debris on lee-side laminae, range from less than 10° to about 42°. Southwest transport directions dominate this facies.

Siltstone bedsets in the A2 interval (Fig. 3) vary from 1.4–16 cm and account for about 65% of the PBL facies thickness. These sedimentation units progressively thicken and thin through the interval (Fig. 5C). Single sedimentation units within the thicker strata of this unit show a repeated vertical sedimentary structure sequence (Lander et al., 1993) consisting of a basal massive- to normally graded interval, parallel lamination, with an abrupt transition to Type A, erosional-stoss-

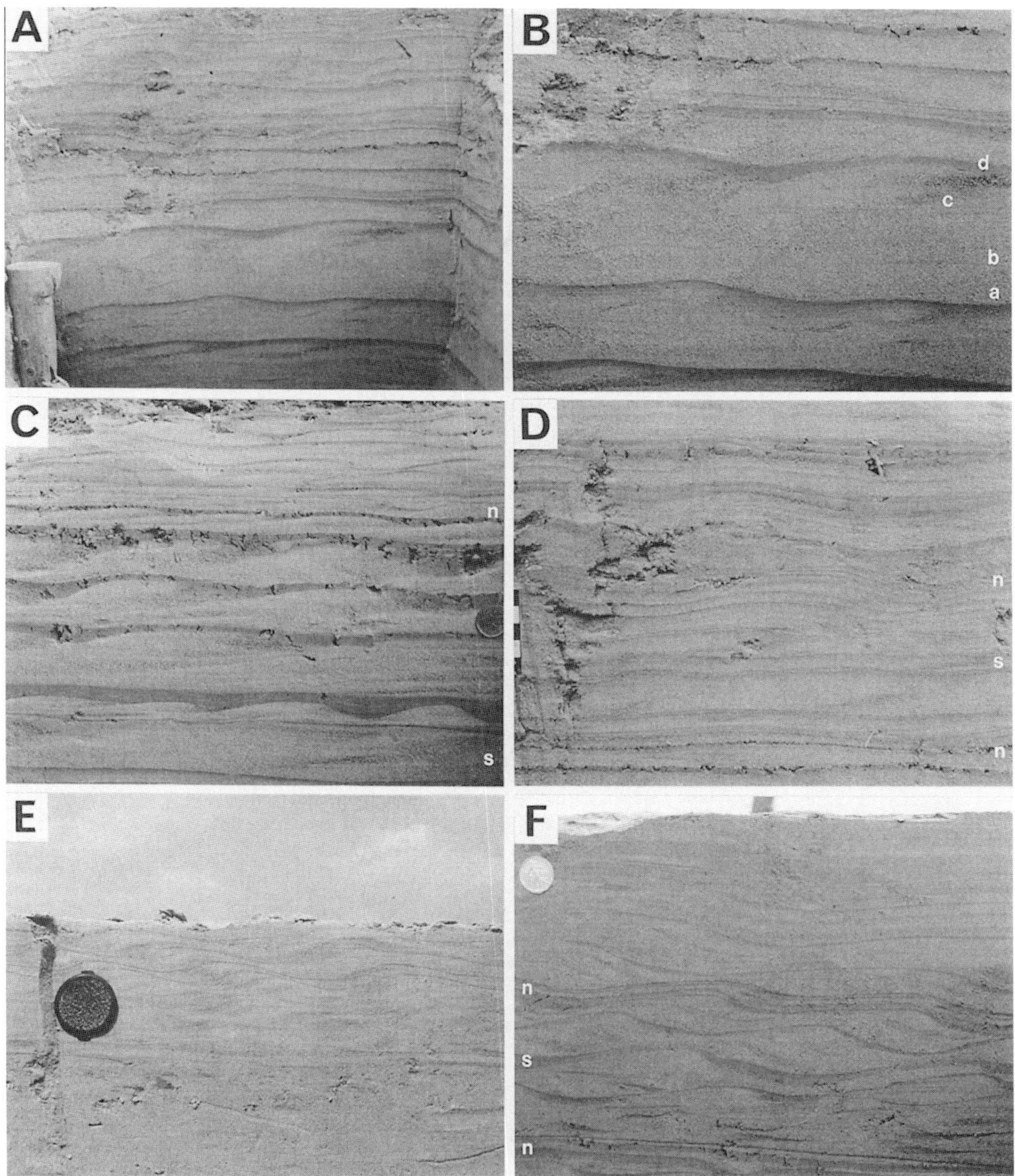

FIG. 4.—Climbing-ripple facies (CRF) in the Mont-Saint-Michel Estuary: (A, B, C and D). Flood-dominated CRF: (A) Succession of thick climbing units. Knife handle (lower left corner) is 12 cm.; (B) Detail of (A) showing the evolution from base to top in a thick climbing unit, with almost structureless stratification (a), to planar laminae (b), through a transition to climbing laminae (c), and a mud drape (d); (C) Climbing units with thick mud drapes. The thinning of units reflects the decreasing energy from spring (s) to neap (n). Coin is 12 mm; (D) A complete neap (n)—spring (s)—neap (n) cycle record in flood CRF. Scale bar in centimeters. (E and F) Ebb-dominated CRF: (E) Typical ebb CRF showing Type B climbing lamination. Mud drapes are absent. Lens mask is 55 mm; (F) a neap (n)—spring (s)—neap (n) cycle record in ebb CRF. Coin is 24 mm.

climbing ripples (Jopling and Walker, 1968), overlain by Type B (depositional-stoss) climbing ripples (Fig. 5D). Organic matter, mostly detrital plant materials and fines, are concentrated on the lee of the ripple crests. Climb angles increase from Type A through the Type B interval at the tops of these single sedimentation units. Bed surfaces are generally flat through this transition and thick mud drapes are lacking. Silty laminae within the Type B interval are frequently convoluted, oversteepened, and may show water escape structures (Lanier et al., 1993, see Fig. 9.). Transport directions are towards the southwest (Fig. 3).

Thicker sedimentation units in the B1 interval of the channel/levee facies (Fig. 3) of the Buildex Quarry section are also characterized by climbing ripples. The most common vertical sequence of sedimentary structures within single sedimentation units consists of a basal cross-laminated interval abruptly overlain by Type B climbing-ripple stratification (Fig. 5E) with angles of climb uniform or gradually increasing upward. Some bedsets consist entirely of Type B climbing ripples. This facies also has linguoid ripples that climb at subcritical angles (Hunter, 1977). As in the PBL facies, the morphology and internal structure of each sedimentation unit is strongly controlled

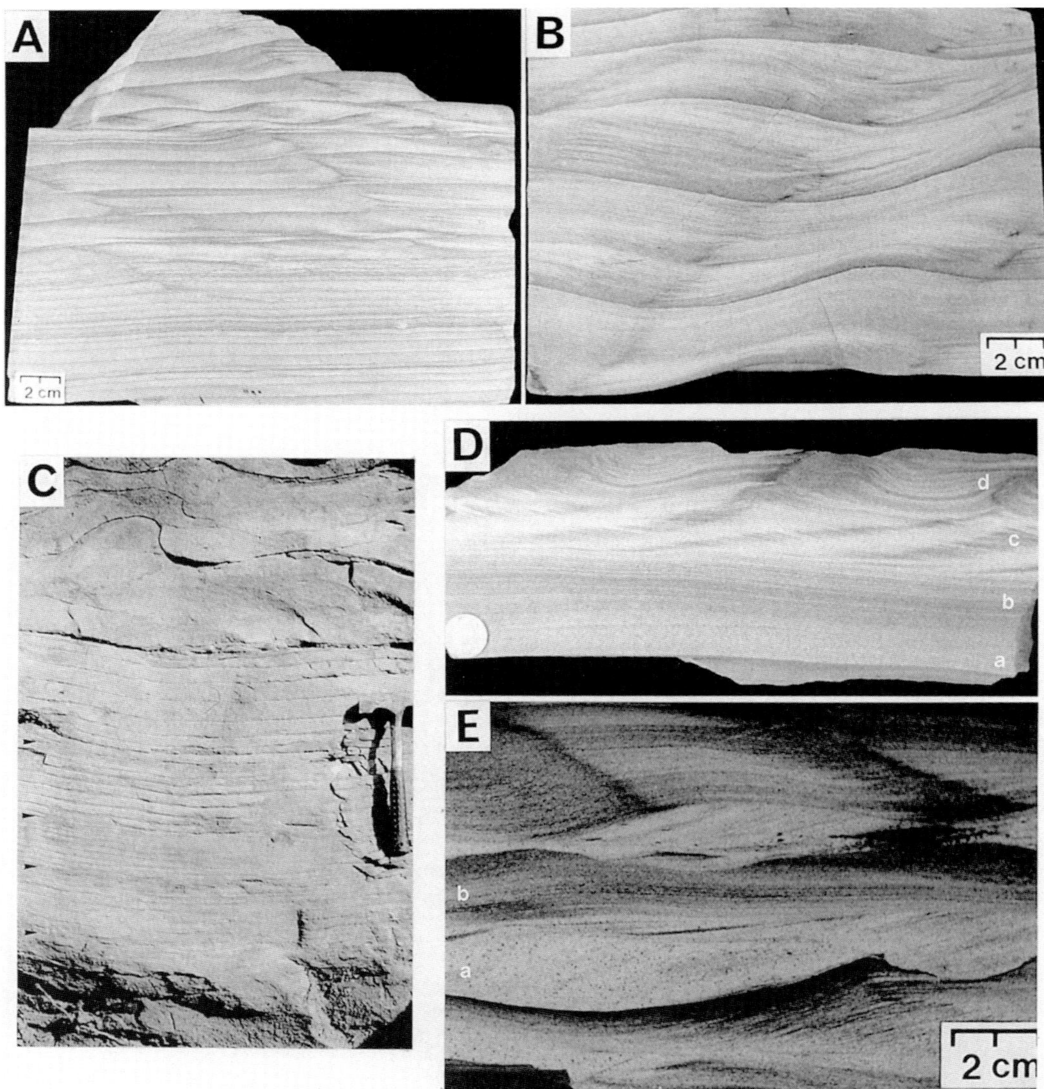

FIG. 5.—Climbing-ripple facies in the Tonganoxie Sandstone at Buildex Quarry (Lanier et al., 1993): (A) A variety of Type B climbing ripples from sedimentation units in a lower part of the PLB facies (transition from A1 to A2, Fig. 3). Mud drapes are absent; (B) Well-developed climbing silt waves from unit A2 (Fig. 3) with Type B climbing-ripple lamination; (C) Successive packages of thickening and thinning units (PLB facies, unit A2, Fig. 3). Arrows indicate where strata are thinnest; (D) Thicker strata in unit A2 (PLB facies, Fig.3) showing a sequence from a normally graded interval (a), to parallel lamination (b), with an abrupt transition to Type A climbing ripples (c), overlain by Type B climbing ripples (d). Coin is 17 mm. (E) Typical climbing strata of B1 unit (CL facies, Fig. 3) with a lower cross-laminated interval (a), abruptly overlain by Type B climbing ripples (b).

by the geometry of the preceeding bed form. In many cases, the bases of thicker bedsets show erosion of clay drapes. Consequently, many of the thicker beds near the axis of the channel consist of amalgamated packages of sedimentation units. Paleocurrents are uniformly towards the southwest.

DISCUSSION

Within a fluvial transition to the open estuarine environment, climbing ripple facies could be assumed to be controlled primarily by fluvial dynamics. However, in the estuary of Mont-Saint-Michel, the occurrence of flood-dominated CRF, the low riverine discharge, and the very low fluvial sediment loads that characterize this setting indicate that tidal currents play the dominant role in the deposition of these sedimentary facies. Moreover, the association of climbing-ripple facies with tidal rhythmites in the modern setting is compelling evidence that the sedimentation of these facies within the fluvio-tidal domain is closely related to tidal current dynamics. This correspondence also demonstrates that each climbing unit in the flood-dominated type and each ripple train in the ebb-dominated facies is the consequence of a single sedimentation event during flood and ebb stages, respectively. Similarly, stratification thicknesses in the ancient example from the Buildex Quarry also show tidal rhythmites (Fig. 5C) and stratification thicknesses display a cyclic thickening and thinning that is indicative of a tidal influence on sedimentation (Lanier et al., 1993).

The deposition of climbing-ripple bedding/stratification has been related to the rate of suspension sedimentation versus the aggradation rate of the traction load (Jopling and Walker, 1989). It has been established that the suspended load/bed load ratio increases from Type A to Type B climbing-ripple stratification (Jopling and Walker, 1968).

In general, the formation of all types of climbing ripples for very-fine sands to silt-dominated sediment is favored by a very high suspended sediment load and a rapidly decelerating flow velocity such that deposition from suspension exceeds that of the bedload. The ratio of suspended silt to the traction load is also significant in this regard. In the Mont-Saint Michel Estuary, these hydrodynamic and sedimentation conditions are achieved during almost each single flood and ebb stage of the high spring portion of the tidal cycle. Measurements of suspended sediment load and current velocities in the Couesnon channel (Giraud, 1996), located in the southern part of the estuary (see location on Fig. 1A), demonstrate that during flood stage of high spring tides, very high suspended sediment loads, reaching up to 10 g/l, are induced by the passage of the tidal bore. Following this, current velocities decrease very rapidly. Although ebb current velocities are generally lower than those associated with flood flow, these measurements indicate that maximum ebb current velocities at mid flow are responsible for significant sediment reworking and thus very high suspended sediment loads (1000 mg/l to 2000 mg/l). Such measurements have not been performed in the channels of the Sée and Sélune Rivers (Fig. 1), where detailed observations of climbing-ripple facies have been made. However, because of their comparable size and morphosedimentary characteristics compared with the Couesnon River, it is assumed that very similar processes occur in these channels as well.

Contrary to common tidal bedding, formed by planar-, lenticular-, wavy-, and flaser bedding, which are widespread within the fluvio-tidal transitional zone, flood- and ebb-dominated climbing-ripple facies appear to be associated with very specific depositional facies in the Mont-Saint-Michel Estuary. The flood-type forms part of the levee facies in the inner/straight part of the fluvial-tidal zone, and the ebb-type characterizes chute bar deposits in the meandering portion (Fig. 1B).

Along the fluvio-tidal channels of the Mont-Saint-Michel Estuary, natural levee aggradation can only occur during spring flood tides when the rising water overtops the channel shortly after the passage of the tidal bore. Fluvial levees formed by river flooding do not exist within this zone of the system. Moreover, these tidal flood levees are only found in the inner/straight channel portion of the transitional zone. Channels within the meandering zone of the system are apparently too wide to allow levee construction. In the inner/straight zone, however, the flow is concentrated in a narrower channel cross-section, limited on each side by marshes, and the intertidal flats are only a few meters wide. Channel overbanking occurs suddenly and remains restricted to the immediate channel margins so that levees can aggrade. Flood-dominated climbing-ripple units are thought to be deposited at this time, and probably within a relatively short time interval insofar as flow reversal, from flood to ebb, is very short. Flood flow velocity decelerates and suspended silt-sized sediment load decreases rapidly during this transition. Mud drapes can be deposited during the relatively brief slack-water interval between flood and ebb.

Ebb-oriented climbing-ripple facies have been identified within the meandering zone of the fluvio-tidal transition of the Mont-Saint-Michel Estuary. These sedimentary structures are found at mid- to upper levels of point bars immediately downstream from the crossover reach of the meander loop. They are somewhat less common, however, because average ebb current velocities are generally too low to erode and transport a significant sediment load. The occurrence of ebb-dominated CRF in chute channel/bar facies is, however, consistent with the fact that ebb current velocities are herein dominant and thus can potentially lead to significant sedimentation. High sediment loads are transported from the channel, and current velocities decelerate in the chute channel as flow velocities wain. In purely fluvial environment, climbing-ripple laminations are present in chute bar facies sequences, but these fluvial chute channel sequences are deposited only during high river flood conditions and are capped by mud drapes (Levey, 1978). The sedimentation of ebb-oriented climbing ripples found in Mont-Saint-Michel is not controlled by fluvial processes, although the hydrodynamic/sedimentologic conditions of their formation may be quite similar, but is related to current dynamics and the falling water stage during the ebb phase of the tidal cycle. High concentrations of silt-sized sediments are derived from erosion within the inner zone of the fluvio-tidal transition during maximum spring tides, and subsequently deposited as CRF during waning stages of ebb flow on the upstream zones of point bars and within chute channel/bar facies. The lack of a slack-water phase during this interval of the tidal cycle explains the fact that these CRF do not commonly show mud drapes except in low-standing areas of the chute channel.

These observations of climbing ripples along the fluvio-tidal channels of the Mont-Saint-Michel Estuary demonstrate that CRF deposition is closely related to the hydrodynamic and morphosedimentary characteristics that are specific to this setting. Thus, the distribution of flood CRF and ebb CRF as described herein may not necessarily define a general pattern that can be applied to all other macrotidal estuaries. For instance, in the Cobequid Bay-Salmon River estuary in Canada, flood CRF are found in the outer/straight to meandering part of the fluvio-tidal domain (Dalrymple and Makino, 1989; Dalrymple et al., 1991). This is never the case in the Mont-Saint-Michel Estuary, where flood CRF are restricted to the inner/straight channel zones of the system. Given this fact, flood-dominated climbing ripple facies may not be specific sedimentary facies indicators within the depositional setting along the transitional estuarine domain. Ebb-dominated CRF, however, remain strong evidence of deposition in the meandering zone of these systems and are specifically associated with sediments deposited at the upstream ends of point bars and within chute channel/bar systems.

Climbing ripple facies from the Tonganoxie Sandstone at the Buildex Quarry in eastern Kansas are also interpreted to have been deposited within the fluvial-tidal transitional zone of an estuarine system (Lanier et al., 1993; Archer et al., 1995). Similarities between climbing ripple stratification from the modern and the ancient are compelling. Both are or were silt-dominated systems and should thus be comparable on a hydrodynamic basis. Vertical stratification successions consist of single sedimentation units deposited under conditions of decelerating flow, and which thicken and thin in a manner consistent with tidal influence on sediment deposition. Both show comparable climbing ripple bedform morphologies with strikingly similar internal structuring within these single sedimentation units.

Other features which are comparable between the ancient and modern include additional physical sedimentary structures which clearly indicate periodic/episodic emergence of bed surfaces in association with tidal processes and similar biogenic sedimentary structures (Lanier et al., 1993, Tessier et al., 1995).

Another similarity between the ancient and modern relates to the filling of available accommodation space in the tidal estuary. Deeply rooted horizons and a thin, silty coal are found at the top of the sedimentary succession at the Buildex Quarry (Fig. 3), which indicates that approximately 5 m of available accomodation space had been filled, perhaps within a relatively short period of time (Lanier et al., 1993). Rooted horizons are also associated with modern soils in the Mont-Saint-Michel Estuary, where these tidal sediments have also aggraded within their accomodation space.

Differences between ancient and modern climbing-ripple facies are also noteworthy. As demonstrated by the paleogeographical reconstruction for the Tonganoxie Sandstone at the Buildex Quarry (Lanier et al., 1993), most of the CRF appear to be ebb-dominated. This fact is consistent with modern ebb-dominated CRF, however, insofar as the Buildex climbing ripples also lack mud drapes. Most of the finest-grained sediments must have been transported further down-basin during the ebb phase of tidal deposition. The fluvial dynamics and sediment sources in the ancient estuarine setting may also have been quite different than those that characterize the present-day estuaries of the Mont-Saint-Michel and Salmon River. Whereas the fluvial contribution to both modern macrotidal settings is rather insignificant compared to sediment loads derived from tidal erosion, the drainage basin and fluvial source for suspended sediment may have been quite large (Lanier et al., 1993). However, the well-defined association of these Carboniferous climbing-ripple facies with tidal rhythmites and evidence of frequent and short-term emergence clearly demonstrates that their deposition cannot be related to a seasonal fluvial overbank discharge, but rather to deceleration of ebb currents with very high suspended sediment loads. These processes are similar to those occurring within the meandering zone of the fluvial-tidal transition in the Mont-Saint-Michel Estuary.

CONCLUSIONS

Climbing-ripple bedding characterizes a variety of modern depositional environments in which suspension sedimentation exceeds the rate of deposition from traction transport, but is poorly documented from tidal environments. Studies within the modern macrotidal estuary of Mont-Saint-Michel (France), in comparison with an ancient estuarine analog from the Carboniferous of eastern Kansas, clearly demonstrate that these sedimentary structures and their vertical stratigaphic sequences may be far more common than previously thought. The following general conclusions are relevant to this research:

1. Both the modern and the ancient are silt-dominated and are comparable on a hydrodynamic basis, exhibiting well-developed tidal rhythmites, and a cyclic thickening and thinning of strata, which indicate strong tidal influences on sediment deposition.
2. Climbing-ripple sedimentary structures that have been studied within the modern estuary of Mont-Saint-Michel are restricted to the fluvio-tidal zones of the estuarine transition.
3. In the modern case presented herein, flood-dominated climbing-ripple facies are associated with natural levee systems, which only characterize the inner, straight channels of the transitional zone, whereas ebb-dominated facies are found in the meandering channel portion of this macrotidal system.
4. Ebb-dominated climbing-ripple strata from the ancient show similar vertical structure sequences within single sedimentation units as compared to the modern; they also lack or show very thin clay drapes, which would indicate prolonged slack-water periods for the deposition of fines. This is also a very important feature of the modern facies analog.
5. Both the ancient and the modern show remarkably similar physical and biogenic sedimentary structures on bedding surfaces, which are unambiguous indicators of periodic/episodic emergence of these sediments.

ACKNOWLEDGMENTS

Partial funding for W. P. Lanier was provided by NSF grant EAR-9405123. Funding to support travel and field work in France was provided through a grant from the Research and Creativity Committee at Emporia State University. W. P. Lanier also gratefully acknowledges the enthusiastic participation of numerous undergraduate students at Emporia State University who assisted in the collection of samples and, through their undergraduate research projects, contributed to the overall success of the research program.

REFERENCES

Archer, A. W., Lanier, W. P., and Feldman, H. R., 1994, Stratigraphy and depositional history within incised-paleovalley fills and related facies, Douglas Group (Missourian/Virgilian; Upper Carboniferous) of Kansas, USA, *in* Dalrymple, R. W., Boyd, R., and Zaitlin, B. A., eds., Incised-Valley Systems: Origin and Sedimentary Sequences: Tulsa, Society of Economic Paleontologists and Mineralogists Special Publication 51, p. 175–190.

Bourcart, J., and Charlier, R., 1959, The tangue: A "nonconforming" sediment: Geologic Society of America Bulletin, v. 70, p. 565–568.

Dalrymple, R. W., and Makino, Y., 1989, Description and genesis of tidal bedding in the Cobequid Bay-Salmon River Estuary, Bay of Fundy, Canada, *in* Taira, A., and Masuda, F., eds., Sedimentary Facies in the Active Plate Margin: Tokyo, Terra Scientific Publishing Company, p. 151–177.

Dalrymple, R. W., Makino, Y., and Zaitlin, B. A., 1991, Temporal and spatial patterns of rhythmite deposition on mud flats in the macrotidal Cobequid Bay-Salmon River, Bay of Fundy, Canada, *in* Smith, D. G., Reinson, G. E., Zaitlin, B. A., and Rahmani, R. A., eds., Clastic Tidal Sedimentology: Calgary, Canadian Society of Petroleum Geologists Memoir 16, p. 137–160.

Dalrymple, R. W., Zaitlin, B. A., and Boyd, R., 1992, Estuarine Facies models: Conceptual basis and stratigraphic implications: Journal Sedimentary Petrology, v. 62, p. 1130–1146.

Feldman, H. R., Gibling, M. R., Archer, A. W., Wightman, W. G., and Lanier, W. P., 1995, Stratigraphic architecture of the Tonganoxie paleovalley-fill (Lower Virgilian) in Northeastern Kansas: American Association of Petroleum Geologists Bulletin, v. 79, p. 1019–1043.

Giraud, A., 1996, Le site du Mont-Saint-Michel: étude hydrosédimentaire et dynamique estuarienne du Couesnon. Unpublished M.S. Thesis, University of Caen, France, 50 p.

Harms, J. C., Southard, J. B., and Walker, R. G., 1982, Structures and Sequences in Clastic Rocks: Tulsa, Lecture notes for Society of Economic Paleontologists and Mineralogists Short Course 9, 394 p.

Hunter, R. E., 1977, Terminology of cross-stratified sedimentary layers and climbing ripple structures: Journal of Sedimentary Petrology, v. 47, p. 697–706.

Jopling, A. V., and Walker, R. G., 1968, Morphology and origin of ripple-drift cross-lamination, with examples from the Pleistocene of Massachusetts: Journal of Sedimentary Petrology, v. 38, p. 971–984.

Lanier, W. P., Feldman, H. R., and Archer, A. W., 1993, Tidal sedimentation from a fluvial to estuarine transition, Douglas Group, Misourian-Virgilian, Kansas: Journal of Sedimentary Petrology, v. 63, p. 860–873.

Larsonneur, C., 1989, La baie du Mont-Saint-Michel: Bulletin de l'Institut Géologique du Bassin d'Aquitaine, v. 46, p. 5–74.

LARSONNEUR, C., 1994, The Bay of Mont-Saint-Michel: A sedimentation model in a temperate macrotidal environment: Senckenbergiana maritime, v. 24, p. 3–63.

LEVEY, R. A., 1978, Bedform distribution and internal stratification of coarse-grained point bars, Upper Congaree River, S.C., *in* Miall, A. D., ed., Fluvial Sedimentology: Calgary, Canadian Society of Petroleum Geologists Memoir 5, p. 105–128.

TESSIER, B., 1993, Upper intertidal rhythmites in the Mont-Saint-Michel Bay (NW France): Perspectives for paleoreconstruction: Marine Geology, v. 110, p. 355–367.

TESSIER, B., ARCHER, A. W., LANIER, W. P., AND FELDMAN, H. R., 1995, Comparison of modern analogues (The Bay of Mont-Saint-Michel) with ancient tidal rhythmites (Carboniferous of Kansas and Indiana, U.S.A.), *in* Flemming, B. W., and Bartholomä, A., eds., Tidal Signatures in Modern and Ancient Sediments: Oxford, International Association of Sedimentologists Special Publication 24, p. 259–271.

WUNDERLICH, F., 1969, Studien zur Sedimentbewegung. 1. Transportformen und Schichtbildung im Gebiet der Jade: Senckenbergiana maritime, v. 1, p. 107–146.

YOKOKAWA, M., KISHI, M., MASUDA, F., AND YAMANAKA, M., 1995, Climbing ripples recording the change of tidal current condition in the middle Pleistocene Shimosa Group. Japan, *in* Flemming, B. W., and Bartholomä, A., eds., Tidal Signatures in Modern and Ancient Sediments: Oxford, International Association of Sedimentologists Special Publication 24, p. 301–311.

MODELS FOR THE HOLOCENE VALLEY-FILL HISTORY OF ALBEMARLE SOUND, NORTH CAROLINA, USA

ERIC D. SAGER AND STANLEY R. RIGGS
Department of Geology, East Carolina University, Greenville, North Carolina, USA 27858

ABSTRACT: Previous lithostratigraphic studies of incised, valley-fill systems on the mid-Atlantic coast of the United States indicate a continuous Holocene sea-level rise. Likewise, the lithostratigraphic evaluation of vibracores from Albemarle Sound, North Carolina, can be interpreted as a simple infill history resulting from a continuous Holocene transgression. However, an integrated litho-, high-resolution seismic-, and chronostratigraphic approach suggests that the infill record of Albemarle Sound may be the result of several relative sea-level oscillations during the Holocene.

The Holocene section of Albemarle Sound contains at least three depositional sequences (ASDS-1 to ASDS-3), each consisting of lithologically similar successions of depositional environments. Basal sediments from ASDS-1 were deposited in a restricted-estuarine environment behind a continuous barrier island system between 8.1 ka and 5.9 ka. For some interval between 5.9 ka and 5.5 ka sea level dropped (<6 m), producing an extensive erosional surface throughout the middle- and outer-estuarine zone. Deposition resumed as sea level rose to form ASDS-2. Basal sediments of ASDS-2 were deposited in an open-estuarine system between 5.5 ka and 2.9 ka, suggesting that the previous barrier island complex was extensively breached. ASDS-2 is capped with closed-estuarine sediments that date from 2.9 ka to 1.5 ka, suggesting the reformation of a continuous barrier island complex. Another small drop of sea level occurred for some interval between 1.6 ka and 0.4 ka, which truncated ASDS-1 and ASDS-2. This drop was only between 1–2 m based on the absence of basal open-estuarine sediments in the overlying depositional sequence (ASDS-3), suggesting that the barrier island system was preserved and active during this interval. ASDS-3 consists of closed-estuarine sediments that are being deposited during the ongoing sea-level transgression and behind the modern, well developed barrier island system.

INTRODUCTION

Traditionally, Quaternary formations have been defined utilizing standard litho- and biostratigraphic techniques based on the assumption that deposits are laterally continuous and generally flat lying. The terraced geomorphology of the Atlantic Coastal Plain (Cooke, 1930; Doering, 1960; Winker and Howard, 1977) demonstrated that major sea-level fluctuations have occurred throughout the Quaternary. Development of oxygen isotope records of deep-sea sediments (Shackleton and Opdyke, 1973) suggested that these eustatic sea-level fluctuations reflected episodes of glaciation and deglaciation. Sea-level oscillations caused coastal environments to migrate across the low sloping continental margin and produced a complex succession of lithologically indistinguishable deposits that have been highly dissected and only partially preserved (Belknap and Kraft, 1981; Hine and Snyder, 1985, and Riggs and Belknap, 1988). Therefore, similar lithologies may not define time-stratigraphic units. However, integration of high-resolution seismic stratigraphy and age dating are now aiding in the recognition and interpretation of these complex Quaternary depositional units (Riggs, et al., 1992). A seismic stratigraphic section provides a relative chronostratigraphic framework that can be combined with age dating techniques and other geologic data to help ascertain the detailed geologic history for a given region (Vail and Mitchum, 1977).

Recent studies of Holocene coastal evolution have concentrated on coastal marshes (Fletcher, et al., 1993a, 1993b) and incised valley systems (Dalrymple, et al., 1994). Lithologic studies of coastal marshes reveal a simple and gradual succession of depositional environments resulting from a continuous Holocene sea-level rise (Bloom, 1964; Kraft, 1971; Belknap and Kraft, 1977), but the preservation potential of these deposits is low. However, sediments of incised valley systems have a greater chance of containing a detailed geologic record because they have a higher preservation potential (Belknap and Kraft, 1981; Hine and Snyder, 1985; and Belknap, et al., 1994). Albemarle Sound provides an excellent opportunity to evaluate coastal evolution and sea-level history of this region because it contains a well preserved valley-fill sequence with a detailed geologic database. In addition, the modern depositional systems are analogous to some of the conditions in which the Holocene sediments were deposited. The lithostratigraphic model is reevaluated using a multidisciplinary approach that integrates high-resolution seismic stratigraphy and radiocarbon dating with the lithostratigraphy. This integration has established a detailed infill history of Albemarle Sound that suggests a more complex Holocene sea-level history.

REGIONAL GEOLOGIC SETTING

Albemarle Sound is a large, drowned-river estuarine system occupying the paleo-Roanoke River valley (Fig. 1), which is situated entirely within the Coastal Plain province of northeastern North Carolina, USA. The Albemarle estuarine system occupies a portion of the Albemarle Embayment, a major depositional basin that existed throughout the Plio-Pleistocene (Ward et al., 1991). The Albemarle Embayment occurs on the northern flank of the Carolina Platform High and is bounded to the north by the Norfolk Platform (Ward et al., 1991). Plio-Pleistocene sediments crop out along the perimeter of Albemarle Sound and form the framework for the Holocene infill. The Plio-Pleistocene units were deposited during previous glacial and interglacial periods and contain fluvial, estuarine, barrier-island, and continental shelf sediments (Blackwelder, 1981; Riggs and Belknap, 1988; Ward, et al., 1991) that directly control the morphology of the Albemarle estuarine system (Riggs, et al., 1993; Riggs, 1996).

The Albemarle estuarine system encompasses approximately 2000 km^2 of water and consists of two trunk rivers, two trunk estuaries, and six major tributary streams (Fig. 1). This network forms the largest drainage system in North Carolina (Riggs et al., 1993). The tributary estuaries are the drowned segments of small, black-water streams that drain the swamp forests and poccosins of the sandy Coastal Plain. The riverine zone of the Roanoke and Chowan trunk rivers is bounded by extensive swamp forest floodplains and extends into the Piedmont province of North Carolina and Virginia. The riverine zone passes into a transitional zone characterized by eroding swamp-forest floodplains as the rivers are drowned by the ongoing rise in sea level. The trunk estuaries are the drowned portions of the Roanoke and Chowan Rivers, which increase in width and depth

Tidalites: Processes and Products, SEPM Special Publication No. 61

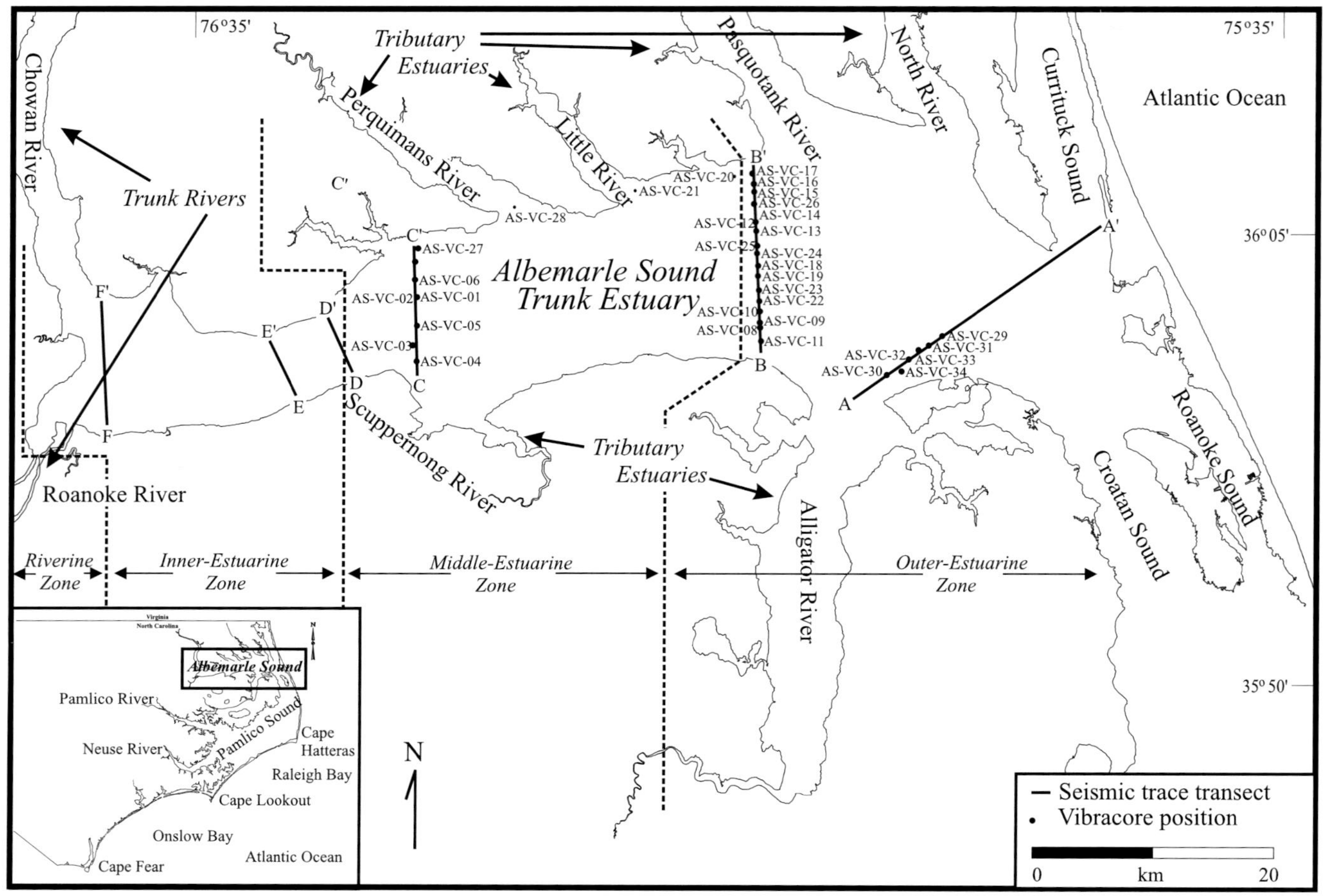

FIG. 1.—Location map of Albemarle Sound, North Carolina, USA, showing the geographic components of the estuarine system and the position of vibracore and seismic sampling transects.

seaward from the transitional zone. The Albemarle trunk estuary is further subdivided into the inner-, middle-, and outer-estuarine zones (Fig. 1). The estuarine zones are characterized by eroding sediment banks resulting from irregular storm tides and high wave energy associated with the increased fetch. The bottom morphology within the estuaries is analogous to a shallow dish. A narrow perimeter platform, composed of Pleistocene sediments with minor amounts of modern surficial sand on top, slopes gently from the shoreline to water depths of 1.5–2 m. At the edge of the perimeter platform, the slope increases abruptly down to a broad, relatively flat, organic-rich mud floor of the central basin that is 4.5–6.5 m below the water surface (Riggs et al., 1993).

METHODS

A single-channel, high-resolution seismic reflection system consisting of an EG&G UNIBOOM and Huntec Sea Otter tow sled belonging to Stephen W. Snyder, North Carolina State University, were utilized to collect seismic data along six cross-sectional transects (Fig. 1). The UNIBOOM system employs a broad band, high frequency (400 Hz to 14 kHz) wave spectrum that allows for enhanced vertical and horizontal resolution in the shallow subsurface, but sacrifices the depth of penetration (Snyder, 1994). Subsurface penetration was less than 0.1 s (85 m), two-way travel time, with the highest resolution obtained in the upper 20 ms (17 m).

Thirty-four vibracores were positioned along three seismic lines within the middle- and outer-estuarine zones (Fig. 1). Each core was described, photographed, and subsampled for lithologic analysis and radiocarbon dating. The major lithologic units infilling Albemarle Sound were identified and characterized according to grain size, sedimentary structures, and paleontological evidence. Accelerator-mass-spectrometer (AMS) and standard radiocarbon analyses were performed by Beta Analytic, Inc., on shell, peat, and organic-rich sediment samples from the major lithofacies. Radiocarbon ages are reported as calibrated calendar years before present (BP, present = 1950 AD) using the Pretoria Calibration Procedure program. Radiocarbon calibration is used to convert BP dates to calendar years. The variations between the two dates are the result of fluctuations in ancient atmospheric ^{14}C concentrations. Calibrated dates are reported in Table 1 with 1 σ probability (age range in parenthesis) and as the intercept of the radiocarbon age with the calibration curve. Marine carbonate samples that have been corrected for $\sigma^{13/12}$C (AMS dates only) have also been corrected for both global and local geographic reservoir effects prior to calibration. Marine carbonates that have not been corrected for $\sigma^{13/12}$C (standard dates) have been adjusted by an assumed value of zero per mil in addition to the reservoir cor-

TABLE 1.—RADIOCARBON AGE DATA FOR ALBEMARLE SOUND, NORTH CAROLINA, USA. SEE FIG. 1 FOR SAMPLE LOCATIONS. SEE TEXT FOR CALIBRATION METHODOLOGY. MBMSL, METERS BELOW MEAN SEA LEVEL; SED, SEDIMENT. ASTERISKS INDICATE ACCELERATOR MASS-SPECTROSCOPY DATES

Sample Number	Elevation (mbmsl)	Location (Lat/Long)	Lithofacies	Material Dated	Conventional C-14 Age BP	Calibrated C-14 Age BP (1 σ)
AS-VC-01	7.15–7.25	36.02.260/76.22.508	LF-6	Organic Sed	3220 +/− 70	3440 (3365–3480)
AS-VC-01	7.45–7.55	36.02.260/76.22.508	LF-3	Organic Sed	5150 +/− 70	5915 (5770–5945)
AS-VC-01	14.20–14.30	36.02.260/76.22.508	LF-3	Organic Sed	5990 +/− 90	6845 (6735–6905)
AS-VC-03	7.92–7.97	36.00.203/76.23.043	Pre-Holo	Peat	21360 +/− 140	N/A
AS-VC-05	6.60–6.70	36.00.953/76.22.799	LF-6	Organic Sed	3210 +/− 70	3400 (3360–3475)
AS-VC-05	6.80–6.90	36.00.953/76.22.799	LF-3	Organic Sed	6130 +/− 100	7000 (6875–7170)
AS-VC-08	9.48–9.53	36.01.099/76.04.508	LF-2	Organic Sed	23,800 +/− 1000	N/A
AS-VC-10	5.65–5.75	36.01.417/76.04.504	LF-6	Organic Sed	2090 +/− 80	2025 (1945–2140)
AS-VC-12	7.63–7.73	36.04.497/76.04.457	LF-3	Organic Sed	4160 +/− 80	4740 (4540–4835)
AS-VC-12	8.92–8.97	36.04.497/76.04.457	LF-5	Organic Sed	5320 +/− 160	6150 (5920–6290)
AS-VC-12	11.90–12.00	36.04.497/76.04.457	LF-5	Organic Sed	4600 +/− 80	5305 (5080–5445)
AS-VC-12	13.63–13.73	36.04.497/76.04.457	LF-3	Organic Sed	5550 +/− 90	6315 (6280–6420)
AS-VC-12	14.55–14.60	36.04.497/76.04.457	LF-3	*Crassostrea*	5730 +/− 80	6570 (6460–6665)
AS-VC-13	6.33–6.43	36.04.043/76.04.505	LF-6	Organic Sed	1630 +/− 70	1525 (1415–1570)
AS-VC-13	7.63–7.73	36.04.043/76.04.505	LF-3	Organic Sed	4750 +/− 80	5525 (5325–5590)
AS-VC-16	5.44–5.54	36.06.607/76.04.493	LF-6	Organic Sed	5530 +/− 110	6305 (6210–6420)
AS-VC-16	7.35–7.45	36.06.607/76.04.493	LF-6	Organic Sed	6470 +/− 260	7370 (7160–7545)
AS-VC-16*	4.02	36.06.607/76.04.493	LF-1B	Wood	6150 +/− 60	6890 (6855–6990)
AS-VC-16	9.27–9.37	36.06.607/76.04.493	Pre-Holo	Organic Sed	>37,480	N/A
AS-VC-18	5.98–6.08	36.02.809/76.04.466	LF-6	Organic Sed	2810 +/− 60	2880 (2840–2965)
AS-VC-18	0.37	36.02.809/76.04.466	LF-5/LF-6	*Crassostrea*	2470 +/− 60	2690 (2570–2720)
AS-VC-18	8.55–8.65	36.02.809/76.04.466	LF-4	*Tagelus*	3790 +/− 90	4270 (4125–4405)
AS-VC-18	14.58–14.68	36.02.809/76.04.466	LF-3	Organic Sed	6580 +/− 70	7415 (7385–7515)
AS-VC-23*	8.88–8.98	36.02.030/76.04.560	LF4	*Cyrtopleura*	4450 +/− 40	5585 (5485–5600)
AS-VC-23	9.76–9.86	36.02.030/76.04.560	Pre-Holo	Organic Sed	>25,950	N/A
AS-VC-25	7.40–7.50	36.03.394/76.04.542	LF-6	Organic Sed	4720 +/− 80	5460 (5315–5580)
AS-VC-25	11.30–11.55	36.03.394/76.04.542	LF-4	*Mercenaria*	3200 +/− 50	4040 (3955–4040)
AS-VC-25	12.95–13.05	36.03.394/76.04.542	LF-4	*Cyrtropleura*	4450 +/− 100	5190 (4975–5290)
AS-VC-25	13.05–13.15	36.03.394/76.04.542	LF-3	Organic Sed	7310 +/− 80	8095 (7970–8140)
AS-VC-25	14.35–14.45	36.03.394/76.04.542	LF-3	Organic Sed	6760 +/− 70	7555 (7530–7630)
AS-VC-27	4.30–4.35	36.04.260/76.22.496	LF-2	Peat	6000 +/− 70	6855 (6755–6900)

rections. Dates are also reported in Table 1 as Conventional ^{14}C Ages, which result after applying C13/C12 correction to the measured radiocarbon age.

RESULTS AND DISCUSSION

Lithostratigraphic Model (LM)

Geologic cross-sections (Fig. 2) were constructed from the vibracores along B-B′ and C-C′ to determine the lithostratigraphy of Albemarle Sound. These sections revealed that the Holocene sediments of Albemarle Sound are comprised of six lithofacies (LF) summarized in Table 2: Quartz-rich sand (LF-1); Peat (LF-2); Laminated mud (LF-3); Fossiliferous, muddy sand (LF-4); Interlaminated mud and sand (LF-5); and Organic-rich mud (LF-6).

A fine- to medium-grained sand lithofacies (LF-1) infills channels along B-B′ (Fig. 2A) and C-C′ (Fig. 2B) at various depths beneath the surface. Variations in grain size, sedimentary structures, and paleontological evidence have been utilized to divide LF-1 into two subfacies: LF-1A and LF-1B. LF-1A is characterized by a slightly muddy, very fine- to fine-grained, quartz sand with mud laminae ranging in thickness from 1–3 mm. The bedding in LF-1A is similar to flaser bedding and is indicative of deposition in response to tidal processes (Reineck and Wunderlich, 1968). Channels backfilled with LF-1A contain high-brackish water microfossils (*Elphidium* and *Campylodiscus*). This supports the interpretation that LF-1A was deposited in tidal channels with salinity conditions ranging from meso- to polyhaline.

LF-1B is characterized by clean, fine- to medium-grained, quartz sand that contains no microfossils. LF-1B contains uniform, planar-type cross-bedding with occasional mud drapes that are indicative of point bar deposits (Reineck and Singh, 1980). LF-1B also contains convoluted laminae that suggest high flow regimes and soft sediment deformation common in fluvial systems. A muddy, dark brown to black peat (LF-2), ranging in thickness from 0.2–0.8 m, occurs beneath channel complexes in two cores taken along profile C-C′ (Fig. 2B). LF-2 is associated with fluvial sediments (LF-1B) and was deposited in swamp-forest floodplains that fringed these fluvial environments.

A laminated mud (LF-3) is the stratigraphically lowermost lithofacies (Figs. 2A and B). LF-3 is characterized by laminae of muddy, very fine-grained sand embedded within an organic-rich mud and is comparable to lenticular bedding as described by Reineck and Wunderlich (1968). X-radiographs reveal that contacts between the sand and mud layers are sharp and individual layers are uniform. The mean thickness of sand and mud layers of LF-3 is 3.0 and 18.3 mm, respectively, and the sand to mud ratio is 0.17 (Table 3). The paleofauna within LF-3 are dominated by *Macoma* and *Mulinia*. Today, these organisms occur extensively throughout North Carolina in environments dominated by fine-grained sands and muds with mesohaline salinity conditions. This suggests that LF-3 was deposited during the initial marine flooding of the innermost Roanoke Valley with mesohaline conditions.

LF-3 is overlain by a fossiliferous, muddy sand lithofacies (LF-4) that is between 0.25–2.0 m thick along profile B-B′ (Fig. 2A). Articulated, disarticulated, and fragmented *Mulinia* are commonly disseminated in a 0.5-m zone along the base of the facies. LF-4 is characterized by burrowing invertebrates *Tagelus* and *Cyrtopleura* that occur in life position. These organ-

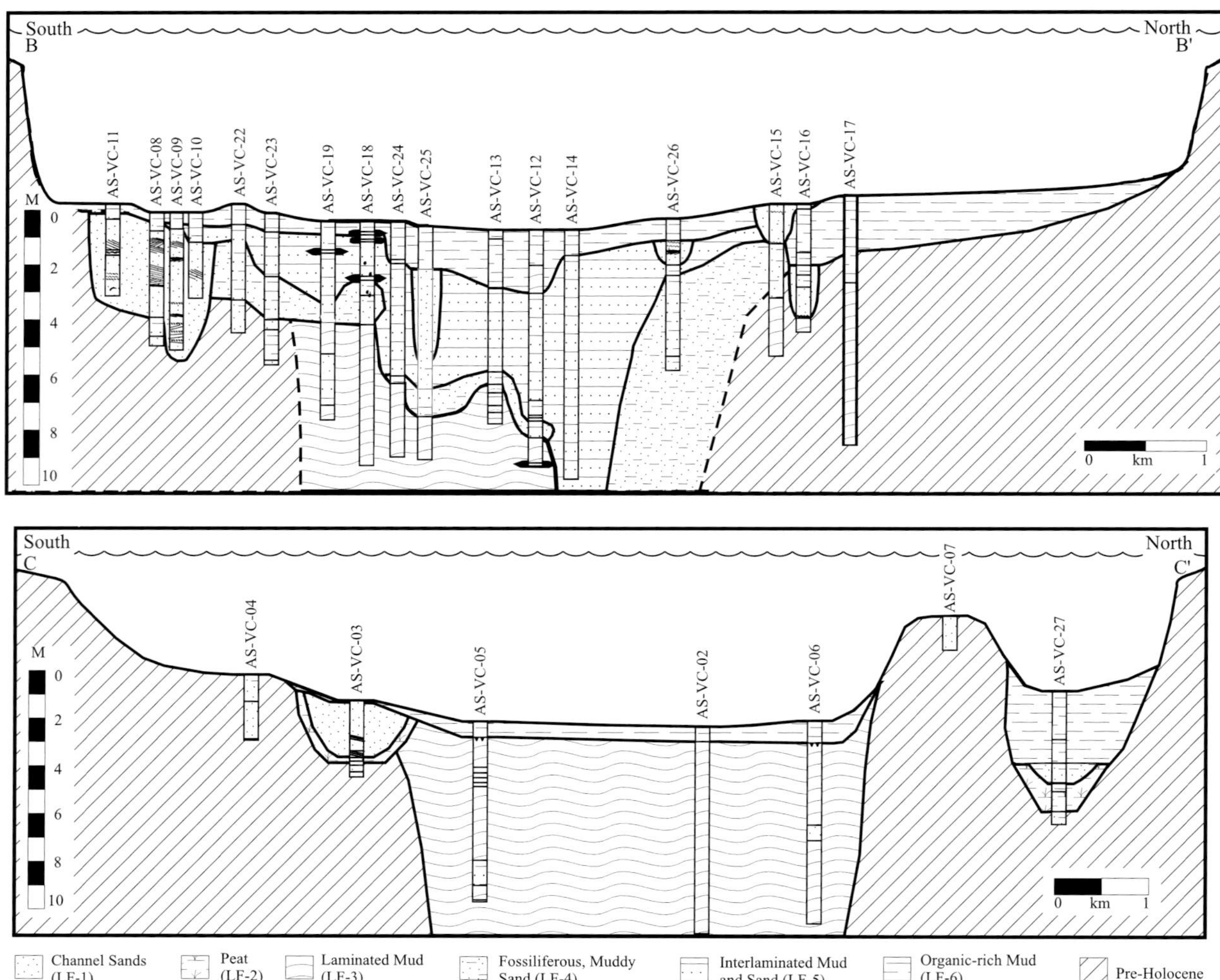

FIG. 2.—Lithologic cross-sections along profiles B-B′ and C-C′. See Fig. 1 for locations. Notice the different scales on each panel.

isms are common in open-estuarine environments with polyhaline conditions. In North Carolina, these organisms occur along the edges of marshes and on muddy sand shoals adjacent to inlets through the barrier island complex. The abundance of *Tagelus* and *Cyrtopleura* within the fossiliferous, muddy sand lithofacies suggests a once extensive shallow-water environment with polyhaline conditions and intertidal habitats along the perimeter of Albemarle Sound.

The fossiliferous, muddy sands of LF-4 grade vertically and laterally into the interlaminated sand and mud lithofacies (LF-5) (Fig. 2A). LF-5 ranges from 0.75 to >8 m in thickness and is confined to the central portion of Albemarle Sound along B-B′. LF-5 is distinguished by alternating layers of very fine- to fine-grained sand and mud that are similar to flaser bedding (Reineck and Wunderlich, 1968). X-radiographs reveal that the contacts between the mud and sand layers are sharp, with sand layers displaying forset laminae. The sand:mud ratio of LF-5 is 0.82, whereas the mean thickness of sand and mud layers is 6.5 mm and 7.9 mm, respectively (Table 3). The fauna within LF-5 are dominated by *Cyrtopleura, Mulinia,* and *Tagelus.* The interlaminated mud and sand of LF-5 occurs adjacent to tidal channel facies (LF-1A) and may represent shoal complexes associated with tidal channels in the central basin. Demarest and Kraft (1987) have shown that tidal shoal and channel lithosomes will occur in a vertical transgressive sequence produced in an open-estuarine environment. Weil (1976) described similar sediment types in shoals adjacent to flood-tidal channels in Delaware Bay, USA. These shoals are 1.5–6.0 m thick, asymmetrical in cross-section, and exhibit well-developed internal cross-beds produced by lateral migration of tidal channels and shoal complexes. Weil (1976) refers to the shoals as tidal current ridges because they are shaped by, and oriented parallel to, tidal currents with flaser bedding resulting from alternating periods of current activity and quiescence. Formation of these shoals may be analogous to the processes responsible for deposition of LF-5 in an open-estuarine environment within the central basin.

Organic-rich mud (LF-6) is the dominant surficial sediment type within the central basin of the Albemarle estuarine system, which constitutes approximately 70% of this system (Riggs et al., 1993; Riggs, 1996). LF-6 is comprised of inorganic sediments delivered to the estuary as suspended load by the trunk rivers during high discharge periods and organic matter derived from erosion of the extensive swamp forests and marshes that fringe the Albemarle estuarine system (Riggs et al., 1993; Riggs, 1996). LF-6 is currently being deposited in a closed

TABLE 2.—SUMMARY OF THE LITHOFACIES THAT COMPRISE THE HOLOCENE SEDIMENTS OF THE MIDDLE- AND OUTER-ALBEMARLE SOUND, NORTH CAROLINA, USA

Lithofacies	Description	Sedimentary Structures	Paleofauna	Environment (Salinity)
Quartz Sand LF-1A	Very fine- to fine-grained, quartz and	Mud laminae	*Campylodiscus Elphidium*	Tidal Channel (Meso- to Polyhaline)
Quartz Sand LF-1B	Fine- to medium-grained quartz sand	Heavy mineral cross laminae	Absent	Fluvial Channel (Infrahaline)
Peat LF-2	Dark brown to black mud with organic detritus composed of plant fragments and wood	None	Absent	Swamp Forest (Infrahaline)
Laminated Mud LF-3	Laminae of very fine- to fine-grained sand within an organic-rich mud	Lenticular bedding	*Crassostrea* *Macoma* *Mulinia*	Restricted/Inner Estuary (Mesohaline)
Fossiliferous, Muddy Sand LF-4	Fosiliferous, muddy, fine- to medium-grained sand	Bioturbated	*Crassostrea* *Cyrtopleura* *Ensis* *Mercenaria* *Mulinia* *Tegalus* *Urosalpinx*	Tidal Flats in an Open Estuary (Polyhaline)
Interlaminated Mud and Sand LF-5	Laminae of mud and very fine- to fine-grained sand	Flaser bedding	*Crassostrea* *Cyrtropleura* *Macoma* *Mulinia* *Tegalus*	Shoal Complex in an Open Estuary (Polyhaline)
Organic-Rich Mud LF-6B	Loose sediment with elevated heavy metal concentrations	Homogenous	*Macoma* *Rangia*	Closed Estuary (Infra- to Oligohaline)
Organic-Rich Mud LF-6A	Dense sediment with low heavy metal concentrations	Homogenous	*Macoma*	Closed Estuary (Infra- to Oligohaline)

estuary under infra- to oligohaline salinity conditions in response to the establishment of a continuous barrier island complex. The organic-rich mud lithofacies (LF-6) caps the entire sequence and rests upon LF-5 along B-B′ (Fig. 2A) and LF-3 along C-C′ (Fig. 2B). Sediment texture, heavy metal concentrations, and paleontological evidence were utilized to divide LF-6 into two subfacies: LF-6A and LF-6B. LF-6B comprises the surficial sediments and rests upon a denser organic-rich mud substrate (LF-6A). LF-6B consists of loose sediment, ranging from 0 m to 2 m in thickness with a mean thickness of <0.5 m (Riggs, 1996), that scuba divers can swim into. Riggs et al. (1993) analyzed 156 short cores for heavy metal contaminants and discovered elevated concentrations within LF-6B, which they attributed to urbanization and industrialization during post-colonial times (<400 BP). The decreased concentration in heavy metals in the underlying LF-6A, in conjunction with the increased degree of compaction, is attributed to a previous depositional episode during pre-colonial times. The environments and processes of the older depositional system (LF-6A) are probably analogous to the modern environment of LF-6B.

Utilizing only the lithostratigraphy, vibracores reveal that the lithologic framework of Albemarle Sound consists of six major lithofacies that appear to display a simple infill history (Fig. 2A). Fletcher et al. (1990, 1992) described a similar sediment record for Delaware Bay and attributed the infill sequence to changing tide versus wave-related processes during a continuous sea-level rise. Initial sedimentation and the onset of the constructive deltaic phase (CDP) (Weil, 1976; Fletcher et al., 1990, 1992) began during the earliest inundation of Delaware Bay. The CDP is characterized by the deposition of fine-grained sediments in an estuarine deltaic sequence at the turbidity maximum zone. As inundation continued, Delaware Bay entered the destructive deltaic phase (DDP) of Weil (1976) and Fletcher et al. (1990, 1992), which was prompted by an increased fetch and higher wind-wave energy. The DDP is characterized by the erosion of the previously deposited sediments by wind-wave activity associated with the increased fetch. The latter stages of development were characterized by open-estuarine sedimentation. Therefore, the DDP is represented as an erosional hiatus

TABLE 3.—MEAN THICKNESS OF SAND AND MUD LAYERS AND SAND-MUD RATIOS FROM THE LAMINATED MUD LITHOFACIES (LF-3) AND THE INTERLAMINATED MUD AND SAND LITHOFACIES (LF-5)

				SAND		MUD	
Lithofacies	Core Number	Thickness (m)	Sand-Mud Ratio	Number of Layers	Average Thickness (mm)	Number of Layers	Average Thickness (mm)
LF-3	AS-VC-01	2.42	0.13	85	3.3	85	25.1
LF-3	AS-VC-06	2.98	0.21	167	3.8	168	18.6
LF-3	AS-VC-24	2.54	0.13	125	2.4	126	17.8
LF-3	AS-VC-25	1.47	0.21	102	2.5	103	11.7
LF-3	MEAN		0.17		3.0		18.3
LF-5	AS-VC-24	2.36	0.99	167	7.0	168	7.1
LF-5	AS-VC-25	1.04	0.65	70	5.9	71	8.7
LF-5	MEAN		0.82		6.5		7.9

that separates sediment of the CDP from the overlying open-estuarine sediments.

Following this model, the Albemarle infill sequence appears to show a similar infill history. The lowermost lithofacies is a laminated mud (LF-3) that could have been deposited during the initial drowning of Albemarle Sound and represents the CDP. Fine-grained sediments were deposited by tidal circulation at the turbidity maximum zone, and as sea level continued to rise, the loci of deposition migrated westward up the valley axis. As the inundation of Albemarle Sound continued, the shoreline would have retreated, and the increased fetch resulted in the initiation of the DDP. This phase was probably characterized by the erosion of the laminated mud (LF-3), deposited during the CDP, as the ever-widening shoreline produced higher wind-wave energy. LF-3 is overlain by a fossiliferous, muddy sand (LF-4) characterized by burrowing invertebrates, which grades into the interlaminated mud and sand lithofacies (LF-5). These sediments would have been deposited by tidal current activity in an open-estuarine environment as Holocene aggradation resumed following the DDP (Weil, 1976; Fletcher et al., 1990, 1992). The entire sequence is capped by an organic-rich mud (LF-6) deposited in a closed-estuarine environment following the formation of an almost continuous barrier island complex, as it occurs today.

Integrated Litho-, Seismic-, and Chronostratigraphic Model (ILSCM)

The LM is based upon the principle of superposition and assumes that deposits are laterally continuous and relatively flat lying. However, Quaternary sea-level fluctuations produced a complex succession of lithologically similar deposits that were dissected and modified during low stands of sea level (Belknap and Kraft, 1981; Hine and Snyder, 1985; Riggs and Belknap, 1988). If similar sea-level oscillations occurred within the Holocene, the lithostratigraphic approach alone would not be sufficient to recognize these events because lithologically similar units may not be geometrically or time correlative.

DePratter and Howard (1981), Colquhoun and Brooks (1986), and Gayes et al. (1992) present data that suggest fluctuations in sea level along the southeast U.S. coast have occurred in the Holocene. Also, oxygen isotope records from ice cores (Dansgaard et al., 1969; Dansgaard et al., 1984) and marine sediments (Fairbanks, 1989) show that climatic conditions have oscillated throughout the Holocene. Because climate influences sea level, it seems reasonable that sea level may have fluctuated during the Holocene. Utilizing a lithostratigraphic approach, these small sea-level events have not been widely recognized in marsh deposits because they are susceptible to compaction, variations in sediment supply, and freshwater discharge that mask regional sea-level trends (Fletcher et al., 1993b). Furthermore, the preservation potential of these depositional sequences is low due to subaerial oxidation and weathering during low stands, erosion from migrating tidal channels and inlets, and shore-face retreat (Belknap and Kraft, 1981; Gayes, et al., 1992).

Assuming that sea-level oscillations have occurred during the Holocene, a reevaluation of the lithostratigraphic model (LM) utilizing a multidisciplinary approach that integrates high-resolution seismic stratigraphy and radiocarbon dating with the lithostratigraphy can result in a very different interpretation than the LM. The lithostratigraphic framework is utilized to interpret the processes and environments of deposition. Seismic stratigraphy provides a continuous record that can trace seismic sequences laterally through a basin permitting correlation of isochronous strata between sample localities, regardless of lithology. Seismic analysis is utilized to define the seismic sequence, a conformable succession of genetically related strata deposited during a given interval of time and bounded by unconformities and their correlative conformities (Vail and Mitchum, 1977). Such a seismic sequence defines a depositional sequence. The concept of sequence analysis and recognition of depositional sequences is important because the unconformities are chronostratigraphic markers that separate older strata from younger strata. Utilizing this concept, a relative chronostratigraphic framework can be established utilizing the basic principles of superposition and cross-cutting relationships.

Radiocarbon dating permits an absolute chronology to be superimposed upon the seismic data and allows each depositional sequence to be temporally positioned. Also, radiocarbon dates provide a critical method for distinguishing between depositional sequences because each sequence consists of similar lithofacies. Integration of these data sets suggests a more complex Holocene sea-level history for Albemarle Sound than indicated by the LM.

Lithologic and seismic data collected along profile B-B′ reveal that the Holocene section of Albemarle Sound consists of three distinct depositional sequences (Fig. 3) that are characterized by numerous channels. Fig. 4 provides a chronostratigraphic summary of the Albemarle Sound valley-fill sequence. The oldest Albemarle Sound Depositional Sequence (ASDS-1) identified in this study has an apparent southward dip, and is incised into Pleistocene sediments along Sequence Boundary-1 (SB-1). ASDS-1 has a maximum thickness of > 8 m along the northern segment of profile B-B′, where it is truncated by both SB-2 and SB-3 and thins to the south, where it is truncated by SB-2. Radiocarbon dates indicate that ASDS-1 was deposited between 8.1 ka and 5.9 ka. Litho- and seismic stratigraphy reveal that Albemarle Sound was a restricted-estuarine environment during this period. This occurred in response to the development of a barrier island with few inlets that produced mesohaline conditions in the outer estuary.

ASDS-1 is truncated by ASDS-2, which ranges in thickness from 5.5 to 6.0 m in the central portion of profile B-B′, from 1.0 to 2.0 m in the north-central section, and is <0.5 m thick along the southern margin. ASDS-2 was deposited between 5.5 ka and 1.5 ka. The sequence boundary (SB-2) separating ASDS-1 and ASDS-2 represents a 400-year hiatus and occurs within 1.0 m of the sediment-water interface (6.75 m below present sea level [MBSL]) within the middle Albemarle Sound and slopes eastward to 12.5 MBSL along profile B-B′. The seismic- and lithostratigraphic data do not reveal fluvial incision, oxidation, or pedogenesis along SB-2. Along C-C′, SB-2 is recognized by a burrowed surface separating LF-6 from LF-3, while along profile B-B′, SB-2 is characterized by an abrupt change in lithology in which sediments of LF-3 are overlain by LF-4 and by a concentration of *Mulinia* disseminated within a 0.5-m zone at the base of LF-4. This erosional surface may represent a drop of sea level between 5.9 ka and 5.5 ka. Sea level could not have fallen more than 6 m during this period because subaerial conditions conducive to fluvial incision and

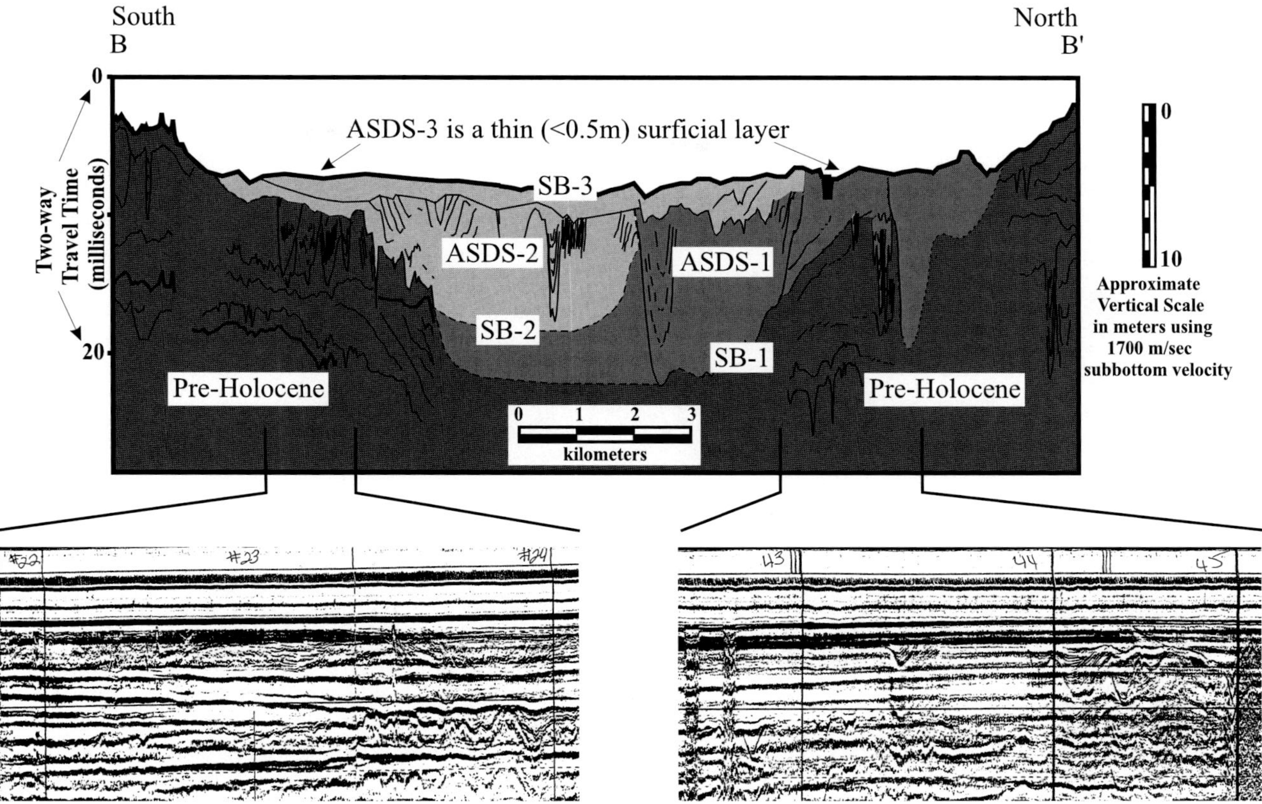

FIG. 3.—Raw and interpreted seismic reflection data collected in Albemarle Sound. The integrated seismic-, litho-, and chronostratigraphy reveal that the infill sequence of Albemarle Sound is characterized by three depositional sequences (ASDS-1, ASDS-2, and ASDS-3). ASDS-3 is not seismically resolvable; but, it is depicted as a heavy line.

soil formation would have been established in the middle-estuarine zone. During the low stand, colder climates probably increased the storminess in the mid-latitudes (Dansgaard et al., 1984), which generated storm surges and waves capable of eroding large portions (up to 6.5 m) of ASDS-1 and creating SB-2.

Deposition resumed in outer Albemarle Sound by 5.5 ka with the formation of ASDS-2 under open-estuarine conditions. This suggests that the barrier island previously developed by 8.1 ka had been either extensively breached or totally eliminated. The shallow portions of the outer-estuarine zone were dominated by intertidal flats while an extensive network of migrating tidal channels and shoal complexes characterized the central basin. ASDS-2 is capped by organic-rich muds of LF-6A that were deposited between 2.9 ka and 1.5 ka in the outer-estuarine zone. LF-6A was deposited by 2.9 ka in response to the formation of a continuous barrier island complex that produced closed-estuarine conditions similar to the modern.

ASDS-3 consists of a thin veneer (average thickness = 0.5 m) of organic-rich mud sediment (LF-6B) that occurs everywhere in the central basin and truncates ASDS-1 and ASDS-2. ASDS-3 cannot be resolved seismically, but the sequence boundary (SB-3 at 6.0 MBSL) separating ASDS-3 from ASDS-1 and ASDS-2 is recognized by a dramatic change in the (1) sediment density and (2) heavy-metal concentration. ASDS-3 is comprised of sediments of LF-6B and represents <400 years of deposition. Radiocarbon dates below SB-3 range from 1525 years BP (ASDS-2) to 6305 years BP (ASDS-1) and indicate that erosion has removed at least 1125 years of the estuarine record. This erosional surface is interpreted to be another slight drop of Holocene sea level for some portion of the interval between 1.5 ka and 0.4 ka. After 0.4 ka, deposition resumed in the outer Albemarle Sound in a closed-estuarine system. This suggests that the barrier island developed by 2.9 ka had not been abandoned by a major drop in sea level and destroyed by erosion. Rather, the drop in sea level was no more than 1–2 m, allowing for a slight recession of the shoreline. Sediments of ASDS-3 are currently being deposited by the ongoing Holocene transgression behind a continuous barrier island system.

Mechanisms for Sea-level Movement and Climatic Evidence

Fletcher et al. (1993a) summarized several possible mechanisms for sea-level movements during the Holocene, which include, (1) thermal expansion of the oceanic mixed layer, (2) decoupling of the Antarctic ice sheet, and (3) changes in the surface circulation in the North Atlantic. However, without good control on the magnitude of the sea-level events in Albemarle Sound, it is difficult to determine which mechanism, if any, may have produced the observed record. The thermal expansion of the oceanic mixed layer may result in a sea-level rise of a few centimeters for a 1° C increase in atmospheric temperature (Wigley and Raper, 1987), which does not seem adequate to produce the record in Albemarle Sound. Decou-

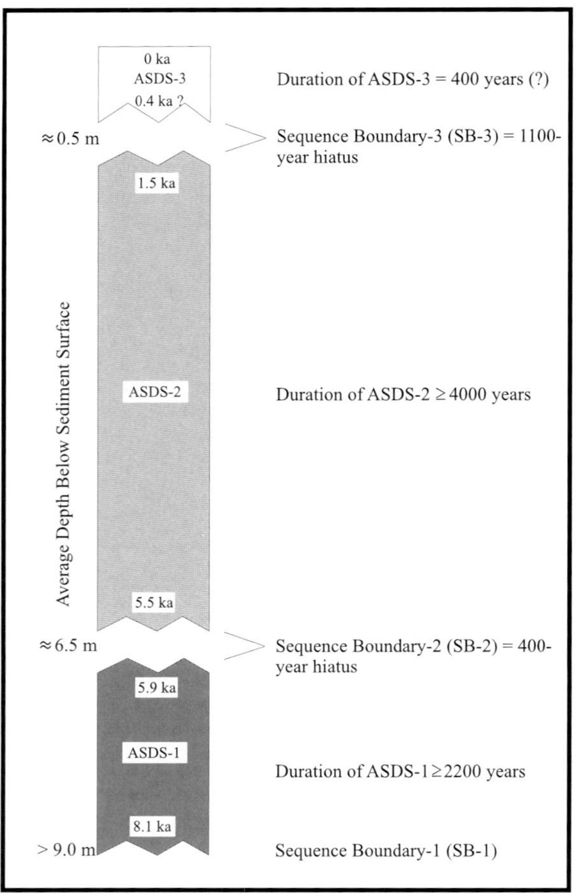

FIG. 4.—Chronostratigraphic summary of the Albemarle Sound Holocene valley-fill sequence. ASDS, Albemarle Sound Depositional Sequence.

pling of the Antarctic ice sheet (Anderson and Thomas, 1991) may produce episodic sea-level movements. However, this mechanism accounts for rapid rises of sea level followed by stillstands, but does not explain a decline in sea level. Fletcher et al. (1993a) attributed changes in the surface circulation of the North Atlantic to sea-level movements observed in marsh sediments in Wolfe Glade, Delaware, USA. Periods of warming and cooling cause changes in the thermal gradient between high and low latitudes. This can result in surges in the Gulf Stream that may regionally alter sea level as much as 1 m. None of these mechanism alone appear to account for the supposed magnitude and duration of sea-level movement needed to create the record in Albemarle Sound. However, because they are all related, they are interdependent and each process probably contributes to the overall sea-level response recorded in the Albemarle Sound sediment record.

Because palynological data do not exist for the Albemarle Sound region, the climatic conditions for this region cannot be directly related with the proposed history for this region. However, the proposed sea-level record for Albemarle Sound can be compared with (1) neoglaciations in the Scandinavian and European Alps (Fletcher et al., 1993a), and (2) estimated degree of glaciation from the Camp Century, Greenland, δ O^{18} record (Dansgaard et al., 1984). The interpreted decline of sea level between 5.9 ka and 5.5 ka in Albemarle Sound correlates with a broad neoglacial maximum observed in the European Alps and a high degree of glaciation from the Camp Century δ O^{18} record. The proposed lowstand between 1.5 ka and 0.4 ka is a small event that is constrained in time. Several small neoglacial maxima occurred in the Scandinavian and European Alps, including the Little Ice Age, during this period that could explain this event. Erosion during such a minor event could have eroded a significant portion of the previously deposited record. Additional age dating may help clarify timing and aid in calibration with the climatic data. This fall of sea level for some portion of the period between 1.5 ka and 0.4 ka could represent the response of sea level to the Little Ice Age.

CONCLUSIONS

Two models have been developed for the Holocene history of Albemarle Sound. The first model was constructed utilizing a standard lithostratigraphic (LM) approach while the second model was based on a multidisciplinary integration of litho-, seismic, and chronostratigraphy (ILSCM). Comparison of these models yields very different interpretations. The lithostratigraphic model reveals that Albemarle Sound consists of six lithofacies that show a gradual succession of depositional environments resulting from a continuous sea-level rise. The basal facies represents the CDP, which included estuarine deltaic sedimentation at the turbidity maximum during the initial marine flooding of the paleo-Roanoke River. As the inundation of Albemarle Sound continued, the fetch increased and initiated the DDP. During the DDP the basal sediments were eroded as wind-wave energy was enhanced. Deposition resumed in an open estuary characterized by tidal channels and shoals in the central basin and tidal flats along the perimeter of Albemarle Sound. Continued sea-level rise lead to the emplacement of the present barrier island system and deposition of the subsequent organic-rich muds in a closed-estuarine system behind the modern barrier islands.

However, the ILSCM reveals that Albemarle Sound may contain a more complex Holocene infill history. Basal sediments of ASDS-1 were deposited in an restricted-estuarine environment behind a continuous barrier island system that existed from at least 8.1 ka until 5.9 ka. Between 5.9 ka and 5.5 ka, a drop of sea level (<6 m) produced an extensive erosional surface that separates ASDS-1 and the overlying ASDS-2. This event correlates with climatic data from the European Alps and Camp Century, Greenland, which suggests a possible climatic control. The breached barrier island produced open-estuarine conditions as sea level began to rise within outer Albemarle Sound. The outer-estuarine zone was characterized by intertidal flats along the perimeter and a network of tidal channels and shoal complexes within the central basin. A closed-estuarine system was established again within outer Albemarle Sound by 2.9 ka with the formation of another continuous barrier island complex. The closed-estuarine environment persisted from 2.9 ka to 1.5 ka with the deposition of organic-rich mud (LF-6A). For some interval between 1.5 ka and 0.4 ka another drop in Holocene sea level (1–2 m) occurred, which dropped wave base and truncated portions of both ASDS-1 and ASDS-2. Organic-rich muds (LF-6B) of ASDS-3 are currently being deposited in a closed-estuarine system by the present Holocene marine transgression that has been in progress for the past several hundred years.

ACKNOWLEDGMENTS

Funding for this study was provided in part by East Carolina University, Sigma Xi, Southeastern Section of the Geological Society of America, and the North Carolina Geological Survey. The authors would like to acknowledge the contribution of Stephen W. Snyder, North Carolina State University, in acquiring the seismic data. Numerous students from East Carolina University aided in the collection of field data. Chip Fletcher provided important editorial comments.

REFERENCES

ANDERSON, J. B., AND THOMAS, M. A., 1991, Marine ice-sheet decoupling as a mechanism for rapid, episodic, sea-level change: The record of such events and their influence on sedimentation, *in* Biddle, K. T., and Schlager, W., eds., The record of sea-level fluctuations: Sedimentary Geology, v. 70, p. 87–104.

BELKNAP, D. F., AND KRAFT, J. C., 1977, Holocene relative sea-level changes and coastal stratigraphic units on the northwest flank of the Baltimore Canyon Trough geosyncline: Journal of Sedimentary Petrology, v. 47, p.610–629.

BELKNAP, D. F., AND KRAFT, J. C., 1981, Preservation potential of transgressive coastal lithosomes on the U. S. Atlantic shelf: Marine Geology, v. 42, p. 429–442.

BELKNAP, D. F., KRAFT, J. C., AND DUNN, R. K., 1994, Transgressive valley-fill lithosomes: Delaware and Maine, *in* Dalrymple, R. W., Boyd, R., and Zaitlin, B. A., eds., Incised-valley systems: Origins and Sedimentary Sequences: Society of Economic Paleontologists and Mineralogists Special Publication 51, p. 303–320.

BLACKWELDER, B. W., 1981, Stratigraphy of Upper Pliocene and Lower Pleistocene marine and estuarine deposits in northeastern North Carolina and southeastern Virginia: United States Geological Survey Bulletin 1502-B, 19 p.

BLOOM, A. L., 1964, Peat accumulation and compaction in a Connecticut coastal marsh: Journal of Sedimentary Petrology, v. 34, p. 599–603.

COOKE, C. W., 1930, Correlation of coastal terraces: Journal of Geology, v. 38, p. 577–589.

COLQUHOUN, D. J., AND BROOKS, M. J., 1986, New evidence from the southeastern United States for eustatic components in the late Holocene sea levels: Geoarcheology, v. 1, p. 275–291.

DALRYMPLE, R. W., BOYD, R., AND ZAITLIN, B. A., 1994, Incised-valley systems: Origins and sedimentary sequences: Society of Economic Paleontologists and Mineralogists Special Publication 51, 380 p.

DANSGAARD, W., JOHNSEN, S. J., CLAUSEN, H. B., DAHL-JENSEN, D., GUNDESTRUP, N., HAMMER, C. U., AND OESCHGER, H., 1984, North Atlantic climatic oscillations revealed by deep Greenland ice cores, *in* Hansen, J. E., and Takahashi, T., eds., Climate Processes and Climate Sensitivity: American Geophysical Union Monograph Series 29, p. 288–298.

DANSGAARD, W., JOHNSEN, S. J., MOLLER, J., OERSTED, H. C., AND LANGWAY, C. C., JR., 1969, One thousand centuries of climatic record from Camp Century on the Greenland ice sheet: Science, v. 166, p. 377–380.

DEMAREST, J. M., AND KRAFT, J. C., 1987, Stratigraphic record of Quaternary sea levels: Implications for more ancient strata, *in* Nummedal, D., Pilkey, O. H., and Howard, J. D., eds., Sea-level Fluctuations and Coastal Evolution: Society of Economic Paleontologists and Mineralogists Special Publication 41, p. 223–229.

DEPRATTER, C. B., AND HOWARD, J. D., 1981, Evidence for a sea-level low-stand between 4500 and 2400 years BP on the southeast coast of the United States: Journal of Sedimentary Petrology, v. 51, p. 1287–1296.

DOERING, J. A., 1960, Quaternary surface formations of southern part of Atlantic Coastal Plain: Journal of Geology, v. 68, p. 182–202.

FAIRBANKS, R. G., 1989, A 17,000-year glacio-eustatic sea level record: Influence of glacial melting rates on the Younger Dryas event and deep-ocean circulation: Nature, v. 342, p. 637–642.

FLETCHER, C. H., KNEBEL, H. J., AND KRAFT, J. C., 1990, Holocene evolution of an estuarine coast and tidal wetlands: Geological Society of America Bulletin, v. 102, p. 283–297.

FLETCHER, C. H., KNEBEL, H. J., AND KRAFT, J. C., 1992, Holocene depocenter migration and sediment accumulation in Delaware Bay: A submerging marginal marine sedimentary basin: Marine Geology, v. 103, p. 165–183.

FLETCHER, C. H., III, VAN PELT, J. E., BRUSH, G. S., AND SHERMAN, J., 1993a, Tidal wetland record of Holocene sea-level movements and climate history: Palaeogeography, Palaeoclimatology, Palaeoecology, v. 102, p. 177–213.

FLETCHER, C. H., III, PIZZUTO, J. E., JOHN, S., AND VAN PELT, J. E., 1993b, Sea-level rise acceleration and drowning of the Delaware Bay coast at 1.8 ka: Geology, v. 21, p. 121–124.

GAYES, P. T., SCOTT, D. B., COLLINS, E. S., AND NELSON, D. D., 1992, A late Holocene sea-level fluctuation in South Carolina, *in* Fletcher C. H. and Wehmiller, J. F., eds., Quaternary coasts of the United States: Marine and lacustrine systems: Society of Economic Paleontologists and Mineralogists Special Publication No. 48, p. 155–160.

HINE, A. C., AND SNYDER, S. W., 1985, Coastal lithosome preservation: Evidence from the shoreface and continental shelf off Bogue Banks, North Carolina: Marine Geology, v. 63, p. 307–330.

KRAFT, J. C., 1971, Sedimentary facies patterns and geologic history of a Holocene marine transgression: Geological Society of America Bulletin, v. 82, p. 2131–2158.

REINECK, H. E., AND SINGH, I. B., 1980, Depositional sedimentary environments, New York: Springer-Verlag, 549 p.

REINECK, H. E., AND WUNDERLICH, F., 1968, Classification and origin of flaser and lenticular bedding: Journal of Sedimentology, v. 11, p. 99–104.

RIGGS, S. R., 1996, Functions of organic-rich mud habitats of the Albemarle Sound estuarine system, North Carolina: Estuaries, v. 19, no. 2A, p. 169–185.

RIGGS, S. R., AND BELKNAP, D. F., 1988, Upper Cenozoic processes and environments of continental margin sedimentation: Eastern United States, *in* Sheridan, R. E., and Grow, J. A., eds., The Atlantic Continental Margin, U.S.: Geological Society of America, The Geology of North America, v. I-2, p. 131–176.

RIGGS, S. R., BRAY, J. T., WYRICK, R. A., KLINGMAN, C. R., AMES, D. V., HAMILTON, J. C., LUECK, K. L., AND WATSON, J. S., 1993, Heavy metals in organic-rich muds of the Albemarle Sound estuarine system: North Carolina: Albemarle-Pamlico Estuarine Study, Raleigh, N.C., U.S. Environmental Protection Agency and N.C. Department of Health, Environment and Natural Resources, Report No. 93-02, 173 p.

RIGGS, S. R., YORK, L. L., WEHMILLER, J. F., AND SNYDER, S. W., 1992, Depositional patterns resulting from high-frequency quaternary sea-level fluctuations in northeastern North Carolina, *in* Fletcher, C. H., and Wehmiller, J. F., eds., Quaternary Coasts of the United States: Marine and Lacustrine Systems: Society of Economic Paleontologists and Mineralogists Society for Sedimentary Geology Special Publication no. 48, p. 141–153.

SHACKLETON, N. J., AND OPDYDE, N. D., 1973, Oxygen isotope and paleomagnetic stratigraphy of equatorial Pacific core V28-238; Oxygen isotope temperatures and ice volumes on a 10^5 and 10^6 year scale: Quaternary Research, v. 3, p. 39–55.

SNYDER, S. W., 1994, Miocene sea-level cyclicity: Frequency and amplitude estimates from the Carolina Platform: Unpublished PhD Dissertation, Department of Marine Science, University of South Florida, St. Petersburg, 688 p.

VAIL, P. R., AND MITCHUM, R. M., JR., 1977, Seismic stratigraphy and global changes of sea level, part 1, *in* Payton, C. E., ed., Seismic stratigraphy-applications to hydrocarbon exploration: American Association of Petroleum Geologists Memoir 26, p. 51–52.

WARD, L. W., BAILEY, R. H., AND CARTER, J. G., 1991, Pliocene and early Pleistocene stratigraphy, depositional history, and molluscan paleobiogeography of the Coastal Plain, *in* Horton, J. W., and Zullo, V. A., eds., The Geology of the Carolinas: University of Tennessee Press, Knoxville, p. 274–289.

WEIL, C., 1976, A model for the distribution, dynamics, and evolution of Holocene sediments and morphological features of Delaware Bay: Unpublished Ph.D. Dissertation, Department of Geology, University of Delaware, Newark, 408 p.

WIGLEY, T. M. L., AND RAPER, S. C. B., 1987, Thermal expansion of sea water associated with global warming: Nature, v. 330, p. 127–132.

WINKER, C. D., AND HOWARD, J. D., 1977, Correlation of tectonically deformed shorelines on the southern Atlantic Coastal Plain: Geology, v. 5, p. 123–127.

UNCONFORMITIES WITHIN A PROGRADATIONAL ESTUARINE SYSTEM: THE UPPER SANTONIAN VIRGELLE MEMBER, MILK RIVER FORMATION, WRITING-ON-STONE PROVINCIAL PARK, ALBERTA, CANADA

RUDI MEYER AND FEDERICO F. KRAUSE
Department of Geology and Geophysics, The University of Calgary, Calgary, Alberta T2N 1N4, Canada
AND
DENNIS R. BRAMAN
Royal Tyrrell Museum of Palaeontology, Box 7500, Drumheller, Alberta T0J 0Y0, Canada

ABSTRACT: The Virgelle Member of the Milk River Formation, Alberta, Canada represents a sandy progradational depositional systems tract that contains linkages between offshore, estuarine, and coastal plain environments. Distinct upward-shoaling depositional successions include regional erosion surfaces that punctuate transitions from storm- and fair-weather-dominated deposition to tidal sand bars and estuarine channel complexes that developed as the systems tract prograded basinward.

The lower part of the Virgelle Member is characterized by hummocky and swaley cross-bedded sandstones depicting the transition from offshore to storm-dominated middle shoreface. A sharp, regionally flat erosion surface separates middle shoreface deposits below from two end-member upper shoreface/foreshore lithofacies associations above: (1) rare fair weather wave-reworked deposits or, (2) common tidally reworked deposits represented by outer estuarine tidal bars. The upper shoreface unconformity is thus dominantly a *tide-cut source diastem* (TSD), overlain by bathymetrically equivalent subtidal to intertidal sand bars typified by planar-bound, herringbone, cross-bedded sandstone.

The erosive base of extensive, laterally accreted estuarine channels (ECh) cuts into middle shoreface deposits and truncates the flat disconformity and the above-mentioned shoreface and estuarine successions. Tidal influence within the channels is recorded by carbonaceous bundles and couplets, reactivation surfaces, and subordinate, flood-directed, three-dimensional dunes. In addition, a restricted ichnofauna documents the influence of an estuarine environment.

The resulting depositional model depicts a west-northwest/east-southeast trending estuarine system, open to the east, that truncates the storm-dominated middle shoreface, and is itself cut by a belt of meandering estuarine channels merged into overlying supratidal coastal plain mudstones. The general distribution of palynomorphs is consistent with the progression of marine dinoflagellate-rich assemblages in outer estuarine tidal bars to mostly terrestrial assemblages in ebb-dominated estuarine channels.

A qualitative analysis of depositional regime variables *Q, M, D* and *R* within the supply-dominated depositional systems tract of the Virgelle Member highlights the critical importance of the sediment dispersal function *D,* in this case controlled by tides and storms. The corresponding relationship may be expressed as $Q \cdot M \geq D \cdot R$. The proposed model for a *progradational estuary* contrasts with sequence stratigraphic models of transgressive estuaries because it is not restricted to specific relative sea-level stages, and because it arises from the linkage of depositional processes along the entire systems tract, from offshore to coastal plain.

INTRODUCTION

Ever since sequence stratigraphy has become widely accepted as a paradigm for depositional models (Van Wagoner et al., 1988), concepts and patterns of sediment stacking have been developed to conform to expectations of sequence-stratigraphic criteria. One such model is that of the transgressive estuary (Dalrymple et al., 1992), which is commonly tied to the origin and development of incised valleys (Posamentier et al., 1992). Estuaries are hence defined as representing exclusively transgressive depositional systems (Reinson, 1992), as geomorphologic features that can only form by the drowning of incised-valley systems, and which "become filled and cease to exist as soon as sea-level rise slows" (Dalrymple et al., 1992, p. 1132). Estuaries and their deposits are thus diagnostic of the sequence-stratigraphic late lowstand to transgressive systems tracts (Zaitlin et al., 1994).

In this context, it is clear that estuaries may commonly be transgressive, however characteristic features of estuaries such as tidal sediment transport paths, salinity fluctuations and distinct biotic communities, do not depend on the history of relative sea level. The risk of restricting the sequence-stratigraphic position of estuaries is even more apparent when we consider that much of the data on the morphodynamics of modern tide-dominated estuaries and tidal flats is obtained from clearly regressive facies successions, for example, the Cobequid Bay-Salmon River estuary, Canada (Dalrymple et al., 1990), the Gironde estuary, France (Allen, 1991), the Ord estuary, Western Australia (Coleman and Wright, 1975), and the German and Dutch tidal flats (Reineck, 1970).

Cooper (1993) proposes an alternative view to the transgressive systems tract model, where not all estuaries are simply ephemeral features waiting to be filled and become non-tidal river mouths, deltas or swamps, but instead, "attain dynamic equilibrium with sediment supply and sea level changes and so retain estuarine characteristics indefinitely" (Cooper, 1993, p. 1009). Similar concepts are expressed by Swift and Thorne (1991), who propose the equilibrium estuary type, capable under certain conditions of bypassing significant volumes of sediment to the shelf, and Perillo (1995), who introduces the so-called eternal estuary, in which sediment supply is balanced or overcome by subsidence.

The variability of estuarine sedimentation and stratigraphy may be approached using the concept of depositional regime proposed by Swift and Thorne (1991) following Sloss (1962). Denoting the dynamic equilibrium at geological time- and space-scales, the depositional regime is defined by sediment input rate *Q,* sediment type or grain size distribution *M,* dispersive sediment transport rate *D,* and the rate and sense of relative sea-level change *R* (Thorne and Swift, 1991). In this context the progradational estuarine system of the Virgelle Member represents a supply-dominated regime that can be characterized qualitatively by the accommodation-supply relationship $Q \cdot M > D \cdot R$ (Swift and Thorne, 1991). Any of the depositional regime variables on either side of this inequality may be evaluated separately, e.g., $Q \cdot M \geq R$ or $Q \cdot M \geq D$. This approach is applied in our study because it allows one to conceptually separate depositional process variables (e.g., *D,* the dispersive sediment transport rate) from those that are only

Tidalites: Processes and Products, SEPM Special Publication No. 61

indirectly and partially dependent on depositional processes, such as *R*, the rate and sense of relative sea level change.

The Upper Santonian Virgelle Member at Writing-on-Stone Provincial Park (WOSPP), southern Alberta, Canada (Fig. 1), is interpreted to represent a progradational tide- and river-dominated estuarine system coupled with a storm-wave-reworked lower and middle shoreface. In this paper we describe the lithofacies framework that characterizes this coastal depositional systems tract as exemplified at WOSPP, including a number of interlacing unconformities across which depositional processes change abruptly. In our analysis we emphasize the role of these regionally significant erosion surfaces that punctuate transitions from storm-dominated deposition to sedimentation in tidal sand bars and estuarine channel complexes. Two key associations are proposed as representative of the general model: (1) the transition from storm-dominated middle shoreface to tidal bar and tidally influenced upper shoreface/foreshore (in contrast to the classic open-marine, wave-dominated beach succession) and (2) the subtidal-to-intertidal estuarine channel. With the Virgelle paleo-estuary at WOSPP as an example, we introduce a new model, namely the progradational estuarine valley, which is analyzed within the framework of depositional regime variables and contrasted with sequence stratigraphic models of transgressive estuaries.

BACKGROUND

Location of Study Area and Stratigraphy

The study area at WOSPP, southern Alberta, is about 9 km north of the international border between Canada and the United States Field work was conducted in Sections 23, 25–26, 35–36, Twp1-Rge13-W4, NTS Map 72 E/4 (Fig. 1). On both sides of the Milk River valley and along several smaller creeks south of the Milk River (Van Cleeve, Police, Davis, and Humphreys' creeks), nearly continuous, variably eroded cliffs and hoodoos provide three-dimensional exposures of sandstones assigned to the Virgelle Member of the Milk River Formation of Late Santonian to Campanian age (Dowling, 1916; Nichols and Sweet, 1993; Leahy and Lerbekmo, 1995) (Fig. 2). In addition we studied outcrops of the Virgelle Member in the following locations: (1) along the Milk River to 2 km east, and about 12 km west, of WOSPP; (2) along Red Creek west of Coutts, Alberta (Sections 5, 8, 9, Twp1-Rge16-W4, NTS Map 82 H/1); and (3) along Buckley Coulee, southwest of Sweetgrass, Montana (Toole County, Hillside Colony and Sunburst Quadrangles, USGS 7 1/2-minute Series, Sections 14–17, E 1/2 18, W 1/2 8, R3W, T37N). Reconnaissance-level visits were made to outcrops in the vicinity of the Sweetgrass Hills, just south of the Canada/U.S.A. border (Toole and Liberty County Quadrangles, USGS 7 1/2-minute Series, R1-4E, T36-37N).

The study area is located on the western margin of the Alberta Foreland Basin and on the axis of a north-northeast-northeast trending cratonic structural high, the Sweetgrass Arch. This feature is proposed to have been intermittently active since the Precambrian and in the Jurassic-Cretaceous (Lorenz, 1982). The Milk River Valley and its tributaries are considered to be the type area for the Milk River Formation even though the lower boundary of the formation is not exposed there (Tovell, 1956; Meijer-Drees, 1990). The formal and informal stratigraphic subdivisions pertaining to the study are shown in Fig. 3.

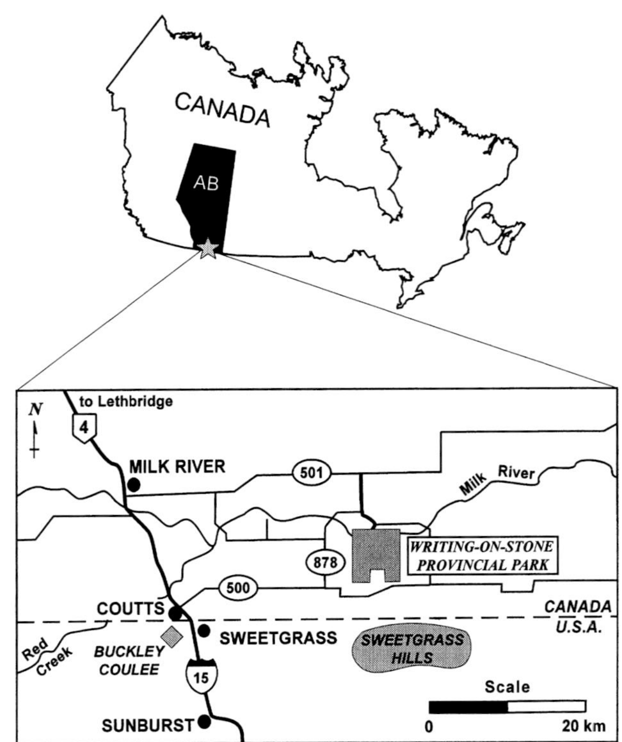

FIG. 1.—Access and location map of the study area in south-central Alberta (AB), Canada, near the Canadian/U.S. border. Writing-on-Stone Provincial Park is shown in gray as well as the locations of Buckley Coulee and Sweetgrass Hills.

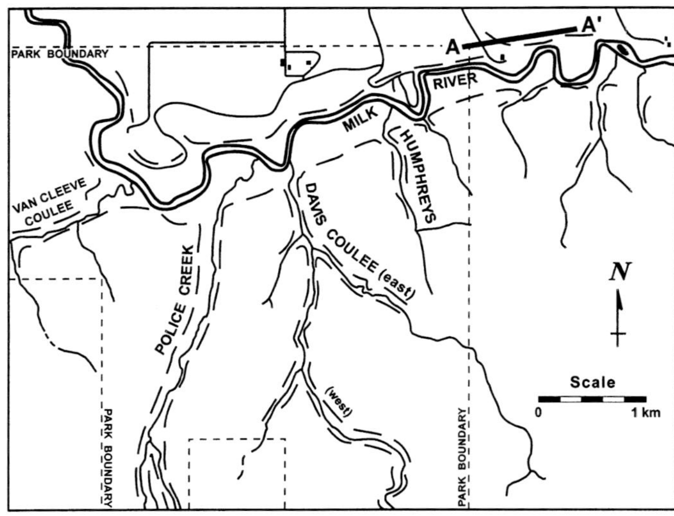

FIG. 2.—Map of Writing-on-Stone Provincial Park. Long dashed lines trace the edge of cliff exposures along the Milk River Valley and coulees. The location of cross-section A-A′ is shown.

Work and Objectives

This account forms part of a study in which sedimentologic and petrographic interpretations are integrated with permeability data to derive a reservoir model for fluid flow. Previous sedimentologic studies were restricted to selected outcrops along the Milk River at WOSPP, thereby severely limiting re-

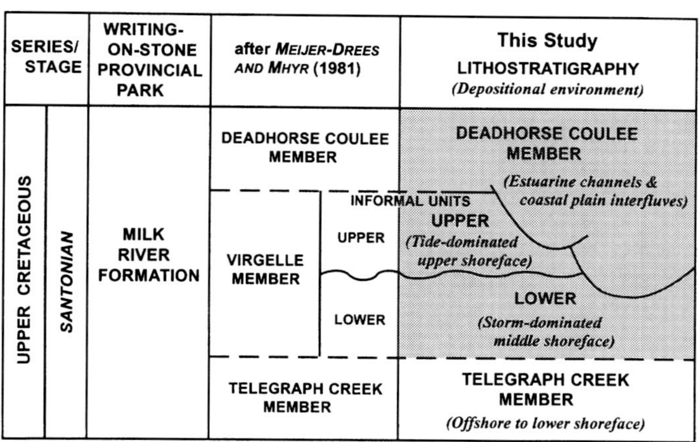

FIG. 3.—Stratigraphy applicable to Writing-on-Stone Provincial Park. Depositional environments are also indicated.

gional application of proposed models (McCrory and Walker, 1986; Leckie and Rosenthal, 1986; Cheel and Leckie, 1990). To generate a depositional model for the Virgelle Member, detailed sedimentologic data were collected continuously along outcrops within WOSPP and adjacent areas. The work included descriptions and measurements of current and bar-form orientations in 71 stratigraphic sections. Additional data were collected in the vicinity of Coutts and Sweetgrass to obtain a more regional perspective of the depositional system (Fig. 1). The field work also included accurate mapping of the top of the middle shoreface in an area of approximately 20 km^2, using a micrometer theodolite.

LITHOFACIES FRAMEWORK

Lithofacies Occurrence and Interpretation

A summary of lithofacies, their characteristics and interpretations of depositional subenvironments is given in Table 1. Due to the limited lithologic differences in the interval, the subdivision into nine lithofacies is based primarily on physical and biogenic sedimentary structures. The term dune is used to describe large-scale, subaqueous flow-transverse bedforms, modified by shape descriptors two-dimensional (2-D) and three-dimensional (3-D) following the classification proposed in Ashley (1990). In this section we briefly discuss the depositional significance of each lithofacies listed in Table 1.

Lithofacies 1 at WOSPP it is well exposed only in Police Coulee, where it is known as the upper part of the Telegraph Creek Member (Russell and Landes, 1940; Meijer-Drees and Mhyr, 1981). However, good, complete exposures of the equivalent unit are present in Buckley Coulee, Montana, U.S.A. Lithofacies 1 is thought to represent the transition from offshore to lower shoreface, mostly below fair-weather wave-base, but is influenced by occasional storm waves.

Lithofacies 2 makes up the prominent cliffs along the Milk River Valley and creeks in the study area. Hummocky and swaley cross-bedding of this lithofacies are interpreted to indicate high shear-stress storm-wave erosion of the seafloor and deposition within the middle shoreface (Fig. 4). Storm-generated beds may grade into wave-rippled tops, possibly representing waning-storm flow conditions, or be truncated at the top by fair-weather wave ripples. The trace fossil assemblage may be assigned to a mixed *Skolithos-Cruziana* ichnofacies indicative of shifting, unconsolidated substrates under conditions of intermittently rapid currents (Frey and Pemberton, 1984). Palynomorphs from six samples of mudstone lenses from both *Lithofacies 1* and *2* include diverse dinoflagellate assemblages, consistent with a marine origin for these facies (D. R. Braman, Unpublished Manuscript, 1993, 1995).

The 3-D and 2-D dunes of *Subfacies 3a* are clear evidence of an environment of high bottom-shear stress under the influence of strong, unidirectional currents. In addition, various diagnostic structures document the influence of tidal currents: 1) inclined-to-subhorizontal reactivation surfaces, in some instances overlain by thin, subordinate, reversely oriented, foresets; 2) thick, carbonaceous- and quartzose zones containing 10–12 pairs of quartzose/sideritic laminae (depicting neap/spring cyclicity); and 3) carbonaceous laminae separating alternating thicker (10–15 mm) and thinner (1–2 mm) quartzose laminae, resembling tidal bundles produced by semidiurnal tides (Fig. 5). Wave-rippled tops of dunes illustrate the effects of marine wave reworking. Among the 21 palynologic samples collected from interbedded mudstone lenses, 18 contain exclusively nonmarine palynomorphs, whereas in the remaining three only one dinoflagellate taxon is present (Braman, 1993, 1994). These results are interpreted to indicate ebb-dominated deposition in tide- and wave-influenced brackish-water channels.

Herringbone cross-bedded units (HCB) of *Subfacies 3b* are interpreted as representing subtidal to intertidal sand bars subject to strong bidirectional currents. Four palynologic samples collected from very thin interbedded mudstone lenses have all yielded common dinoflagellate assemblages, thus supporting a marine-influenced depositional environment (D. R. Braman, Unpublished Manuscript, 1993). In contrast with the tide-dominated channels of Subfacies 3a, the nearly equal thickness of oppositely directed HCB demonstrates the relative balance between ebb and flood tidal currents.

Lithofacies 4, mudstone-clast conglomerate, commonly overlies the erosional basal contacts of the cross-bedded channel facies. The composition and internal structures of the clasts indicate that these were derived primarily from an adjacent, coeval estuarine coastal plain, from horizons that had been subject to near-surface weathering processes of varying intensity (e.g., compaction, pedoturbation, and leaching).

Occurrences of *Lithofacies 5* are interpreted as remnants of wave-dominated, upper shoreface-beach foreshore sequences in which low-angle 3-D dunes may represent shoreface-attached bars, and planar parallel-laminated sandstone the swash zone (Fig. 6). A palynologic sample of rare, interlaminated, wave-rippled mudstone yielded a relatively abundant and diverse dinoflagellate assemblage consistent with a marine depositional environment (D. R. Braman, Unpublished Manuscript, 1993). Strongly leached and/or mottled bedform tops are consistent with shallow-water or subaerial diagenetic conditions.

Current-rippled sandstone of *Lithofacies 6* is interpreted as the result of relatively shallow unidirectional flow on estuarine channel margins, flooding nearly filled and/or abandoned channels and the adjacent muddy coastal flat.

Mudstones of *Lithofacies 7* are representative of several sub-

TABLE 1.—SUMMARY OF LITHOFACIES FOR THE MILK RIVER FORMATION IN WRITING-ON-STONE PROVINCIAL PARK, ALBERTA, CANADA: DIAGNOSTIC SEDIMENTOLOGIC FEATURES.

Lithofacies number and description *interpretation*	Physical structures (bedding, lamination, occurrence)	Biogenic structures (degree of bioturbation, traces)
1 Burrowed, interbedded very-fine grained, grey, wave-rippled and thinly laminated hummocky cross-bedded sandstone and dark grey mudstone. *Offshore to lower shoreface.*	Asymmetric wave ripples and very low angle inclined, slightly undulating laminae; HCS. Mudstone in beds and lenses of 1 to 15 cm in thickness.	Variable: slight to extremely churned. *Anconichnus isp., Helminthopsis isp., Chondrites isp., Zoophycos isp., Paleophycus isp., Planolites isp., Ophiomorpha nodosa, Terebellina isp.*
2 Medium to thickly bedded, laminated, very-fine to fine-grained, light brown, hummocky and swaley cross-bedded sandstone. *Middle shoreface.*	Hummocky and swaley cross-bedding; very rare, thin (< 3 cm) mudstone interbeds. Some bed tops reworked by wave ripples. Pillows, load casts near basal contact.	Variable: slight to highly bioturbated. Burrowed intervals w/ gradational base and sharp, truncated tops. *Ophiomorpha nodosa, Rosselia isp., Terebellina isp., Conichnus isp., Paleophycus isp., Planolites isp.*
3a Fine- to coarse-grained, medium-to-very thickly bedded 3-D and 2-D cross-bedded sandstone. *Estuarine channel.*	3-D and 2-D compound cross-bedded dunes, 10 to 150 cm in thickness, up to 15 m in exposed width. Alternating carbonaceous (1–3 mm) and quartzose laminae (5–30 mm), reactivation surfaces, tidal bundles.	Locally very abundant *Skolithos isp.* (1–2 mm dia. w/thin mud lining) and allocthonous *Teredolites isp.* Common microbioturbate texture.
3b Fine- to medium-grained, medium-to-thickly bedded herringbone cross-bedded sandstone. *Subtidal/intertidal distal tidal bar.*	Herringbone cross-bedding within subhoriz, bounding surfaces; 10–18 degree bedding dips. Wave-modified, reactivation surfaces. Interbedded with Lithofacies 5 and 8, and occasional individual HCS-dunes.	Absence of macro-scale traces. Common microbioturbade texture.
4 White-gray-brown, subrounded to angular, pebble-to boulder-sized, mudstone- and sandstone-clast conglomerate. *Estuarine channel basal dunes.*	Thin-medium bedded. Clasts with variety of internal structures: massive, bioturbated, thinly interlaminated w/lt. brown, bleached sandstone, pedogenic structures, fractured and lenticular. Coal fragments and siderite concretions.	Absence of biogenic structures.
5 Fine- to medium-grained, white-gray-brown, planar parallel-laminated and low-angle 3-D cross-bedded sandstone. *Wave-influenced upper shoreface to foreshore.*	Planar-parallel lamination inclined 2 to 6 degrees, overlying or interbedded w/medium-bedded, low-angle (5–15 degrees) trough cross-beds. Individual dunes interbedded w/ Lithofacies 2 of the upper middle shoreface.	Absence of macro-scale traces. Common microbioturbate texture.
6 Fine-grained, very thin to thinly bedded, laminated current-rippled sandstone *Estuarine channel margin, abandoned channels, coastal flat.*	Current ripples of amplitude 2–15 cm; alternating quartzose/carbonaceous laminae. Occurs as 15–40 cm bedsets at top of channel Lithofacies 3a or interbedded w/mudstones of Lithofacies 7 within heterolithic abandoned channel fills.	Absence of biogenic structures.
7 Gray-brown-black, laminated mudstone and shale, and varicolored, pedoturbated mudstone. *IHS, abandoned channel fill, coastal flat paleosols.*	1–30 cm interlaminated beds and lenses; carbonaceous shales w/ flaky-crumbly texture (possibly pedoturbated). Thick intervals of 0.5–4 m, varicolored, structureless-appearing mudstone.	Absence of biogenic structures.
8 Fine-grained, very thin to thinly-bedded, wave-rippled sandstone. *Upper shoreface.*	5–25 cm bedsets of symmetrical wave ripples. Thin, carbonaceous wisps or interlaminated w/1–50 mm lenses of carbonaceous and micaceous mudstone. Occurs w/ Lithofacies 5 and interbedded w/ Lithofacies 3a 2-D dunes.	Rare *?Planolites isp.* near contacts with fine-grained sandstone lenses.
9 Paleo-weathered sandstone. *Intertidal to supratidal estuarine flat.*	Weathering features: mottles, nodules, yellow-orange over lt. gr.-purple horizons. Overprinting Lithofacies 3a and 3b; below muddy paleosols; truncated by channels.	Root traces, 5–60 cm in length, in-filled or with drab haloes.

environments: inclined heterolithic strata (IHS) interbedded with channel sandstones within lateral accretion bar complexes, abandoned channel fill, and thick, muddy paleosols of a shallow, intertidal-supratidal coastal plain adjacent to the proposed estuarine channel/bar system. Six palynomorph samples collected from channel fill and paleosol sequences do not include any marine taxa (D. R. Braman, Unpublished Manuscript, 1993, 1994).

Wave-rippled sandstones of *Lithofacies 8* are interpreted as having been formed by oscillatory currents in relatively shallow water, on the tops of estuarine channel bars and protected beaches.

Lithofacies 9 encompasses all the sandstone units thought to have been affected by subaerial soil formation or gleization processes within the zone of water-table fluctuation (Fig. 7) (C. A. Williams, personal communication, 1994; Retallack, 1990).

Lithofacies Successions

In the following paragraphs we describe three lithofacies successions that characterize typical vertical profiles observed in the field:

1. Wave-dominated shoreface, W-FACE
2. Tide-dominated shoreface, T-FACE
3. Shoreface-to-estuarine channels, CH-FACE

These successions are approximately equivalent in terms of inferred paleobathymetry and stratigraphic position. They are distinguished by the characteristics of their upper intervals corresponding to the upper shoreface and foreshore. A muddy, supratidal flat deposit overprinted by hydromorphic paleosols overlies all three successions, representing the coastal plain or backshore.

FIG. 4.—*Lithofacies 2:* hummocky cross-bedded sandstone (HCS) of the storm-dominated middle shoreface; staff is 1.5 m long.

FIG. 5.—*Lithofacies 3a:* 2-D cross-bedded sandstone with well-developed quartzose/carbonaceous tidal couplets. In lower part of bed, lamination is obscured by microbioturbation texture (*?Macaronichnus isp.*). Scale is 10 cm long.

Historically, the shoreface has been defined as the slope developed by breaking waves, separating the subaerial from the subaqueous plain (Barrell, 1912). Particularly with respect to upper shoreface/foreshore successions this definition is very limited, considering that shorelines extend from open marine beaches into embayments, occupied by estuaries or lagoons, away from the dominance by wave currents. The sedimentologic and stratigraphic significance of bayside beaches in the rock record has been relatively neglected, although the dynamics of bayside beach-forming processes are clearly very different from those on ocean-side beaches (Nordstrom, 1977). The successions described herein represent typical shoreface end-members, from open marine coast into an adjacent, major estuary.

1. Wave-dominated shoreface (W-FACE).—

The succession is composed of lower-to-middle shoreface at the base (*Lithofacies 1* and *2*), and is truncated sharply by a surface (disconformity) that separates it from the overlying fair-weather wave-influenced upper shoreface/foreshore (*Lithofacies 5*).

The lower-to-middle shoreface is 20–25 m thick, and depicts the storm-dominated transition from offshore to upper middle shoreface, coarsening and thickening upwards from interbedded, very fine-grained sandstone and mudstone, to thick, fine-grained sandstones and rare, very thin mudstone lenses. In addition to the upward-increasing mean grain size (from <60 μm to 200 μm), the progradational nature of the succession is recorded by increasing wavelength of hummocks, from about 1 m at the base up to 6 m near the top.

FIG. 6.—*Lithofacies 5* (Buckley Coulee): medium-scale bedded, low-angle 3-D dunes and overlying planar parallel-laminated sandstone, representing wave-dominated upper shoreface and foreshore. Scale is 15 cm long.

FIG. 7.—*Lithofacies 9:* paleo-weathered horizons at the top of tidal bar succession. Note sharp local contact of mottled, nodular, cemented interval with incipiently weathered sandstone below. Scale is 15 cm long.

A sharp, regionally planar unconformity separates middle shoreface from the overlying upper shoreface/foreshore. This unconformity depicts a shift from fine-grained, burrowed *Lithofacies 2* to medium-grained, laminated sandstone of *Lithofacies 5*. In terms of fluid dynamics it may be interpreted to represent the abrupt change from (storm)-friction-dominated, onshore/offshore currents to fair-weather, wave-driven longshore currents of the upper shoreface and surf zone (Hunter et al., 1979; Niedoroda. et al., 1985). Stratigraphically, the surface corresponds to the source diastem of a supply-dominated depositional system, cut by thė seaward-migration of the surf zone within a prograding shoreline (Swift et al., 1991).

The wave-influenced upper shoreface/foreshore sandstones of *Lithofacies 5* have a maximum thickness of 8 m and are similar to the classic, open-marine upper shoreface/foreshore facies sequence (Harms et al., 1975; Hunter at al., 1979; Niedoroda et al., 1985). This lithofacies is rarely present in the study area, exposed in outcrops of only limited lateral extent and commonly as thin, 2–3-m-thick erosional remnants below channel *Lithofacies 3a.*

2. Tide-dominated shoreface (T-FACE).—

The lower part of the T-FACE succession is identical to the W-FACE succession: the lower-to-middle shoreface at the base, truncated by the unconformity described above (Fig. 8). In contrast to the W-FACE, the unconformity is overlain by tidal sand bar deposits of *Subfacies 3b*, assigned to the zone of subtidal to intertidal sand bars and flat of the outer estuary.

We define the unconformity within the T-FACE as the *tide-cut source diastem* (TSD) of a supply-dominated depositional system, cut by offshore migration of estuarine tidal bars. The depth at which the proposed tide-cut source diastem occurs must be dependent on the relative magnitude of tides and position across the width of the estuary. Because of the importance of the unconformity in our model, we briefly describe this erosion surface and recount some theoretical considerations that help elucidate its origin and stratigraphic position.

Characteristically, cross-bedded sandstone of *Lithofacies 3b* above the unconformity sharply truncates burrows in the bioturbated middle shoreface below (Fig. 9). Near an estuary mouth, conjunction of high fluvial discharge and tides would dampen the effects of shoaling waves and longshore currents, thus inhibiting the buildup of spits and barrier islands across the estuary. Over a long period of time, storm deposits may dominate the preserved record at middle shoreface depths, but as tides approach an estuary mouth, tidal amplitude may be expected to increase due to landward convergence (Nichols and Biggs, 1985), and erosion/deposition is regulated by tidal currents. Near the estuary mouth, tidal deposits would have a relatively high preservation potential because the recurrence time of tidal cycles is several orders of magnitude smaller than that of major storms or floods (Swift and Niedoroda, 1985; Einsele et al., 1991).

The tidal sand bars above the TSD commonly have a thickness of 5–8 m but reach a maximum of 14 m at the south end of Police Coulee, occupying the entire interval between the middle shoreface deposits below, and the thick, muddy coastal plain paleosols of *Lithofacies 7* above. The bars are made-up of planar-bound herringbone crossbeds of *Lithofacies 3b*, with rare, thin interbedded mudstone lenses (Fig. 10). At lateral contacts with the wave-influenced upper shoreface/foreshore, the interfingering nature of the transition make it difficult to iden-

FIG. 8.—Tide-cut source diastem: planar unconformity between middle shoreface below and tide-dominated, upper shoreface above. Surface is approximately coincident with change in coloration from brown-ochre (due to siderite cement) to light gray-white sandstone above. Exposed section is about 7 m thick.

FIG. 9.—Detail of tide-cut source diastem in Fig. 8. Note the sharp erosional surface (indicated by the black triangle) separating fine-grained, burrowed sandstone (middle shoreface) from the overlying medium-grained sandstone lacking meso-scale burrows (tidal bars). Scale bar is 15 cm long.

FIG. 10.—Example of tidal sandbar succession (large herringbone cross-bedded 2-D dunes). T-FACE unconformity is visible just below undercut zone at bottom of photograph (indicated by black triangle). Staff is 1.5 m long.

tify tidal- and wave-dominated end-members. Also, in at least one locality (Buckley Coulee) tidal bar deposits grade into *Lithofacies 5* at the top. Similar to the W-FACE deposits, those of tidal bars have sharp, subplanar-to-slightly undulating basal contacts on the middle shoreface beds, and are commonly modified by top and lateral channel truncation.

The tidal sand bar deposits are regionally extensive with consistently oriented, approximately equally significant current directions: east-southeast and west-northwest at Writing-on-Stone Provincial Park and southeast and northwest at Buckley Coulee. They are interpreted to represent deposition at the mouth of a southeast to east-southeast-trending estuary. The absence of evidence for deep infaunal burrowers and common presence of dinoflagellates are consistent with subtidal to intertidal deposition from high bottom shear stress, marine-influenced tidal currents.

3. Shoreface-to-estuarine channels (CH-FACE).—

The lower part of this succession is the same as the basal parts of successions W-FACE and T-FACE, that is, the storm-dominated lower to middle shoreface. This interval is cut at its top by an undulating, erosional unconformity at the base of estuarine channel complexes (*Lithofacies 3a* and *4*). From field relationships it is clear that the estuarine channel deposits are stratigraphic equivalents of muddy paleosols representing the updip coastal plain (*Lithofacies 7*), which are formally assigned to the overlying Deadhorse Coulee Member (Fig. 3). The unconformity is thus a compound, diachronous surface made-up, in a landward direction, of the conformable basal contact of supratidal, coastal plain mudstones with underlying T-FACE successions, and seaward, by interlacing erosional contacts at the bases of estuarine channels. The channel complex deposits are prominent and characteristic of many outcrops at WOSPP and west of the Park. To the south, channel successions at the level of the upper Virgelle Member are replaced by complete T-FACE successions (south end of Police Coulee).

Individual channels are about 200–400 m wide, 3–10 m thick and have coalesced to form a meander belt at least 5 km wide at WOSPP. The meandering nature of these channels is well illustrated in Fig. 11A, showing the elevation of contours at the truncated top of the middle shoreface in the area east of Police Creek. The outlines of tightly curved, overlapping channel bottoms can be identified, as well as intervening relatively flat plateaus representing truncation of the middle shoreface by the TSD. Channel bottoms are slightly undulating with flat margins separated by 1–2 m steep steps. Local scours 30–60 m in width have margins dipping more steeply at 20–30 degrees (Fig. 12).

Most of the channels contain both sandy, active channel bar deposits, consisting of sets of *Subfacies 3a* 2-D and 3-D dunes, commonly with basal mudstone/sandstone-clast conglomerates (*Lithofacies 4*), as well as muddy, heterolithic, abandoned channel fills (*Lithofacies 7*). A typical channel sequence begins with very thick (1–1.5 m) tabular dunes at the base, grading upwards into 50–80 cm-thick 3-D dunes that become progressively smaller and are sharply overlain at the top by current-rippled sandstone, often incipiently weathered (e.g., mottled, rooted). Other channels appear to be composed entirely of 3-D dunes, very large at bottom, diminishing in size to 3-D current ripples at the top. Both types of successions illustrate the progressive shallowing and infilling of subtidal to supratidal channels; roots and muddy paleosols record the onset of supratidal conditions.

Lateral accretion surfaces within the channels are planar or irregularly curved, as modified by the erosive bases of 3-D dunes (Fig. 13). Some of the clinoforms include relatively continuous, interbedded mudstone lenses as expected for tidally influenced meanders (Smith, 1987); (Fig. 14). The characteristic low diversity and locally high abundance of *Skolithos isp.* in the lower parts of channels is very similar to that documented for other estuarine environments, for example the sandy channel fills described by Sellwood (1972); (Fig. 15).

Within any given channel-form one general, dominant paleocurrent direction is found with minor reversely oriented paleocurrents. Overall, the dominant, ebb-directed paleocurrents range from northwest to south-southwest, thus spanning the eastern quadrants. Lateral accretion surfaces in contrast, dip in various directions, making high or slightly oblique angles to the local paleocurrents. The above relationships are consistent with a west-southwest-to-east-northeast-trending estuarine meander-belt, prograding across more distal, underlying tidal bars and sand flat (Fig. 11B).

DISCUSSION

Depositional Model

The depositional model proposed for the Virgelle and basal Deadhorse Coulee Members is illustrated in the block diagram of Fig. 16A. Fig. 17 is a cross-section east of WOSPP exemplifying typical relationships among successions. The system is clearly progradational as evidenced by a depositional systems tract representing environments from marine offshore through shoreface and foreshore, subtidal to intertidal estuarine channels, and capped by thick units of terrestrial, supratidal, muddy hydromorphic paleosols. The lower and middle shoreface intervals are clearly storm-dominated, but an open marine, wave-dominated upper shoreface-foreshore succession (W-FACE) is rarely present. Instead, regional, bathymetrically equivalent, subtidal to intertidal sand bar deposits are found (T-FACE), characterized by bipolar, east-west-oriented paleocurrent modes.

Based on the comparable thickness of oppositely directed current deposits and the presence of a diverse dinoflagellate assemblage, the tidal bars may correspond to the outer sand bars of a broad, tide-dominated estuary. Figs. 16B–D represent the dissected model showing the major unconformities described above: (1) the TSD, separating storm- and fairweather-wave deposition from tide-dominated sedimentation near the estuary mouth, (Fig. 16B); and, (2) the ECh at the base of the estuarine channel complex (Figs. 16C–D). The estuarine channels have overall paleocurrent directions that are consistent with the east–west trend of the interpreted tidal bars. The strong ebb-dominance of channel paleocurrents, the limited thickness of stacked channel deposits (maximum of 10 m), and the lack of marine palynomorphs, are features which, in combination, support a river-fed estuarine system, rather than a barrier island/lagoon system intercepted and eroded by tidal inlets.

Tide-cut source diastem (TSD).—

The TSD is a regional surface of very low relief that truncates the wave-dominated shoreface at the estuary mouth. In a landward direction the TSD should ultimately pinchout against the

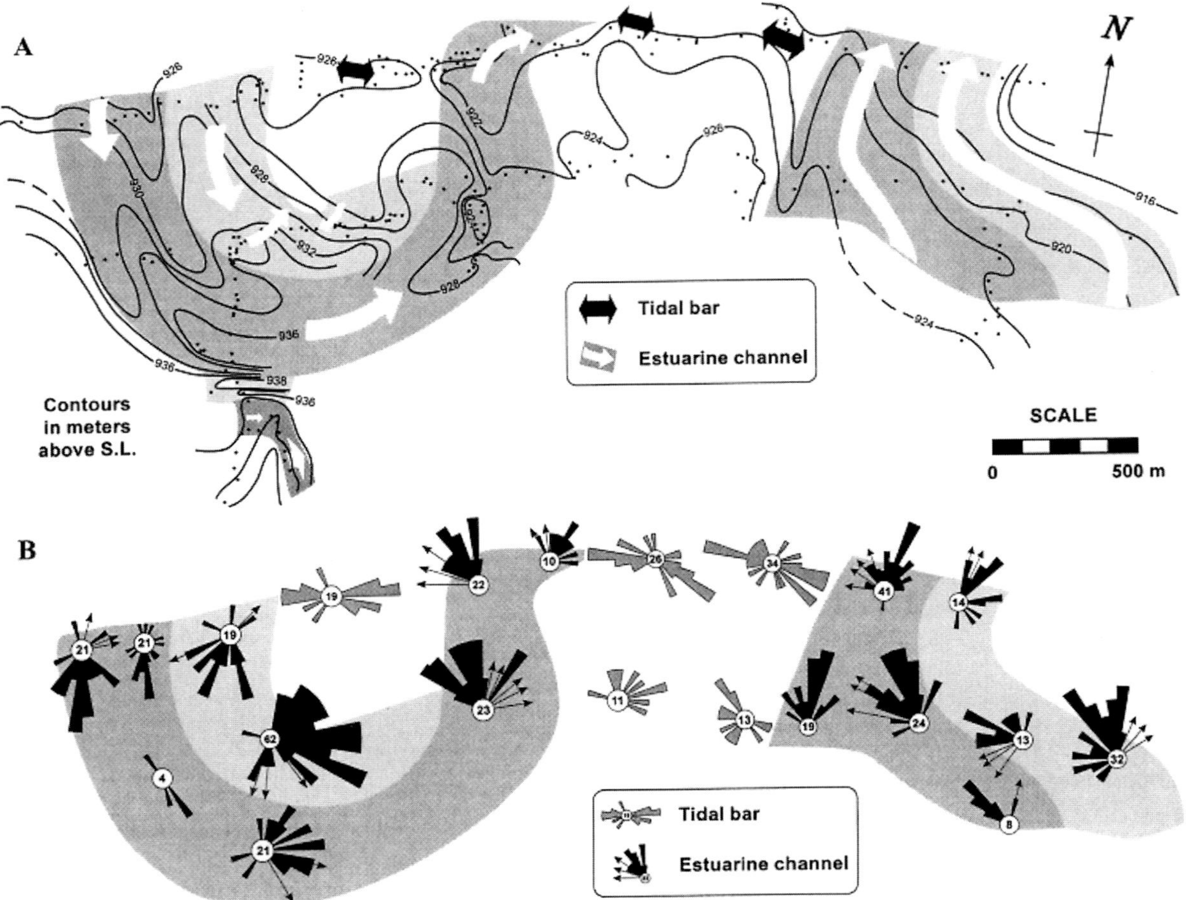

FIG. 11.—A) Contour map of elevation above sea level at the top of the middle shoreface. Outlines of inferred channel complexes are highlighted and general flow directions for both tidal bars and estuarine channels. Note that the map includes a northerly regional structural dip estimated at 1.5°. B). Current roses superimposed on map pattern in A). Current directions are those derived from macro-scale dunes; arrows represent the dip of large-scale bar forms (mostly lateral accretion surfaces). Elevation contours measured with a micrometer theodolite at the top of the truncated middle shoreface.

overlying channels; along-strike (sub-parallel to the coast) it should rise to meet the supratidal coastal plain. The location of these plane intersections remains hypothetical because the TSD is present in both the extreme western and southern outcrops, and dips into the subsurface north of the study area. Based on the relationship of the TSD to adjacent lithofacies, we suggest that it represents a surface of progradation formed when the overlying tidal bars build a subtidal platform at the estuary mouth, upon truncating the storm-dominated shoreface. Having thus defined the TSD, we propose the term estuarine valley, to describe the morphology of the surface. Because the estuarine valley forms at the widest point of the typical conical embayment of estuaries, it establishes the lateral limits of migration and progradation of the corresponding estuarine channel complex.

Estuarine channel-base unconformity (ECh).—

The base of the estuarine channels is interpreted to define a progradational subtidal to supratidal meander belt that progressively eroded the underlying T-FACE down to a depth of 9 m into the middle shoreface. The base of the meander belt therefore constitutes an unconformity, which we term Estuarine channel-base unconformity (ECh). The exact width of the ECh is unknown; the southern margin cuts through the tidal bars in Davis Coulee (west) and Police Creek, but the northern limit is again predicted to be present in the subsurface. Estuarine channel outcrops are found about 15 km west of WOSPP but are thought to be above present-day erosional levels farther west (e.g., in Buckley Coulee).

Other models.—

Several different depositional models for the Virgelle Member at WOSPP have been proposed previously. Based on very limited paleocurrent data, McCrory and Walker (1986) proposed a southwest-trending progradational strandplain, dominated by west-directed longshore drift and cut by distal, fluvio-estuarine channels with dominant northwestward flow. Leckie and others instead proposed a tidal-inlet complex model (Leckie and Rosenthal, 1986; Cheel and Leckie, 1990). Their model is based on locally complex paleocurrent distributions, interpreted as flood-, ebb- or mixed-flow-dominated, and various associated sedimentary structures thought to arise from tidal processes (e.g., shale/sandstone couplets and tidal bundles, flood ramps, and reactivation surfaces). They ascribe the apparent lack of

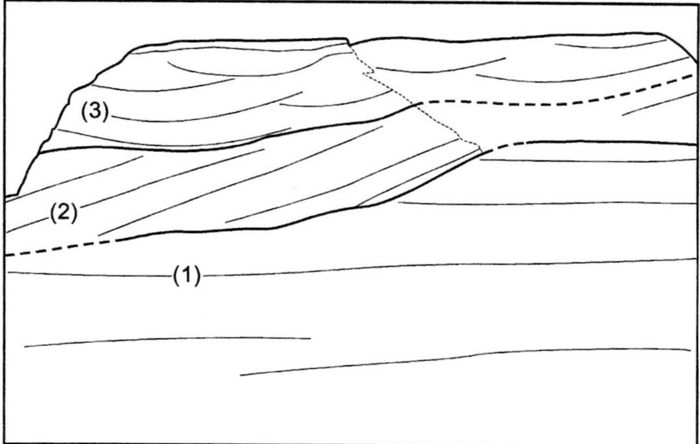

FIG. 12.—Photograph and line diagram of channel succession above sharp, curved, scoured contact with underlying uppermost middle shoreface (1). Note the change in bar- and bed-forms from steep, planar clinoforms containing 2-D-dunes in lower half of sequence (2), and shallow-dipping, undulating lateral accretion surfaces above, containing 3-D-dunes (3). Channel succession is 8 m in thickness.

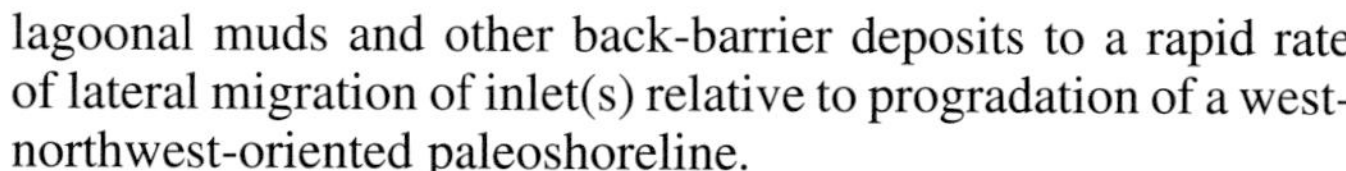

lagoonal muds and other back-barrier deposits to a rapid rate of lateral migration of inlet(s) relative to progradation of a west-northwest-oriented paleoshoreline.

Our model differs fundamentally from the others because of the interpretation of two key elements: the thick, regionally extensive, east-west-oriented estuarine tidal bars, and the stacked, meandering estuarine channels that can be traced landward into the overlying coastal plain mudstones. Thus, we suggest that channels prograded as a meander belt, along and within the estuarine valley, rather than as tidal inlet(s) migrating parallel to the coast.

Depositional Regime

Sloss (1962) proposed that changes in the external geometry of a body of sedimentary rocks through time are a function of process variables *Q, M, D, R.* Swift and Thorne (1991) and Thorne and Swift (1991) developed the concept of depositional regime based on a very similar set of parameters, herein referred to as regime variables. In this section we present a qualitative

FIG. 14.—View of lateral accretion surfaces with thin, relatively continuous interbedded mudstone drapes described as inclined heterolithic strata (IHS). Mudstone layers within active channel dunes such as these are interpreted to have formed by deposition of suspended load during slackwater tide conditions. Outcrop is approximately 9 m thick.

FIG. 13.—Large 3-D-dunes with paleoflow directed toward the viewer (north), contained within lateral accretion surfaces dipping left (east). Scale is 15 cm long.

FIG. 15.—Example of localized dense population of *Skolithos isp.* in trough cross-bedded sandstone of estuarine channel succession. Scale is 15 cm long.

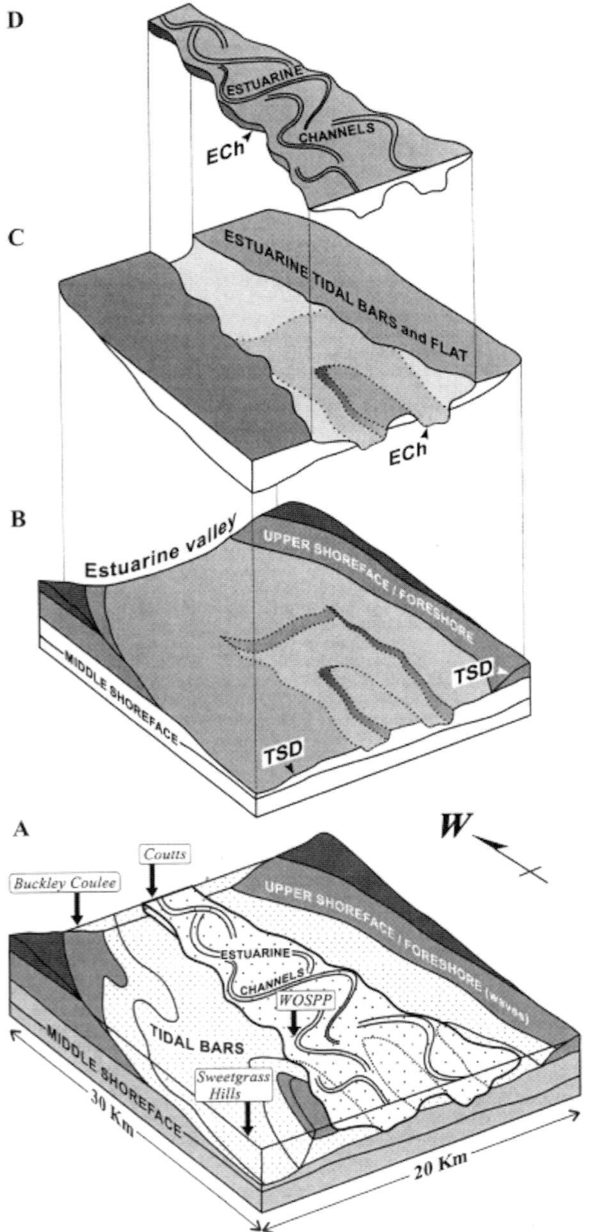

FIG. 16.—Depositional model. (A) Block diagram based on distribution of lithofacies successions. (B)—(D) Exploded views of block model exposing unconformities. TSD = tide-cut source diastem; ECh = estuarine channel base unconformity.

evaluation of the relative magnitudes of these parameters in order to gain some insight into the depositional regime of the Virgelle Member at WOSPP. As a simplification we consider a 2-D profile along the axis of the estuary, from the estuarine channels to the offshore.

The sediment input rate Q can be estimated to have been high, based on the progradational nature of a very thick shoreface. A distinct tectonic regime of right-lateral transpression that dominated the evolution of the southern Canadian Cordillera during the Late Cretaceous to Paleocene triggered significant subsidence and deposition in the southern Alberta foreland basin, providing the active source for the Virgelle Member sediments (Price, 1994). Regarding the bulk grain-size distribution within the system: (1) the Virgelle Member is very sand-rich and mud deposits are restricted to the offshore, thin tidal mud drapes on lateral accretion surfaces, and flood, spring-tide or storm flood overbank deposits on the adjacent coastal plain; (2) analysis of a limited number of samples (n = 15) yielded mean grain sizes of 200 μm for the middle shoreface and 300 μm for upper shoreface and channel intervals; (3) thick, coarse-grained subtidal dunes are very common at the base of estuarine channel successions; and, (4) on a basin-scale, the east-east-southeast dispersal direction for the Virgelle Member is transverse to the strike of the Alberta foreland basin, indicating that long-distance longitudinal transport and sorting parallel to the orogenic belt was not a significant factor in retaining the coarser fraction of the sediment load. The above observations are consistent with a high value for the type of sediment input, M, as is typical for convergent margin settings.

The rate of sediment dispersal D represents the time-averaged transport of sediments in response to the fluid power of waves and currents (Swift and Thorne, 1991). In the case of the Virgelle Member the sedimentary record indicates that storm waves and storm and tidal currents exerted dominant controls on sediment dispersal. The presence of HCS bedforms of large wavelengths interbedded with bioturbated offshore mudstones may be interpreted as indicative of high-intensity storms. The intervals of waning-storm wave ripples, though rarely preserved, are relatively thick (15–20 cm), representing about 30% of the preserved dune form, possibly reflecting the large amount of resuspended load during storms. Regarding the tidal range, our estimate of a high mesotidal range (4 m) is based on the following considerations: (1) the maximum erosional paleo-relief of estuarine channels relative to the top of the middle shoreface is 4–5 m (in a macrotidal regime, channels would have eroded most if not all of the middle shoreface); (2) the maximum thickness of active channel fill is 9 m, comparable to that of modern mesotidal estuaries, for example, Oosterschelde in the Netherlands (Yang and Nio, 1989); (3) a microtidal regime is not consistent with the paucity of wave-reworked upper

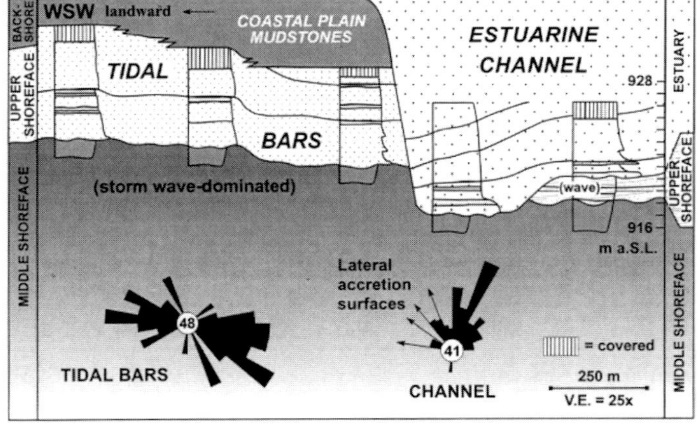

FIG. 17.—Cross-section A-A′ along the Milk River Valley east of Writing-on-Stone Provincial Park showing typical relationships among successions. Note that the top of the middle shoreface corresponds to the surface contoured in Fig. 11A. See Fig. 2 for location. V.E. = vertical exaggeration; m.a.S.L. = meters above sea level.

shoreface deposits. Based on the above, we infer a relatively high sediment transport rate D.

If the mentioned regime variables cannot be evaluated confidently, the rate and sense of relative sea-level change R remains extremely uncertain. The position of the magnetic polarity Chron 34-33r boundary in the middle of the Deadhorse Coulee Member (representing the coastal plain of the Virgelle Member shoreface), can be used to define an age of about 84 Ma for the Virgelle Member (Leahy and Lerbekmo, 1995). This Late Santonian age coincides with a third- or fourth-order transition from maximum transgression to eustatic highstand on the eustatic-cycle charts of Haq et al. (1987) and the correlative transgressive-regressive curves from the Cretaceous Western Interior Basin of Kauffman and Caldwell (1993). Our estimate of R as being very low positive or about zero is only based on the eustatic component of R as inferred above, because we have not evaluated other relevant effects such as the degree of compaction-induced and/or tectonic subsidence.

By combining the regressive character of the depositional system with inferred relative magnitudes of the regime variables (albeit uncertain), we can derive the general, qualitative relationship of a supply-dominated system, i.e., $Q \cdot M > D \cdot R$. Because the supply terms Q and M are inferred to be much larger than R, $Q \cdot M \gg R$, the dispersal rate D may be larger than the supply terms, i.e., $D > Q \cdot M$. This inequality highlights the fundamental importance of tides and storms as dominant coastal processes. If D is high, as was the case for the Virgelle Member, the depositional system may be predominantly estuarine as proposed for equilibrium estuaries by Swift et al. (1991) and Cooper (1993). Conversely, if D should be low the depositional system would be deltaic. With regard to the fundamental supply-accommodation relationship, $Q \cdot M/D \cdot R$, it is interesting to note that once the progradational nature of a coarse-grained estuarine depositional system is established, the relative sea-level change R is required to be low simply to satisfy $D > Q \cdot M$, without making use of third- or fourth-order eustatic cycle charts.

Comparison with Transgressive Estuarine Models

The evolution of estuaries has long been considered to begin with the drowning of river valleys or coastal embayments (e.g., Nichols and Biggs, 1985), a view that derives strong support from the large number of studies of modern estuaries, all of which originated during the most recent eustatic rise in sea-level. It is therefore not surprising that recent sequence-stratigraphic models have emphasized the role of incised fluvial valleys, and their subsequent drowning and infill, to the extent of being considered requisite phases implicit in the definition of estuaries (Reinson, 1992; Dalrymple et al., 1992; Zaitlin et al., 1994). On the other hand, in a recent review on tidal depositional systems, Dalrymple (1992) questioned whether our lack of knowledge of regressive tidal settings is "due to a real scarcity of such deposits, or to the absence of appropriate models?" (Dalrymple, 1992, p. 216). Based on our study of the Virgelle Member at WOSPP we suggest that actively prograding estuarine systems should be another variant of sedimentologic/stratigraphic models, rather than simply assigning regressive estuaries to the passive filling of previously established highstand estuaries.

Our depositional model entails several important differences with respect to the transgressive estuary. Although the river-fed estuary clearly requires an alluvial valley connected to the open sea, subsequent development is not dependent and restricted to a precursor incised-valley/transgressive depositional sequence. The progradation of a regressive estuarine system is not limited to deltas, beach ridges, or open-coast tidal flats (Dalrymple et al., 1992), but in addition may occur by seaward migration of the subtidal tide-cut source diastem overlain by outer estuarine tidal bars. The seaward migration of the TSD is favored within a supply-dominated depositional system, but is more critically dependent on the balance of wave- (particularly storm waves) and tidal-power as components of the coastal sediment dispersal D. The importance of the tide/storm balance is similarly exemplified by the development and stability of ebb-tidal deltas built by relatively small mesotidal estuaries with marginal supply components, as for example Oosterschelde (Yang and Nio, 1989).

The progradational estuarine system has to be considered as part of the associated regressive shelf system, thus constituting a large-scale estuarine-to-offshore depositional systems tract. Analogously, based on an example from the Devonian-age Appalachian foreland basin, Prave et al. (1996) highlighted the proximal to distal linkage of tide- to storm-dominated sedimentation during progressive progradation of coastal facies belts. This contrasts with the role of the regression phase of transgressive-model estuaries that appear effectively decoupled from shelf depositional regimes.

We have illustrated above the coupled estuary-storm shelf depositional systems tract as recorded along the axis of the Virgelle Member estuary. Along strike the system is contained within the subtidal to intertidal estuarine valley (instead of a fluvially incised valley), marked by the transition from a tide- to wave-dominated upper shoreface. Due to the significant longshore components of a fair-weather- and storm-wave-dominated regime, further consideration should be made of the along-shore coupling of the estuarine systems tract and the adjacent open marine, wave-dominated shoreline. Characterization of the corresponding regime variables hinges particularly on an improved distinction of the variety of upper shoreface/foreshore successions present along coasts.

SUMMARY OF CONCLUSIONS

1. The Upper Santonian Virgelle Member at Writing-on-Stone Provincial Park, Alberta, Canada, is represented by a progradational depositional systems tract from marine offshore through shoreface and foreshore, subtidal to intertidal estuarine channels, and capped by supratidal muddy paleosols. The qualitative distribution of palynomorphs is consistent with the progression of a mixed marine dinoflagellate-terrestrial assemblage in outer estuarine tidal bars to nonmarine assemblages in ebb-dominated estuarine channels.
2. The depositional model highlights the role of regionally significant unconformities within the progradational systems tract: (1) the relatively flat, low-relief, tide-cut source diastem within the tidal bar succession, that truncates the storm-dominated middle shoreface; and, (2) the undulating base of estuarine channels that erode and truncate the underlying tidal bar succession and shoreface.
3. The proposed model for a progradational estuary relies on the seaward migration of the tidal source diastem (or estu-

arine valley). Development of the progradational estuary is favored within a supply-dominated depositional regime, $Q \cdot M > D \cdot R$, dependent to a large extent on combined tide and storm sediment dispersal D.

ACKNOWLEDGMENTS

This study is part of the work in progress toward the senior author's doctoral dissertation at the University of Calgary, Canada. For financial assistance we thank Lagoven S. A., affiliate of Petróleos de Venezuela (PDVSA), Norcen Energy Resources Limited for project funds via the Norcen Energy Geology and Geophysics Research Grant in memory of Debbie Olsen (awarded to F. F. K.), and the Department of Geology and Geophysics of the University of Calgary. D. B. gives thanks to Kevin Aulenback for processing palynologic samples; discussions with J. F. Lerbekmo have contributed to D. B.'s work on the Milk River Formation. Special thanks to Alberta Parks and Recreation and the staff at Writing-on-Stone Provincial Park for permission to study the above-mentioned rocks and assistance during the study. We thank G. Els for his thorough review that contributed greatly to the improvement of this paper.

REFERENCES

ALLEN, G. P., 1991, Sedimentary processes and facies in the Gironde estuary: A recent model of macrotidal estuarine systems, *in* Smith, D. G., Reinson, G. E., Zaitlin, B. A., and Rahmani, R., eds., Clastic Tidal Sedimentology: Calgary, Canadian Society of Petroleum Geologists Memoir 16, p. 29–39.

ASHLEY, G. M., 1990, Classification of large-scale subaqueous bedforms: A new look at an old problem: Journal of Sedimentary Petrology, v. 60, p. 160–172.

BARRELL, J., 1912, Criteria for the recognition of ancient delta deposits: Bulletin of the Geological Society of America, v. 23, p. 377–446.

CHEEL, R. J., AND LECKIE, D. A., 1990, A tidal-inlet complex in the Cretaceous epeiric sea of North America: Virgelle Member, Milk River Formation, southern Alberta, Canada: Sedimentology, v. 37, p. 67–81.

COLEMAN, J. M., AND WRIGHT, L. D., 1975, Modern river deltas: Variability of processes and sand bodies, *in* Broussard, M. L., ed., Deltas: Models for Exploration: Houston, Houston Geological Society, p. 99–149.

COOPER, J. A. G., 1993, Sedimentation in a river dominated estuary: Sedimentology, v. 40, p. 979–1017.

DALRYMPLE, R. W., 1992, Tidal depositional systems, *in* Walker, R. G., and James, N. P., eds., Facies Models: Toronto, Geological Association of Canada, p. 195–218.

DALRYMPLE, R. W., KNIGHT, R. J., ZAITLIN, B. A., AND MIDDLETON, G. V., 1990, Dynamics and facies model of a macrotidal sand-bar complex, Cobequid Bay-Salmon River estuary (Bay of Fundy): Sedimentology, v. 37, p. 577–612.

DALRYMPLE, R. W., ZAITLIN, B. A., AND BOYD, R., 1992, Estuarine facies models: Conceptual basis and stratigraphic implications: Journal of Sedimentary Petrology, v. 62, p. 1130–1146.

DOWLING, D. B., 1916, Water-supply, southeastern Alberta: Geological Survey of Canada, Summary Report, 26, p. 102–110.

EINSELE, G., RICKEN, W., AND SEILACHER, A., 1991, Cycles and events in stratigraphy—Basic concepts and terms, *in* Einsele, G., Ricken, W., and Seilacher, A., eds., Cycles and Events in Stratigraphy: New York, Springer Verlag, p. 1–19.

FREY, R. W., AND PEMBERTON, S. G., 1984, Trace fossil facies models, *in* Walker, R. G., ed., Facies Models: Geoscience Canada Reprint Series 1, Toronto, Geological Association of Canada, p. 189–207.

HAQ, B. U., HARDENBOL, J., AND VAIL, P. R., 1987, The chronology of fluctuating sea level since the Triassic: Science, v. 235, p. 1156–67.

HARMS, J. C., SOUTHARD, J. B., SPEARING, D. R., AND WALKER, R. G., 1975, Depositional Structures as Interpreted from Primary Sedimentary Structures and Stratification Sequences: Short Course No. 2, Tulsa, Society of Economic Paleontologists and Mineralogists, 161 p.

HUNTER, R. E., CLIFTON, H. E., AND PHILLIPS, R. L., 1979, Depositional processes, sedimentary structures, and predicted vertical sequences in barred nearshore systems, southern Oregon coast: Journal of Sedimentary Petrology, v. 49, p. 711–26.

KAUFFMAN, E. G., AND CALDWELL, W. G. E., 1993, The Western Interior Basin in space and time, *in* Caldwell, W. G. E., and Kauffman, E. G., eds., Evolution of the Western Interior Basin: Toronto, Geological Association of Canada, Special Paper No. 39, p. 1–30.

LEAHY, G. D., AND LERBEKMO, J. F., 1995, Macrofossil magnetobiostratigraphy for the upper Santonian—lower Campanian interval in the Western Interior of North America: Comparisons with European stage boundaries and planktonic foraminiferal boundaries: Canadian Journal of Earth Sciences, v. 32, p. 247–260.

LECKIE, D. A., AND ROSENTHAL, L., 1986, Cretaceous depositional facies in the western interior: The southern Alberta transect: Field Trip Guide Book, Calgary, Canadian Society of Petroleum Geologists, 70 p.

LORENZ, J. C., 1982, Lithospheric flexure and the history of the Sweetgrass Arch, northwestern Montana, *in* Powers, R. B., ed., Geological Studies of the Cordilleran Thrust Belt: 1982 Symposium—Rocky Mountain Association of Geologists, p. 77–89.

MCCRORY, V. L. C., AND WALKER, R. G., 1986, A storm and tidally-influenced prograding shoreline -Upper Cretaceous Milk River Formation of southern Alberta: Sedimentology, v. 33, p. 47–60.

MEIJER-DREES, N. C., 1990, Milk River Formation, *in* Glass, D. J., ed., Lexicon of Canadian Stratigraphy, Volume 4: Western Canada: Calgary, Canadian Society of Petroleum Geologists, p. 408–9.

MEIJER-DREES, N. C., AND MHYR, D. W., 1981, The Upper Cretaceous Milk River and Lea Park Formations in southeastern Alberta: Bulletin of Canadian Petroleum Geology, v. 29, p. 42–74.

NICHOLS, D. J., AND SWEET, A. R., 1993, Biostratigraphy of Upper Cretaceous nonmarine palynofloras in a north-south transect of the Western Interior Basin, *in* Caldwell, W. G. E., and Kauffman, E. G., eds., Evolution of the Western Interior Basin: Special Paper No. 39, Toronto, Geological Association of Canada, p. 539–84.

NICHOLS, M. M., AND BIGGS, R. B., 1985, Estuaries, *in* Davis, Jr., R. A., ed., Coastal Sedimentary Environments: New York, Springer-Verlag, p. 77–186.

NIEDORODA, A. W., SWIFT, D. J. P., AND HOPKINS, T. S., 1985, The shoreface, *in* Davis, Jr., R. A., ed., Coastal Sedimentary Environments: New York, Springer-Verlag, p. 533–624.

NORDSTROM, K. F., 1977, Bayside beach dynamics: Implications for simulation modeling on eroding sheltered tidal beaches: Marine Geology, v. 25, p. 333–342.

PERILLO, G. M. E., 1995, Geomorphology and sedimentology of estuaries: An introduction, *in* Perillo, G. M. E., ed., Geomorphology and Sedimentology of Estuaries: Developments in Sedimentology Series 53, Amsterdam, Elsevier Science Publishers B.V., p. 1–16.

POSAMENTIER, H. W., ALLEN, G. P., JAMES, D. P., AND TESSON, M., 1992, Forced regressions in a sequence stratigraphic framework: Concepts, examples, and exploration significance: American Association of Petroleum Geologists Bulletin, v. 76, p. 1687–709.

PRAVE, A. R., DUKE, W. L., AND SLATTERY, W., 1996, A depositional model for storm- and tide-influenced prograding siliciclastic shorelines from the Middle Devonian of the central Appalachian foreland basin, USA: Sedimentology, v. 43, p. 611–29.

PRICE, R. A., 1994, Cordilleran Tectonics and the Evolution of the Western Canadian Sedimentary Basin, *in* Mossop, G., and Shetsen, I., eds., Geological Atlas of the Western Canadian Sedimentary Basin: Calgary, Canadian Society of Petroleum Geologists and Alberta Research Council, p. 13–24.

REINECK, H.-E., 1970, Marine Sandkörper, rezent und fossil: Geologische Rundschau, v. 60, p. 302–321.

REINSON, G. E., 1992, Transgressive barrier island and estuarine systems, *in* Walker, R. G., and James, N. P., eds., Facies Models: Toronto, Geological Association of Canada, p. 179–94.

RETALLACK, G. J., 1990, Soils of the Past—An Introduction to Paleopedology: New York, Harper Collins Academic, 520 p.

RUSSELL, L. S., AND LANDES, R. W., 1940, Geology of the Southern Alberta Plains: Memoir 221, Geological Survey of Canada, 223 p.

SELLWOOD, B. W., 1972, Tidal-flat sedimentation in the Lower Jurassic of Bornholm, Denmark: Palaeogeography, Palaeoclimatology and Palaeoecology, v. 11, p. 93–106.

SLOSS, L. L., 1962, Stratigraphic models in exploration: Journal of Sedimentary Petrology, v. 32, p. 415–22.

SMITH, D. G., 1987, Meandering river point bar lithofacies models: Modern and ancient examples compared, *in* Ethridge, F. G., Flores, R. M., and Harvey, M. D., eds., Recent Developments in Fluvial Sedimentology: Tulsa, Society of Economic Paleontologists and Mineralogists Special Publication 39, p. 83–91.

Swift, D. J. P., and Niedoroda, A. W., 1985, Fluid and sediment dynamics on continental shelves, *in* Tillman, R. W., Swift, D. J. P., and Walker, R. G., eds., Shelf Sands and Sandstone Reservoirs: Tulsa, Society of Economic Paleontologists and Mineralogists, SEPM Short Course Notes No. 13, 47–133.

Swift, D. J. P., and Thorne, J. A., 1991, Sedimentation on continental margins, I: A general model for shelf sedimentation, *in* Swift, D. J. P., Oertel G. F., Tillman, R. W., and Thorne, J. A., eds., Shelf Sand and Sandstone Bodies—Geometry, Facies and Sequence Stratigraphy: International Association of Sedimentologists Special Publication 14, London, Blackwell Scientific Publications, p. 3–31.

Swift, D. J. P., Phillips, S., and Thorne, J. A., 1991, Sedimentation on continental margins, IV: Lithofacies and depositional systems, *in* Swift, D. J. P., Oertel, G. F., Tillman, R. W., and Thorne, J. A., eds., Shelf Sand and Sandstone Bodies—Geometry, Facies and Sequence Stratigraphy: International Association of Sedimentologists Special Publication 14, London, Blackwell Scientific Publications, p. 89–152.

Thorne, J.A., and Swift, D.J.P., 1991, Sedimentation on continental margins, II: Application of the regime concept, *in* Swift, D. J. P., Oertel, G. F., Tillman, R. W., and Thorne, J. A., eds., Shelf Sand and Sandstone Bodies—Geometry, Facies and Sequence Stratigraphy: International Association of Sedimentologists Special Publication 14, London, Blackwell Scientific Publications, p. 33–58.

Tovell, W. M., 1956, Some aspects of the geology of the Milk River and Pakowki Formations (southern Alberta): Unpublished Ph.D. Dissertation, University of Toronto, 129 p.

Van Wagoner, J. C., Posamentier, H. W., Mitchum Jr., R. M., Vail, P. R., Sarg, J. F., Loutit, T. S., and Hardenbol, J., 1988, An overview of the fundamentals of sequence stratigraphy and key definitions, *in* Wilgus, C. K., Hastings, B. S., Kendall, C. G. St. C., Posamentier, H., Ross, C. A., and Van Wagoner, J., eds., Sea-Level Changes: An Integrated Approach: Special Publication No. 42, Tulsa, Society of Economic Paleontologists and Mineralogists, p. 39–45.

Yang, C.-S., and Nio, S.-D., 1989, An ebb-tide delta depositional model—A comparison between the modern eastern Scheldt tidal basin (southwest Netherlands) and the Lower Eocene Roda Sandstone in the southern Pyrenees (Spain): Sedimentary Geology, v. 64, p. 175–96.

Zaitlin, B. A., Dalrymple, R. W., and Boyd, R., 1994, The stratigraphic organization of incised-valley systems associated with relative sea-level change, *in* Dalrymple, R. W., Boyd, R., and Zaitlin, B. A., eds., Incised-Valley Systems: Origin and Sedimentary Sequences: Tulsa, Society for Sedimentary Geology Special Publication No. 51, p. 45–60.

TRANSGRESSIVE STRATIGRAPHY AND DEPOSITIONAL FRAMEWORK OF CAMBRIAN TIDAL DUNE DEPOSITS, PEERLESS FORMATION, CENTRAL COLORADO, U.S.A.

PAUL M. MYROW

Department of Geology, The Colorado College, Colorado Springs, Colorado 80903

ABSTRACT: Upper Cambrian Sawatch and Peerless formations exposed in the vicinity of Manitou Springs, Colorado, U.S.A., are part of the Cambrian inner detrital belt of North America. Transgressive systems tract (TST) deposits composed of quartz-rich shoreline and nearshore sandstone (Sawatch Formation) are overlain by coarse-grained, glauconite-rich, tidal dune deposits at the base of the Peerless Formation. These compound cross-bedded sandstone units include spectacular complete formsets up to 3.5 m thick and 3- to 5-m-thick co-sets, the latter composed of two or three stacked-to-shingled, cross-stratified sets. The formsets have near-symmetrical cross-sectional shapes and low stoss and foreset dips, which, in conjunction with their internal structure, indicate deposition under strong but nearly symmetrical tidal currents. Slight asymmetry in current strength is, however, indicated by the dip directions of large-scale foresets, which are consistently from north-northeast to northwest. Centimeter- to decimeter-scale internal cross-bedding shows bimodal to polymodal paleocurrent orientations and reflects the migration directions of superimposed dunes. Sedimentological and stratigraphic analysis indicates that the dunes are part of condensed deposits that formed as latest TST deposits.

The underlying Sawatch Formation provides insight into the nature of TST's in epicontinental settings. The lack of valley fill or coastal plain deposits at the base of the Sawatch suggests transgression over a locally unchannelized, low-relief hinterland. This quartz-rich unit presumably formed from reworking of a thick regolith that sat on deeply weathered Precambrian crystalline rock. No remnants of this regolith remain in the Manitou Springs region. The extremely thin nature of the TST and the presence of numerous ravinement and marine erosion surfaces is the expected signature of TST deposits in epicratonic settings.

INTRODUCTION

Thick, coarse-grained, glauconitic compound cross-bed sets from the Upper Cambrian Peerless Formation exposed in the vicinity of Manitou Springs, Colorado, U.S.A., (Fig. 1) are interpreted as ancient, tide-generated, subaqueous dune deposits of the Cambrian inner detrital belt of North America (Lochman-Balk, 1956). The Peerless dune deposits include spectacular formsets and cosets up to 5 m thick. Tidal dunes are an important class of large marine bedforms that are found in regions of strong tidal currents. The term sandwave was commonly applied to these large tidally formed flow-transverse dunes with heights of less than 10 m and spacings of a few hundred meters or less (e.g., Allen, 1980).

Although tidal deposits have been described for a variety of environments, "there are *remarkably few* well described examples of ancient shelf/shallow marine subtidal sand bodies" (Walker, 1985, p. 303; see also Stride, 1988). Most effort on ancient tidal dunes has been limited to deposits with angle-of-repose cross-bedding and well-developed tidal bundles that formed in environments characterized by considerable tidal asymmetry and large accommodation space (e.g., tidal inlets). Such ancient examples are highly atypical relative to the bulk of dunes in Holocene tidally influenced environments that occur in areas of less extreme conditions of tidal asymmetry (e.g., Fenster et al., 1990). These dunes, like those preserved in the Peerless, have low-angle slipfaces and near-symmetrical profiles.

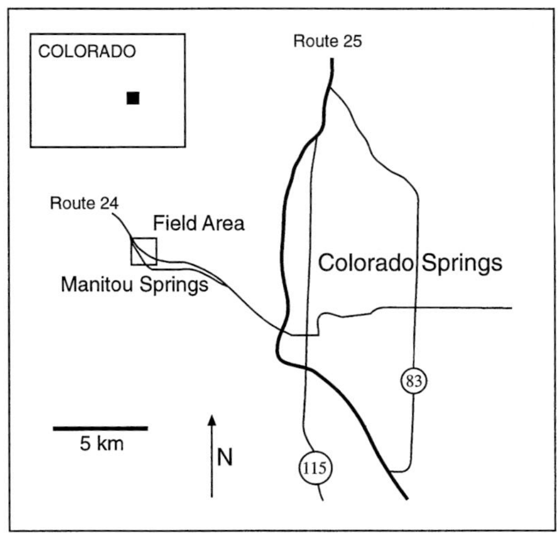

FIG. 1.—Location map of field area, Manitou Springs, Colorado, USA.

Glauconite-rich dune deposits are developed at similar stratigraphic positions in the Riley and Hickory formations of central Texas (e.g., Chafetz, 1978, Fig. 6) and other localities in Cambrian inner detrital belt deposits. This suggests a similarity in the sedimentological and geochemical response to transgression in cratonal settings during the Cambrian. Understanding the dynamics of such a system not only has implications for regional stratigraphy and paleobiogeography, but because these coarse-grained features may be highly permeable, they may therefore be significant as hydrocarbon traps. Such studies may lead to predictive models that hold great potential for exploration of oil and natural gas.

The purpose of this paper is to reconstruct the depositional setting and history of the spectacular tidal dune deposits at the base of the Peerless Formation. Reconstruction of the depositional history will elucidate some of the major controls on dune development within this ancient epicratonal environment. Detailed interpretation of dune geometry and stratification will lead to an understanding of bedform dynamics and the nature of sediment transport on these ancient dunes, but this is beyond the scope of this paper and will therefore be given in a future report. A sedimentological analysis of the Sawatch and Peerless formations is presented with emphasis on the tidal dune deposits. Particular emphasis is given to understanding the evolution of conditions that led to the development of strong tidal currents and the stability of coarse-grained dunes, and finally for conditions that promoted the development of glauconitic and mixed siliciclastic-carbonate facies.

LOCATION AND GEOLOGIC SETTING

The Cambrian-Ordovician rocks of Colorado were deposited in a northwest-southeast-oriented trough (Fig. 2), the Colorado

Tidalites: Processes and Products, SEPM Special Publication No. 61

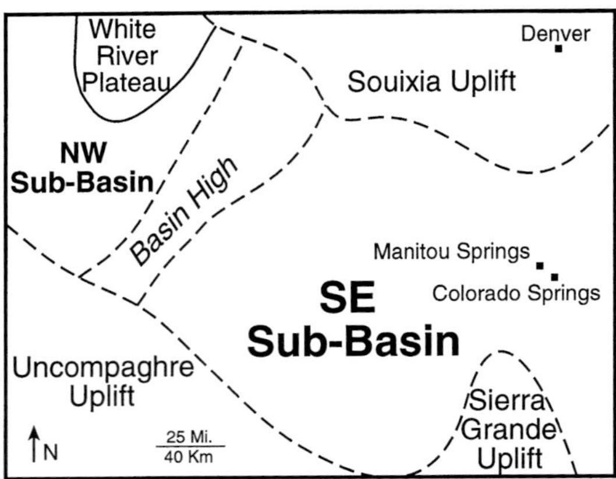

FIG. 2.—Cambrian-Ordovician paleogeography of Colorado. Modified from Gerhard (1972).

Sag (Lochman-Balk, 1956), which represents the only region of major breaching along the entire length of the Transcontinental Arch during this time. Cratonal deposits of the inner detrital belt, one of the facies belts that rimmed the North American continent during the Cambrian (Palmer, 1960; Robison, 1960), consist of quartz- and glauconite-rich nearshore and shoreline sandstone and inner shelf lagoon shale and flat-pebble limestone beds (Lochman-Balk, 1956). The Colorado Sag has been portrayed as consisting of two sub-basins, separated by a subdued arch, with a northwestern sub-basin connected to the paleo-Pacific margin, and a southeastern sub-basin connected to the Midcontinental Sea (Gerhard, 1972; Gerhard, 1974; Fig. 2). This view of the Colorado Sag is criticized by Ross (1996) who believes that all Cambrian-Ordovician rocks in Colorado were on the east side of the Transcontinental arch. Ongoing research (e.g., Myrow et al. 1995) is aimed at resolving these paleogeographic issues. There is no disagreement, however, that the Front Range exposures described in this report were deposited along the western edge of the Midcontinental Sea.

Lower Paleozoic rocks of Colorado rest nonconformably on Precambrian basement consisting of 1.75-Ga-old metamorphic rocks of the Idaho Springs Formation and younger (1030 Ma) intrusive rocks of the Pikes Peak Granite. The Upper Cambrian Sawatch Formation occurs at the base of this cover sequence and consists of white-weathering sandstone and pebbly sandstone that ranges from 4 m to 160 m in thickness across the entire state (Bass and Northrop, 1953; Ross and Tweto, 1980). These shoreline deposits are overlain along the Front Range by nearshore and inner shelf glauconite-rich sandstone and dolostone of the Peerless Formation, which is also of Upper Cambrian age. The Peerless is 12–13 m in thickness in the Colorado Springs area but is over 30 m thick in the subsurface to the east (Stevens, 1961). Tidal dune deposits at the base of the Peerless Formation are found almost exclusively in the area around the northeastern part of Manitou Springs. Numerous exposures occur along small canyons and buttes in a 1-square-mile area (Fig. 1). A detailed measured stratigraphic section for the Sawatch and Peerless formations is given in Fig. 3 along with sedimentological and paleoenvironmental interpretations.

The Peerless is disconformably overlain by limestone and dolostone of the Lower Ordovician Manitou Formation. Trilobite biostratigraphy (Berg and Ross, 1959) shows that the Manitou Formation onlaps onto a pre-Manitou erosion surface that progressively cuts out underlying Cambrian strata south from Colorado Springs along the Front Range so that the Manitou rests successively on the Peerless, Sawatch, and Precambrian basement.

SAWATCH FORMATION

Description

The Sawatch Formation consists of 4–4.5 m of white-weathering coarse and very coarse sandstone and pebbly sandstone (Fig. 4). The sandstone is dominantly quartz arenite that generally contains <5% feldspar (Lewis, 1965). The Sawatch sits on a nonconformity that is remarkably flat at the outcrop scale (Fig. 4). One notable exception occurs in a roadcut along the State Route 24 exit ramp for Manitou Springs (Fig. 1) where large 0.5–3 m-diameter corestones occur throughout the underlying granite, and at the nonconformity surface several of the uppermost corestones project up to 40 cm into the overlying Sawatch (Fig. 5). Stratification in the Sawatch clearly abuts against the corestone, indicating an onlapping relationship of the sedimentary unit with these Precambrian weathering features.

The lower 2.3 m of the Sawatch Formation consists of white and brown weathering, thin- to medium-bedded (5–20 cm thick), coarse to very coarse sandstone and granular to pebbly coarse sandstone. The basal bed consists of 12–15 cm of granular to pebbly very coarse sandstone with angular quartz and feldspar grains and a lag of widely spaced large clasts (up to 10 cm in diameter) of vein quartz that sit directly on the nonconformity. Overlying coarse sandstone beds are dominated by parallel lamination (Fig. 6) but contain widely spaced, small-scale (<8 cm thick), trough cross-stratified beds. Well-preserved polygonal cracks filled with pebbly sandstone occur at 0.95 m and directly overlie a 3–4-cm thick recessively weathered fine sandstone.

The surface at 2.3 m (Fig. 3) represents a fundamental shift from laminated beds below to dominantly massive and bioturbated beds above. The surface is marked by a partially dispersed pebble lag and is overlain by a 60-cm-thick massive sandstone bed. The base of the successive bed (at 2.9 m) contains well-preserved horizontal burrows (Fig. 7). The succeeding 40 cm interval consists of coarse sandstone with widely dispersed pebbles and is marked at its base and top by partially dispersed lags (Fig. 8). The pebbles are generally <2 cm in diameter, but range up to 5 cm across. A 10-cm-diameter sandstone intraclast occurs locally within one of the lags. Bedding is fairly tabular within this upper part of the Sawatch except for the lags, which show clear evidence of post-depositional disruption and dispersion of the clasts into the sandstone as a result of bioturbation.

In most locations, the upper 60 cm of the Sawatch is a distinctive white-weathering, bioturbated, pebbly coarse sandstone (Fig. 9). Blocky remnants of stratified sandstone occur locally at the base of the bed (Fig. 10). These blocks, which appear to have survived the nearly pervasive bioturbation of the bed, have irregular margins and overhanging sides. The bed contains abundant burrows of *Teichichnus* and additional poorly preserved burrow forms including 1-cm-diameter horizontal bur-

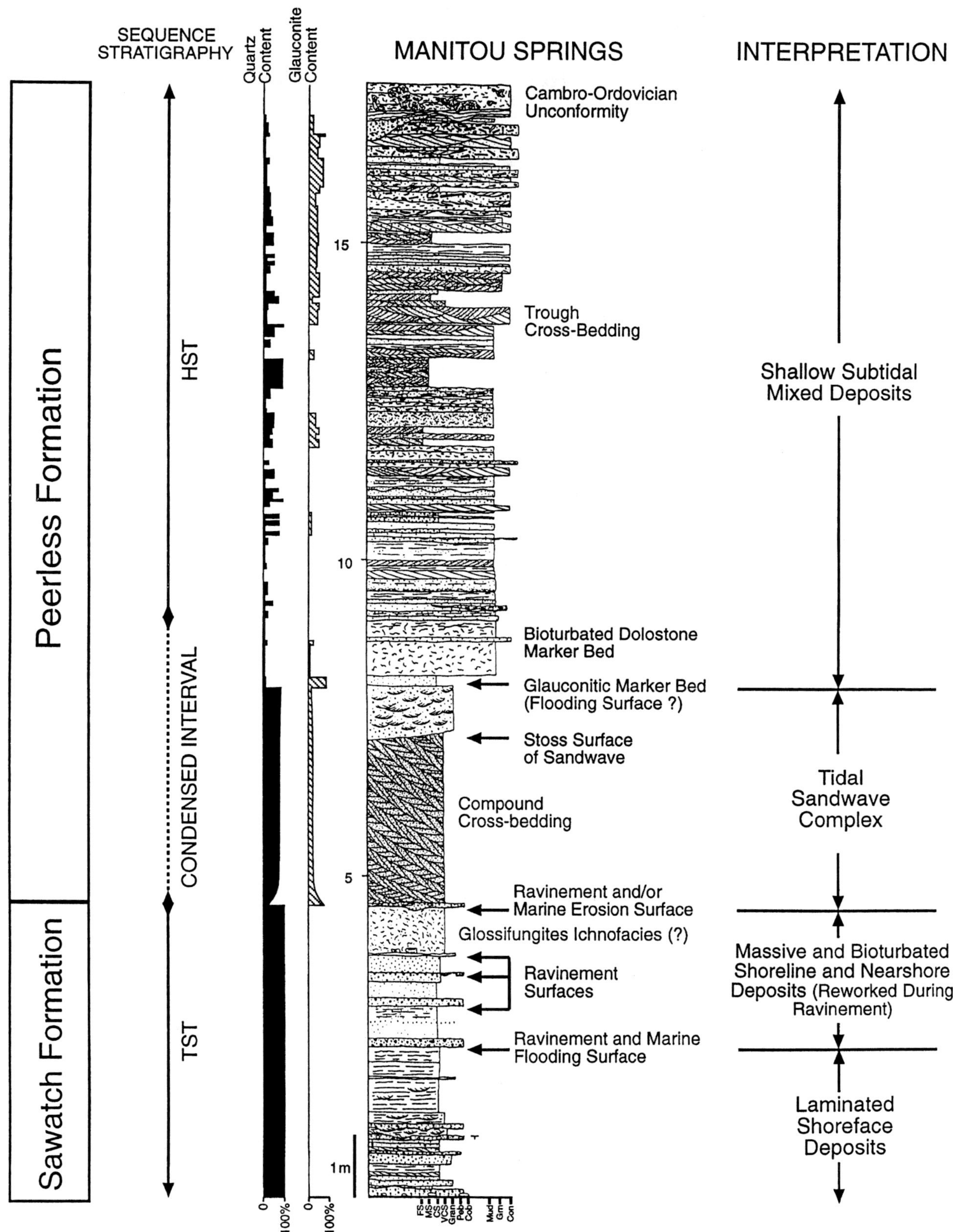

FIG. 3.—Detailed measured section showing lithostratigraphy; stratigraphic changes in quartz and glauconite; and sedimentological, sequence stratigraphic, and paleoenvironmental interpretations. The left side of the grain-size scale shows fine sand (FS) to Cobble (Cob) for siliciclastic sediment. The right side shows mudstone, grainstone and flat pebble conglomerate in carbonate facies. Measured section from outcrops in the vicinity of the first eastbound exit ramp of Route 24 for Manitou Springs (southeast part of Sec. 31, T. 13 S., R. 67 W., Manitou Springs 7.5′ quadrangle, El Paso County, Colorado).

FIG. 4.—Evenly bedded coarse sandstone of Sawatch Formation. Note planar nonconformity surface on granite and dark basal beds of Peerless at top of section. Note: 15-cm scale circled at lower left.

rows and thinner lined burrows of cf. *Paleophycus*. The uppermost 10 cm of the bed consists of a lag with abundant quartz pebbles and large intraclasts of quartz sandstone (Fig. 9). The lag contains pyritic steinkerns (internal moulds) and pyrite-replaced cephalopod, gastropod, trilobite, and brachiopod shells. Many of the fossils, large pebbles, and intraclasts are red-stained with a coating of hematite, and some of the bivalve and gastropod fossils are silicified. In thin section, small, millimeter-diameter burrows are seen projecting inwards radially around rounded (<1 cm diameter) intraclasts of mudstone. The intraclasts must have been firm enough early on to withstand transport and rounding. The burrowing took place while the lag was forming, probably between periods of episodic transport. Directly overlying the lag surface are the glauconite-rich, dune deposits of the Peerless Formation. No glauconite has been observed in the underlying Sawatch.

Interpretation

The sandstone of the Sawatch Formation directly overlies Precambrian basement and contains no evidence of multi-cycle grains (e.g., with rounded overgrowths) (Lewis, 1965), yet it is remarkably mature compositionally, even in poorly sorted pebbly sandstone beds. This suggests that the underlying basement was subjected to an extensive period of intense chemical weathering that produced a highly chemically mature regolith that was a primary source of sediment for the Sawatch. The extensive, nearly flat nature of the nonconformity is consistent with this idea, in that it would reflect removal of weathered granite regolith to the weathering front and the nearly unweathered and erosion-resistant rock below. Erosion of the regolith was uniform to that depth, except where resistant corestones projected up into the weathered profile (Fig. 5). A lack of small-scale bedrock channeling suggests a long period of erosion of the landscape to a low-relief hinterland capped by a deep regolith. This interpretation may explain the paradox of highly compositionally mature deposits that are coarse-grained and directly overlie an extensive basement that should have readily provided abundant feldspar and rock fragments.

Interpretations of depositional processes and paleoenvironments for the Sawatch are difficult because of the paucity of diagnostic sedimentary structures. This is in part a function of the uniformly coarse grain sizes. The presence of bioturbation fabrics and burrows such as *Teichichnus* in the upper part of the formation indicate marine conditions at that time. The stratigraphic position of the first significant marine flooding surface appears to be at 2.3 m, at the sudden shift from laminated beds to massive and bioturbated units.

The coarser and well-laminated units below 2.3 m do not contain structures diagnostic of fluvial braided stream deposition, such as lenticular beds or abundant dune-scale trough cross-stratified beds. Cross-strata occur in thin sets and isolated troughs, but they are rare, and thicker cross-stratified sets are completely absent. This is unusual because dunes are common bedforms in sediment of this grain size under most conditions in fluvial and marginal marine environments. The dominantly horizontal to sub-horizontal stratification, the presence of des-

FIG. 5.—Large corestones in Precambrian granite. The basal beds of the Sawatch Formation onlap onto projecting corestones along the nonconformity in center of photograph. The Sawatch is approximately 4 m thick.

FIG. 6.—Crude horizontal lamination and thin pebble lags in the lower Sawatch Formation. Isolated small-scale trough cross-stratification occurs in this facies, but is not shown in this photograph. Mud-cracked horizon is the recessively weathered zone at the top. Scale bar is 15 cm.

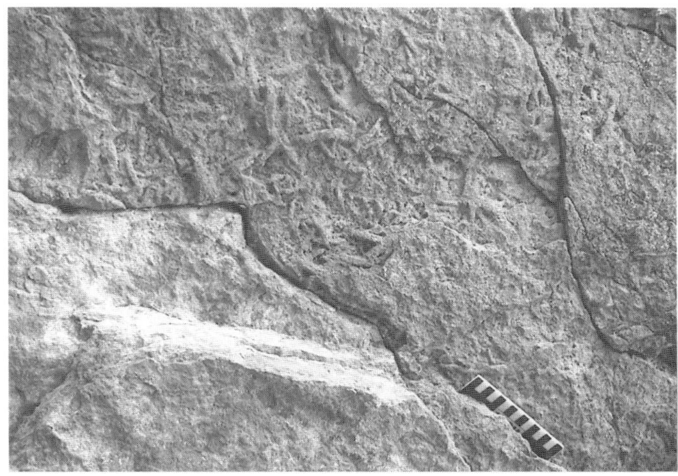

FIG. 7.—Burrows on base of bed at 2.9 m (Fig. 3) within Sawatch Formation. Scale = 15 cm.

FIG. 8.—Pebble lags (just below scale and 30 cm above) partially dispersed in sandstone by bioturbation. Large, rounded sandstone intraclast is at top of photo. Also note widely dispersed pebbles in massive sandstone layer 10 cm above scale that probably represents a thinner, more biogenically disrupted lag bed.

iccation cracks at 0.9 m, and a stratigraphic position directly above basement rocks suggest that these may be foreshore deposits.

The presence of trace fossils and bioturbation fabrics within the upper Sawatch indicates general upward deepening, which is consistent with flooding of the craton at this time. The height of tidal dune deposits in the overlying Peerless Formation provide some constraint on the magnitude of the upward shift in paleobathymetry (discussion below). Beds in the upper part of the Sawatch are marked at their bases by pebble lag horizons, each of which may potentially represent a ravinement or marine erosion surface. The origin of these and other features in the Sawatch are explored below.

Transgressive stratigraphy.—

A wide variety of erosional surfaces are recognized within transgressive systems tracts (TSTs) ranging from sequence-bounding unconformities to more short-term diastems (Nummedal and Swift, 1987). The latter include shoreface ravinement surfaces and offshore deflation surfaces. Ravinement surfaces (Stamp, 1922) result from shoreface erosion by waves and currents during shoreline retreat (Bruun, 1962; Swift, 1968). Numerous studies on the effects of Holocene sea level rise on the Atlantic coast of the United States (Fischer, 1961; Kraft, 1971; Belknap and Kraft, 1981, 1985) demonstrate that shoreface retreat bevels barrier islands and marginal marine (paralic) deposits and deposits a thin layer of sand above the ravinement surface a short distance offshore. Offshore marine erosion surfaces (Nummedal and Swift, 1987) form during transgression due to the combination of vigorous wave/current energy and low sediment input, the latter resulting from trapping of sediment in estuarine and fluvial systems. These surfaces may form regionally due to increased wave and tidal energy, or locally between large sand ridges (Nummedal and

FIG. 9.—Bioturbated white sandstone bed at top of Sawatch Formation and overlying lag with large sandstone intraclasts. The lag lies directly on white bioturbated sandstone bed of the uppermost Sawatch Formation. Scale = 15 cm.

FIG. 10.—Bioturbated white sandstone bed at top of Sawatch Formation. Blocky remnants of undisturbed lamination at the base represent sediment that escaped bioturbation, presumably due to early cementation.

Swift, 1987). They are either coincident with sequence boundaries, or more commonly, rest stratigraphically a short distance above. The thickness of paralic deposits that intervene between these erosion surfaces is a function of a number of factors including depth of shoreface erosion, rate of relative sea-level rise, and the initial thickness of coastal deposits that overlie the sequence boundary (Fischer, 1961; Belknap and Kraft, 1981). Thin transgressive successions occur along shorelines with headlands or highland beaches, and in such cases, if the surface being beveled is underlain by sandy sediment it will be transported seaward to sit upon the ravinement surface.

In the case of the Sawatch, sediment input was extremely low, so very thin transgressive deposits are preserved. As envisioned, the Sawatch records transgression over a landscape lacking significant fluvial systems and consisting instead of a low relief bedrock surface mantled with a thick regolith. Soils would have been stripped and reworked, as documented for interfluves of Pleistocene erosion surfaces during Holocene transgression (Demarest and Leatherman, 1985). Such interfluve areas generally preserve thin transgressive deposits in the rock record (Nummedal and Swift, 1987), because the accommodation space is low prior to ravinement.

Ravinement and offshore erosion surfaces are commonly directly overlain by lags as well as sand sheets up to 10 m thick that consist of sediment eroded from the shoreline during ravinement (Swift et al., 1972; Hollister, 1973; Pilkey et al., 1981). The upper 2 m of the Sawatch is interpreted to record several shifts in shoreline position with ravinement and marine erosion surfaces marked by lags and thin sandstone units representing redistributed sediment and thin remnants of any number of possible nearshore and shoreline subenvironments (Fig. 3); the deposits lack diagnostic criteria for identification of specific paleoenvironments. This is, however, the stratigraphic record one would expect in transgressive deposits in a cratonal setting with low sediment input, low accommodation space, and low equilibrium bottom slopes. In such an environment small changes in relative sea level will be expressed by significant shifts in shoreline position, resulting in numerous erosion surfaces and also in thin intervening deposits.

The bioturbated bed below the uppermost Sawatch surface may contain an assemblage of trace fossils of the *Glossifungites* ichnofacies, which form in firm but unlithified marine sediment in shoreline deposits, particularly above omission surfaces (Frey and Seilacher, 1980; Pemberton and Frey, 1985). Along Holocene shorelines this ichnofacies occurs in cohesive, dewatered salt marsh deposits that, during transgression, were initially covered with beach ridges and then exhumed by wave or tidal erosion (e.g., Frey and Basan, 1981). A distinctive fauna of firmground burrowers colonize the exhumed sediment that was previously dewatered and semi-consolidated during burial. Although *Teichichnus* and cf. *Paleophycus* are not restricted to *Glossifungites* ichnofacies, several lines of evidence point to firmground conditions for the uppermost Sawatch. The presence of sandstone intraclasts in the uppermost lag is evidence for early dewatering and at least partial lithification of sand. The remnants of laminated sand at the base of the burrowed bed indicate not only that the burrowing significantly postdated deposition, but that some parts of the bed may have been too consolidated or lithified to be bioturbated. Finally, burrows in the small mudstone intraclasts in the uppermost lag indicate firmground burrowing of the clasts while they were still actively being rolled episodically, prior to final burial.

Ravinement and marine erosion surfaces are commonly marked by gravel lag pavements with shell accumulations and large intraclasts of shale and early-cemented sandstone (Swift, 1968; Demarest and Kraft, 1987; Nummedal and Swift, 1987; Riggs et al., 1996). The uppermost lag of the Sawatch contains similar features and is interpreted to represent a ravinement surface that was likely modified by strong tidal currents. The ravinement surface may have initially formed by tidal currents, a process generally attributed to scour within tidal inlets (Allen, 1992; Allen and Posamentier, 1993). Regardless of the marine process, resulting facies transitions are commonly very sharp across ravinement surfaces, particularly when zones of active erosion are wide (onshore-offshore) so that more distal marine sediment is deposited above the ravinement surface (Belknap and Kraft, 1981, 1985). This appears to be the case with the transition from the quartz arenitic sandstone of the Sawatch to the glauconitic and dolomitic dune deposits of the Peerless. The lag that marks this surface is interpreted to represent a condensed deposit in which slow sedimentation rates and unusual geochemical conditions produced pyritic and silicic filling and replacement of marine fossils and hematitic coating of a variety of clasts and skeletal fragments.

PEERLESS FORMATION

Description

The Peerless Formation consists of 13 m of dolostone, sandy dolostone, and dolomitic sandstone with abundant glauconite. The base of the formation is marked by the sudden stratigraphic appearance of thick, coarse-grained, glauconite-rich, compound cross-bed sets that rest directly on the uppermost lag of the Sawatch Formation (Fig. 3, 11, 12). Individual cross-bed sets range up to 3.5 m in thickness and consist of complete or near-complete formsets. Thick co-sets between 3–5 m in thickness consist of two to three stacked to shingled sets of cross-strata. The compound cross-bedded formsets contain subdued, near-

FIG. 11.—Large dune deposits at the base of the Peerless Formation directly overlie prominent white-weathered; bioturbated sandstone bed at top of Sawatch Formation. Precambrian granite is in foreground and nearly complete section of the Sawatch and Peerless formations is shown. Cross-bed set is 3 m thick on the left.

FIG. 12.—Photomosaic of dune deposit showing large-scale cross-bedding. Shingled co-sets reach up to 5 m on left side.

symmetrical cross-sectional shapes and low stoss (3–8°) and foreset dips (10–15°). Foreset dip directions are fairly consistent and range from north-northeast to northwest (Fig. 13A). Small-scale cross-bedding is at the angle-of-repose and generally consists of centimeter- to decimeter-scale sets (Fig. 14, 15) that show bimodal to polymodal paleocurrent orientations (Fig. 13B). In some cases these sets show paleocurrent orientations that are dominantly opposite to the dip of the large-scale foresets for a wide area of a dune exposure (Fig. 14). This scale of cross-bedding resulted from the migration of smaller secondary dunes across the lee (and presumably the stoss) side of the dune.

The dunes consist of dolomite-cemented admixtures of detrital grains of quartz and glauconite with minor feldspar, dolomite and rock fragments. White-weathering inarticulate brachiopod shell fragments are common, particularly in association with glauconite. Cross-bed sets exhibit slight upward-coarsening and higher concentrations of glauconite within toesets (up to 80% or more) and along some foresets. Thin sections reveal that glauconite grains are well-rounded and significantly smaller on average than the stable framework grains. The high concentrations of glauconite in the Peerless, particularly in the dune deposits at its base (Fig. 3), is a localized phenomenon, for these southern Front Range exposures are much richer in glauconite than other Peerless deposits in the state, which rarely contain more than a few percent of the mineral (Lewis 1965, Table 5).

When dune deposition ceased, the sea floor had several meters of relief, both in areas with individual formsets (up to 3 m thick) and those in which several sets were locally stacked and shingled. Some of this relief was subsequently filled in with glauconitic and dolomitic coarse sandstone similar to that of the dunes, and locally with beds of granule- and pebble-rich very coarse sandstone.

A regionally extensive, glauconitic (35%) and dolomitic (15%) recessively weathered marker bed of fine to coarse sandstone occurs generally 2.5 m above the base of the dunes (Fig. 2), but climbs stratigraphically where it overlies thicker parts of dunes. In the thickest dune deposits (cosets > 4 m) this bed rises by nearly 2 m and pinches out against the flanks of the dunes. This glauconite-rich sandstone bed marks a sharp transition from coarse sandstone of the dune complex to finer and more dolomitic deposits of the middle and upper Peerless above. The latter are complexly mixed siliciclastic and carbonate strata that range from nearly pure micritic dolostone to quartz arenite. Mixing occurs both within and between beds; the most common lithology is fine dolostone with significant amounts of glauconite and quartz-rich sand. Beds average 10 cm in thickness and are almost exclusively less than 30 cm thick. Sandstone beds are medium- to coarse-grained and range from <6 cm to 50 cm in thickness with planar lamination and trough cross-bedding. The sandstone generally contains admixtures of glauconite grains (up to 50%), dolostone, and larger dolostone intraclasts.

Bioturbation is common in the middle Peerless, particularly in an extensive, bright red, 75- to 80-cm-thick interval of quartz- and glauconite-rich dolostone that directly overlies the glauconite-rich marker bed described above. Burrows in this interval include *Planolites, Chondrites, Curvolithes,* and possible *Thalassinoides horizontalis;* the latter was recently described from the overlying Ordovician Manitou Formation at

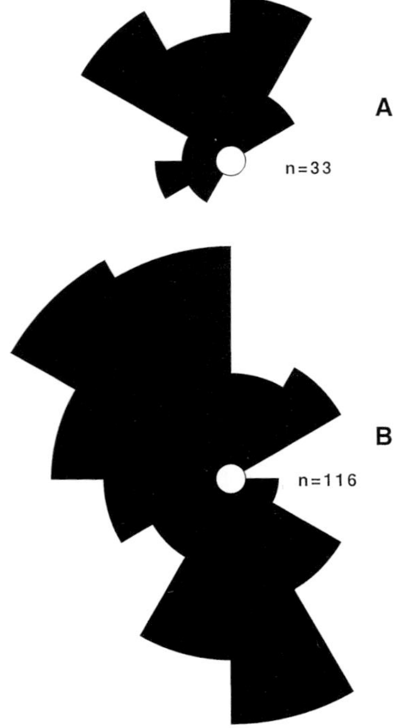

FIG. 13.—(A) Equal-area rose diagram showing dip directions of large-scale forsets of tidal dunes. (B) Rose diagram showing dip directions of superimposed small-scale cross-beds. North toward top of page.

FIG. 14.—Compound cross-bedding in the dune deposits of the Peerless Formation. Large-scale foresets dip from right to left. Small-scale cross-beds show oblique festoon views and a large number of sets showing transport largely up the slipface of the sand wave (left to right). Scale = 15 cm.

this locality (Myrow, 1995). There is a decrease in bioturbation and an increase in the abundance of trough cross-stratified and intraclast-rich beds towards the top of the Peerless.

The contact between the Peerless and overlying Manitou Formation is marked by a distinctive, resistant-weathering, red, coarse dolostone unit that varies from 50 cm to 100 cm in thickness. The base of the resistant bed is highly irregular, with as much as 40 cm of relief. The bed is extremely complex sedimentologically and shows clear evidence of karstification, including collapse breccias and cement-filled vugs. This bed locally marks the Cambrian-Ordovician boundary unconformity, which along the Front Range of Colorado represents a considerable hiatus interpreted as a depositional sequence boundary within Palmer's (1981) Sauk III subsequence (Landing, 1993). The lowermost Manitou consists of thin-bedded ribbon and nodular limestone that is generally in sharp depositional contact with the karsted dolostone bed below.

Interpretation

The shift from quartz arenitic sandstone and pebbly sandstone of the underlying Sawatch Formation to the glauconitic tidal dune deposits of the lowermost Peerless represents a significant lithologic and paleoenvironmental shift across a tidal (?) ravinement surface. Large tidal dunes in many locations globally rest directly on ravinement surfaces produced during Holocene transgression (e.g., Davis et al., 1993). Marine transgression has generally been implicated in the formation of dune complexes (e.g., Hine, 1977), and in deposition of glauconite (Brasier, 1980; Odin and Fullagar, 1988), although transgression is not universally a causal factor for either. The Lower Greensands of England and the Miocene of Belgium contain glauconite-rich tidal dune deposits, and both have been interpreted as transgressive deposits (Nio, 1976; Houbolt, 1982).

A minimum estimate of the magnitude of the transgression, or change in paleobathymetry, recorded in the transition from

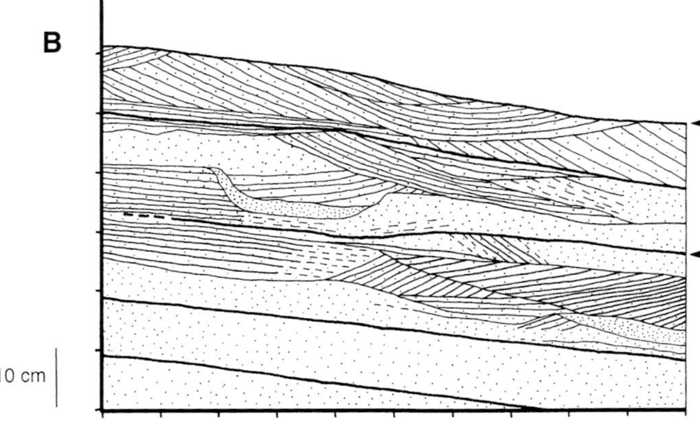

FIG. 15.—(A) Photograph of small-scale cross-stratification of the dune deposits of the Peerless Formation. (B) Sketch corresponding to A; arrows added for reference points. Large-scale foresets dip from right to left and are represented by dark lines in (B). Note complexity and variability of internal structure of the large-scale forsets.

the Sawatch through lower Peerless formation is possible using the thicknesses of tidal dune cross-stratified sets. The size and geometry of bedforms are determined by a number of factors including water depth, velocity, and grain size (Middleton and Southard, 1984). Water depth is the dominant limiting factor on dune height, as it is for all subaqueous dunes because height is limited by erosion due to accelerated flow over bedform crests in shallow water. Yalin (1964) showed that bedform height is generally less than or equal to one-sixth of water depth. In a study of tidal dunes from San Francisco Bay, U.S.A., Rubin and McCulloch (1980) demonstrate that the ratio of bedform height to water depth generally ranges from 1:10 and 1:50, with a maximum of 1:6. They noted larger values of this ratio—1:10 or more—in sediment larger than 0.5 mm in diameter. Dalrymple et al. (1978) found a similar dependence on grain size in their study of Bay of Fundy, Canada, dunes.

An absolute minimum estimate of water depth for the Peerless dunes can be calculated by assuming a maximum height-to-depth ratio of 1:6. The thickest cross-bed set in the Peerless is 3.5 m. This represents a minimum value for the height of the largest bedform, and hence a minimum water depth for the Peerless deposits of 21 m. Using a height-to-depth ratio of 1: 10 would indicate a depth of 35 m, which is probably a more realistic minimum depth, given that (1) the relationship between such factors as grain-size and current velocity are likely to be less than maximizing with regard to bedform height, (2) the thickness of cross-bed sets are some fraction of bedform heights, and (3) water depths are commonly much greater than those representing the minimum for bedform stability. Transgression was potentially fairly rapid; although depositional rates are not possible to calculate, there is only 3.1 m of section between the surface with desiccation cracks in the underlying Sawatch and the base of the dune deposits.

Modern tidal dunes form in a variety of grain sizes, including those with mean grain sizes of coarse sand (e.g., Jones et al., 1965; Harvey, 1966; Smith, 1969), but the Peerless dunes, which consist of medium to coarse sand with quartz clasts up to pebble size, are one of the few recognized ancient coarse-grained tidal dune deposits. The nearly symmetrical geometry and low dips of stoss and lee sides of the Peerless deposits resemble modern tidal dunes (e.g., Fenster et al., 1990) that form under less extreme conditions of tidal asymmetry. Studies of ancient dunes have tended to focus on what are likely the extremes, namely the extraordinary examples that contain a full record of both diurnal tides (bundles) and full lunar month cycles (bundles with systematic changes in thickness; see review by Nio and Yang, 1992) that formed under extremes of tidal velocity asymmetry—Allen's [1980] classes I and II sandwaves. Ancient dune deposits that formed under less extreme conditions of tidal asymmetry—comparable to Allen's (1980) Class V and VI sandwaves—are poorly documented or understood. The well-studied lower Cretaceous Lower Greensands in England (Narayan, 1963, 1971; Allen and Narayan, 1964; Dike, 1972; Nio, 1976) contain evidence for significant, but not extreme, tidal velocity asymmetry (Allen, 1980). The Peerless contains rare examples of dunes that are dominated by low-angle erosional surfaces and small-scale, bimodal or polymodal, cross-stratification (see Allen's [1980] Class VI). Therefore, little is known about the internal structure of such bedforms, and the relevancy of both modern empirical studies (e.g., Dalrymple, 1984; Berne et al., 1988) and theoretical models (Allen, 1980) to ancient deposits has not been established. Although the details of the internal structures and their implications will be described in a later paper, several aspects of the internal structure of Peerless dunes are worthy of discussion herein. Of particular importance is the uniformity of dip directions of the large-scale foresets and the variability of those in the superimposed cross-bedding.

One might assume that the uniform north-northwest direction of the dip of the large-scale foresets represents the orientation of the dominant tidal current, either flood or ebb, as tidal dunes are generally flow-transverse structures. The difference in tidal strength need not have been very large to produce the uniformity in dip directions. Bedform migration resulted from net accumulation over many tidal cycles, and because sediment transport increases nonlinearly with increasing shear stress (Middleton and Southard, 1984), small tidal asymmetry could produce locally consistent bedform migration directions. It is not unusual for modern dunes to show consistent orientation, as Peerless deposits suggest, for areas as large or larger than the outcrop area of this study. On a larger scale, and in some cases over short distances, modern dunes are not systematic in their orientation and behavior. First, the asymmetry of dunes varies spatially; Terwindt (1971) found that dunes in the north part of his study area in the North Sea had steeper sides towards the north but south of a particular line all bedforms were steeper to the south. Fenster et al. (1990) demonstrate that dunes in Long Island Sound, U.S.A., are highly variable in asymmetry and that some show flexing, whereby different sections of crest move in opposite directions relative to a nodal point. They also note variability in migration direction such that a nearly equal percentage of dunes migrated with the ebb direction as with the flood. In addition, many asymmetrical dunes migrated opposite to the direction of the steeper slipfaces. There is no evidence, however, in the Peerless deposits for highly variable migration directions or variable bedform asymmetry.

The Peerless dune deposits do show scatter in the paleocurrent data of small-scale cross-bedding that represents the superimposed dunes. Modern tidal dunes exhibit considerable variation in the size and orientation of superimposed bedforms on and between individual dunes (Houbolt, 1982; Fenster et al., 1990; Davis et al., 1993). This is a reflection of the highly unsteady and nonuniform nature of the currents at the base of tidal flows. Superimposed dunes, particularly in the troughs of large dunes, will respond to the secondary currents set up at the base of the tidal boundary layer—essentially the entire water column—caused by the large-scale bedform roughness. Hence, during the dominant tide the stoss-side migration of superimposed bedforms deliver sand to the advancing crest, yet because of highly irregular and commonly oblique flow, smaller bedform migration may be at a high angle to the crestline orientation of the large dune.

There is some inter-location variability in the inferred transport directions of superimposed dunes from the Peerless, but a large percentage of these dunes migrated either directly up or down from the dip of the large-scale foresets. The scatter in migration directions appear to be lower than for many modern tidal dunes. This may indicate that secondary currents set up at the base of the tidal boundary layer were particularly weak because of lower large-scale bedform roughness due to the subdued near-symmetrical bedform geometries of the Peerless dunes. Alternatively, the high threshold velocities required to move the coarse sediment of these dunes may have only been exceeded during peak flow conditions, during which time secondary flow may have been oriented more close to perpendicular to the dune crests.

The dune deposits are considered to be condensed deposits, for they are rich in glauconite and formed in response to deepening above the ravinement surface at the top of the Sawatch (Fig. 3). Concentration of glauconite in dune toesets, along with upward coarsening within sets, reflects hydraulic segregation during bedform migration. However, the genesis of glauconite and its incorporation into tidal dunes is not well understood, although two possibilities seem likely. First, the dunes were established immediately upon transgression and the glauconite was forming contemporaneously—possibly in the toughs of the bedforms—and was transported and inmixed by large- and small-scale bedform migration. Secondly, a glauconitic layer may have formed following ravinement and deepening, but prior to dune initiation. The dunes would have then reworked and incorporated the glauconite layer. Assuming an average thickness of 3.5 m for the glauconitic dune unit and an average glauconite content of 15%, the hypothetical glauconite bed would have been 50–55 cm thick. There is no sedimentological evidence to suggest one possibility over the other, although the initiation of bedform development is more likely to have occurred immediately as the result of amplification of tidal currents due to transgression. Such amplification would have resulted from the interaction of rising relative sea level and the geometry of the transgressed landscape. The area around Manitou Springs must have been a large embayment in the Cambrian shoreline, which had a funneling effect on tidal currents and possibly also locally created geochemical conditions that favored glauconite formation.

Granule- and pebble-rich very coarse sand that occurs sporadically across the upper part of the glauconitic dune deposits may reflect a minor shoaling prior to a second episode of flooding represented by the beds directly above, namely the glauconite-rich sandstone and red bioturbated dolostone beds. The high glauconite content (40% or more) of the glauconite-rich bed, and the sudden shift to bioturbated, glauconitic dolostone directly above, reflects relative sediment starvation associated with deepening, in which case the base of the glauconite-rich bed may represent a flooding surface. These beds may represent the time of maximum flooding in the Peerless, but regardless, the extensive nature of both beds suggests blanket-like deposition. The overlying 8–9 m of Peerless deposits have evidence for progressive shoaling to the Cambrian-Ordovician unconformity, namely an upward increase in abundance of trough cross-stratification and flat-pebble conglomerate and much less evidence of bioturbation.

Interpretations of depositional environments are difficult for the middle to upper Peerless because these deposits lack obvious small-scale stratigraphic trends, such as shoaling cycles and other highly diagnostic sedimentary structures (e.g., desiccation cracks). There is a distinct lack of wave-diagnostic features such as symmetrical ripples or hummocky cross-stratification, thus suggesting that storm and wave processes were subordinate processes. The abundance of trough cross-stratification indicates that marine currents were important, and their association with underlying tidal dune deposits suggests that tidal currents continued to influence deposition. This part of the formation appears to represent deposits of a moderate energy subtidal environment in which coarse quartz-rich sand was mixed with locally-derived grains of carbonate and glauconite. The quartz fraction of these deposits are generally coarse to very coarse sand, which makes transport by eolian processes unlikely. Sporadic addition of coarse terrigenous sediment was likely from episodic flooding events, and mixing of this sediment with glauconite and carbonate grains that formed *in situ* in the marine environment, resulted from tidal currents, burrowing, and other marine processes.

Shoaling in the upper Peerless culminated in the development of the regionally important sub-Ordovician unconformity surface. The exact position of the unconformity surface with regard to the resistant bed at the top of the Peerless is difficult to pinpoint with accuracy. It may be represented by the irregular surface at the base of the red dolostone bed or by the transition to ribbon bedded limestone at the top. In the former case, the dolomitic unit would be an Ordovician bed that was subjected to karsting shortly after deposition. In the latter case, the sedimentological complexities of the bed would be attributed to late highstand deposition in the Late Cambrian and to subsequent karsting during the Cambrian-Ordovician lowstand.

CONCLUSIONS

1. Coarse-grained, glauconitic tidal dune deposits of the Upper Cambrian Peerless Formation represent a condensed interval at or near the top of thin transgressive systems tract deposits (TST). The TST is thin because of deposition in a low-relief epicratonic setting. The landscape being transgressed at this time consisted of a deeply weathered crystalline basement and a chemically mature, quartz-rich regolith that was the main source of sediment.
2. Foreshore deposits of the lower Sawatch are overlain by thin pebble lags and bioturbated sandstone in the upper part of the formation. The latter represent thin remnants of shallow marine deposits formed by repeated episodes of ravinement, reworking, and redistribution by marine processes. Several lines of evidence indicate that sediment was exhumed during ravinement and that firm sediment in the form of intraclasts, and possibly the underlying substrate as well, was bioturbated and therefore represents a *Glossifungites* ichnofacies. This is supported by evidence for early cementation in the form of sandstone intraclasts and remnants of laminated sandstone at the base of a bioturbated bed.
3. The Peerless dune deposits rest directly on a tidal ravinement surface and are interpreted to record a dramatic change in composition and water depth. These tidal dune deposits are unusually coarse-grained for ancient deposits and are remarkable for their nearly symmetrical profiles and low stoss and lee dips. Such dips are similar to modern dunes (e.g., Fenster et al., 1990) from environments with less extreme conditions of tidal asymmetry.
4. Large-scale foresets of the Peerless dunes dip consistently north-northwest, which presumably represents either the dominant orientation of flood or ebb currents. Superimposed dunes migrated more variably, but much of their movement was generally up or down the slip faces of the dunes.

5. The glauconitic dune deposits are interpreted to be condensed deposits that formed in response to deepening above the ravinement surface at the top of the Sawatch. Deepening in combination with local paleogeographic relief caused amplification of tidal currents and favored the development of glauconite, probably within a large embayment in the shoreline.
6. Beds of glauconite-rich sandstone and red bioturbated dolostone above the dune deposits record blanket-like condensed deposition possibly during maximum flooding. The upper Peerless records deposition in tidally influenced inner shelf environments and upwards shoaling to the Cambrian-Ordovician unconformity.

ACKNOWLEDGMENTS

Acknowledgment is made to the donors of The Petroleum Research Fund, administered by the American Chemical Society, for partial support of this research. This work was also supported by a National Science Foundation grant (EAR-9419141) to the author.

BIBLIOGRAPHY

Allen, G. P., 1992, Sedimentary processes and facies in the Gironde estuary: A recent model for macrotidal estuarine systems, *in* Smith, D. G., Reinson, G. E., Zaitlin, B. A., and Rahmani, R. A., eds.: Calgary, Canadian Society of Petroleum Geologists Memoir 16, p. 29–40.

Allen, G. P., and Posamentier, H. W., 1993, Sequence stratigraphy and facies model of an incised valley fill: The Gironde Estuary, France: Journal of Sedimentary Petrology, v. 63, p. 378–391.

Allen, J. R. L., 1980, Sand waves: A model of origin and internal structure: Sedimentary Geology, v. 26, p. 281–328.

Allen, J. R. L., and Narayan, J., 1964, Cross-stratified units, some with silt bands, in the Folkestone Beds (Lower Greensand) of southeast England: Geologie en Mijnbouw, v. 43, p. 451–461.

Bass, N. W., and Northrop, S. A., 1953, Dotsero and Manitou formations, White River Plateau, Colorado, with special reference to Clinetop Algal Limestone Member of Dotsero Formation: American Association of Petroleum Geologists Bulletin, v. 37, p. 889–912.

Belknap, D. F., and Kraft, J. C., 1981, Preservation potential of transgressive coastal lithosomes on the U.S. Atlantic shelf: Marine Geology, v. 42, p. 429–442.

Belknap, D. F., and Kraft, J. C., 1985, Influence of antecedent geology on stratigraphic preservation potential and evolution of Delaware's barrier system: Marine Geology, v. 63, p. 235–262.

Berg, R. R., and Ross, R. J., Jr., 1959, Trilobites from the Peerless and Manitou formations, Colorado: Journal of Paleontology, v. 33, p. 106–119.

Berne, S., Auffret, J.-P., and Walker, P., 1988, Internal structure of subtidal sandwaves revealed by high-resolution seismic reflection: Sedimentology, v. 35, p. 5–20.

Brasier, M. D., 1980, The Lower Cambrian transgression and glauconite-phosphate facies in western Europe: Journal Geological Society of London, v. 137, p. 695–703.

Bruun, P., 1962, Sea level rise as a cause of shore erosion: Journal of Waterways and Harbors, Division, American Society Proceedings, v. 88, p. 117–130.

Chafetz, H. S., 1978, A trough cross-stratified glaucarenite: A Cambrian tidal inlet accumulation: Sedimentology, v. 25, p. 545–559.

Dalrymple, R. W., 1984, Morphology and internal structure of sandwaves in the Bay of Fundy: Sedimentology, v. 31, p. 365–382.

Dalrymple, R. W., Knight, R. J., and Lambiase, J. J., 1978, Bedforms and their hydraulic stability relationships in a tidal environment, Bay of Fundy, Canada: Nature, v. 275, p. 100–104.

Davis, R. A., Jr., Klay, J., and Jewell, P., IV, 1993, Sedimentology and stratigraphy of tidal sand ridges southwest Florida inner shelf: Journal of Sedimentary Petrology, v. 63, p. 91–104.

Demarest, J. M., and Leatherman, S. P., 1985, Mainland influence on coastal transgression: Delmarva Peninsula: Marine Geology, v. 63, p. 19–33.

Demarest, J. M., and Kraft, J. C., 1987, Stratigraphic record of Quaternary sea levels: implications for more ancient strata, *in* Nummedal, D., Pilkey, O. H., and Howard, J. D., eds., Sea-Level Fluctuation and Coastal Evolution: Society of Economic Paleontologists and Mineralogists Special Publication 41, p. 223–239.

Dike, E. F., 1972, Sedimentology of the Lower Greensand of the Isle of Wight: Unpublished Ph.D. Dissertation, University of Oxford, England.

Fenster, M. S., Fitzgerald, D. M., Bohlen, W. F., Lewis, R. S., and Baldwin, C. T., 1990, Stability of giant sand waves in eastern Long Island Sound, U.S.A.: Marine Geology, v. 91, p. 207–225.

Fischer, A. G., 1961, Stratigraphic record of transgressing seas in the light of sedimentation on the Atlantic coast of New Jersey: American Association of Petroleum Geologists Bulletin, v. 45, p. 1656–1666.

Frey, R. W., and Basan, P. B., 1981, Taphonomy of relict Holocene salt marsh deposits, Cabretta Island, Georgia: Senckenbergiana Maritima, v. 13, p. 111–155.

Frey, R. W., and Seilacher, A., 1980, Uniformity in marine invertebrate ichnology: Lethaia, v. 13, p. 183–207.

Gerhard, L. C., 1972, Canadian depositional environments and paleotectonics, central Colorado, *in* De Voto, R. H., ed., Paleozoic stratigraphy and structural evolution of Colorado: Quarterly of the Colorado School of Mines, v. 67, p. 1–36.

Gerhard, L. C., 1974, Redescription and new nomenclature of Manitou Formation, Colorado: American Association of Petroleum Geologists Bulletin, v. 58, p. 1397–1406.

Harvey, J. G., 1966, Large sand waves in the Irish Sea: Marine Geology, v. 4, p. 49–55.

Hine, A. C., 1977, Lily Bank, Bahamas: History of an active oolite sand shoal: Journal of Sedimentary Petrology, v. 47, p. 1554–1581.

Hollister, C. D., 1973, Atlantic continental shelf and slope of the U.S.: texture of surface sediments from New Jersey to southern Florida: U.S. Geological Survey Professional Paper 529 M, 23 p.

Houbolt, J. J. H. C., 1982, A comparison of recent shallow marine tidal sand ridges with Miocene sand ridges in Belgium, *in* Scrutton, R. A., and Talwani, M., eds., The Ocean Floor: New York, John Wiley and Sons, Ltd., p. 69–80.

Jones, N. S., Kain, J. M., and Stride, A. H., 1965, The movement of sand waves on Warts Bank, Isle of Man: Marine Geology, v. 3, p. 329–336.

Kraft, J. C., 1971, Sedimentary facies patterns and geologic history of a Holocene transgression: Geological Society of America Bulletin, v. 82, p. 2123–2158.

Landing, E., 1993, Cambrian—Ordovician boundary in the Taconic Allochthon, eastern New York, and its interregional correlation: Journal of Paleontology, v. 67, p. 1–19.

Lewis, J. H., 1965, Petrology and diagenesis of Upper Cambrian rocks of central and western Colorado: Unpublished Ph.D. Dissertation, University of Colorado, Boulder, 184 p.

Lochman-Balk, C., 1956, The Cambrian of the Rocky Mountains and Southwest Deserts of the United States and adjoining Sonora Province, Mexico, *in* Rodgers, J., ed., El Sistema Cambrico, su paleogeografia y el problema de su base: Internat. Geol. Congress, 20th, Mexico, D.F., v. 2, pt. 2, p. 521–661.

Middleton, G. V., and Southard, J. B., 1984, Mechanics of sediment movement (2nd ed): Society of Economic Paleontologists and Mineralogists Short Course No. 3, Providence, Rhode Island, 401 p.

Myrow, P. M., Ethington, R. L., and Miller, J. F., 1995, Cambro-Ordovician proximal shelf deposits of Colorado: Short Papers for the Seventh International Symposium on the Ordovician System, Ordovician Odyssey, p 375–379.

Myrow, P. M., 1995, *Thalassinoides* and the enigma of early Paleozoic open-framework burrow systems: Palaios, v. 10, p. 58–74.

Narayan, J., 1963, Cross-stratification and paleogeography of the Lower Greensand of south-east England and Bas-Boulonnais, France: Nature, v. 199, p. 1246–1247.

Narayan, J., 1971, Sedimentary structures in the Lower Greensand of the Weald, England, and Bas-Boulonnais, France: Sedimentary Geology, v. 6, p. 73–109.

Nio, S.-D., 1976, Marine transgressions as a factor in the formation of sandwave complexes: Geologie Mijnbouw, v. 55, p. 18–40.

Nio, S.-D., and Yang, C.-S., 1992, Diagnostic attributes of clastic tidal deposits, *in* Smith, D. G., Reinson, G. E., Zaitlin, B. A., and Rahmani, R. A., eds. Clastic Tidal Sedimentology: Calgary, Canadian Society of Petroleum Geologists Memoir 16, p. 3–28.

Nummedal, D., and Swift, D. J. P., 1987, Transgressive stratigraphy at sequence-bounding unconformities: Some principles derived from Holocene and Cretaceous examples, *in* Nummedal, D., Pilkey, O. H., and Howard, J. D., eds., Sea-Level Fluctuation and Coastal Evolution: Society of Economic Paleontologists and Mineralogists Special Publication 41, p. 241–260.

Odin, G. S., and Fullagar, P. D., 1988, Geological significance of the glaucony facies, *in* Odin, G. S., ed., Green Marine Clays: Developments in Sedimentology, Amsterdam, Elsevier, p. 295–332.

Palmer, A. R., 1960, Some aspects of the early Upper Cambrian stratigraphy of White Pine County, Nevada and vicinity, *in* Geology of East Central Nevada: Intermountain Association of Petroleum Geologists, 11th Annual Field Conference Guidebook, p. 53–58.

Palmer, A. R., 1981, Subdivision of the Sauk Sequence, *in* Taylor, M. E., ed., Short Papers for the Second International Symposium on the Cambrian System: United States Geological Survey Open-File Report 81-743, p. 160–162.

Pemberton, S. G., and Frey, R. W., 1985, The Glossifungites ichnofacies: Modern examples from the Georgia coast, U.S.A., *in* Curran, H. A., ed., Biogenic Structures: Their Use in Interpreting Depositional Environments: Society of Economic Paleontologists and Mineralogists Special Publication 35, p. 237–259.

Pilkey, O. H., Blackwelder, B. W., Knebel, H. J., and Ayers, M. W., 1981, The Georgia Embayment continental shelf. stratigraphy of a submergence: Geological Society of America Bulletin, v. 92, p. 52–63.

Riggs, S. R., Snyder, S. W., Hine, A. C., and Mearns, D. L., 1996, Hardbottom morphology and relationship to the geologic framework: Mid-Atlantic continental shelf: Journal of Sedimentary Research, v. 66, p. 830–846.

Robison, R. A., 1960, Lower and Middle Cambrian stratigraphy of the eastern Great Basin, *in* Geology of East-Central Nevada: Intermountain Association of Petroleum Geologists, 11th Annual Field Conference Guidebook, p. 43–52.

Ross, R. J., Jr., 1996, Quintessence of the Ordovician: From Rocky Mountain beaches to the depths of Nevada, *in* Longman, M. W., and Sonnenfeld, M. D., eds., Paleozoic Systems of the Rocky Mountain Region: Rocky Mountain Section, Society of Economic Paleontologists and Mineralogists, p. 47–62.

Ross, J. R., Jr., and Tweto, O., 1980, Lower Paleozoic sediments and tectonics in Colorado, *in* Kent, H. C., and Porter, K. W., eds., Colorado Geology: Rocky Mountain Association of Geologists—1980 Symposium, p. 47–56.

Rubin, D. M., and McCulloch, D. S., 1980, Single and superimposed bedforms: a synthesis of San Francisco Bay and flume observations: Sedimentary Geology, v. 26, p. 207–231.

Smith, J. D., 1969, Geomorphology of a sand ridge: Journal of Geology, v. 77, p. 39–55.

Stamp, L. D., 1922, An outline of the Tertiary geology of Burma: Geology Magazine, v. 59, 481–501.

Stevens, D. N., 1961, Cambrian and Lower Ordovician stratigraphy of central Colorado, *in* Rocky Mountain Association Geologists Guidebook, 12th Annual Field Conference: p. 7–15.

Stride, A. H., 1988, Preservation of marine sand wave structures, *in* de Boer, P. L., et al., eds., Tide-Influenced Sedimentary Environments and Facies: D. Reidel Publishing Company, p. 13–22.

Swift, D. J. P., 1968, Coastal erosion and transgressive stratigraphy: Journal of Geology, v. 76, p. 444–456.

Swift, D. J. P., Kofoed, J. W., Saulsbury, F. P., and Sears, P., 1972, Holocene evolution of the shelf surface, central and southern Atlantic shelf of North America, *in* Swift, D. J. P., Duane, D. B., and Pilkey, O. H., eds., Shelf Sediment Transport: Process and Pattern: Dowden, Hutchinson and Ross, Stroudsbourg, Pennsylvania, p. 499–574.

Terwindt, J. H. J., 1971, Sand waves in the southern Bight of the North Sea: Marine Geology, v. 10, p. 51–67.

Walker, R. G., 1985, Ancient examples of tidal sand bodies formed in open shallow seas, *in* Tillman, R. W., Swift, D. J. P., and Walker, R. G., eds., Shelf Sands and Sandstone Reservoirs: Tulsa, Society of Economic Paleontologists and Mineralogists Short Course Notes No. 13, p. 303–341.

Yalin, M. S., 1964, Geometrical properties of sand waves: American Society of Civil Engineers Proceedings, Hydraulics Division Journal, v. 90, p. 105–119.

COARSE CLASTIC TIDAL AND FLUVIAL SEDIMENTATION DURING A LARGE LATE ARCHEAN SEA-LEVEL RISE: THE TURFFONTEIN SUBGROUP IN THE VREDEFORT STRUCTURE, SOUTH AFRICA

B. G. ELS AND J. J. MAYER

Geology Division, Potchefstroom University for C.H.E., Private Bag X6001, Potchefstroom 2520, South Africa

ABSTRACT: Most of the economic gold placers of the late Archean Witwatersrand Supergroup of South Africa have been found to be of fluvial origin. A downstream transition from the fluvial paleoenvironment to coastal environments is expected, but such facies changes have not been observed within the previously studied areas. In this paper, the sedimentology of the uppermost formations of the Supergroup in a distal part of the preserved basin, namely the Vredefort structure, is described and evidence for transition from fluvial through tidal to shallow marine depositional conditions is submitted. The study area comprises the exposed arcuate northeastern to southwestern sector of the Vredefort structure, where overturned strata dip radially towards its center. In the Vredefort structure, the upper subgroup of the Witwatersrand Supergroup, the Turffontein has been subdivided into five formations, for which the codes Rt1 through Rt5 are used. The succession between formations Rt1 through Rt3 is the subject of this study.

Only the upper part of the basal formation, Rt1, is exposed in the study area and comprises coarse-grained, trough and planar cross-bedded arenites, which are texturally moderately mature. Paleocurrent distributions are mostly unimodal, except in the extreme eastern region, where the distribution is polymodal. Formation Rt2 disconformably overlies formation Rt1 and consists of interbedded conglomerate and coarse-grained, cross-bedded arenites. Pebble size is largest in the northwestern parts of the study area and decreases towards the east. Paleocurrent distributions are unimodal in the western parts of the study area, but bimodal to polymodal in the central and eastern parts. Formation Rt3 conformably overlies formation Rt2 and comprises predominantly coarse-grained, moderately mature, trough and planar cross-bedded quartz wackes, with scattered small pebbles and thin conglomerates. Paleocurrent distributions of this formation, too, are unimodal in the western part of the study area and mostly bimodal in the remainder. A number of planar cross-bed sets with foreset packages separated by mudstone or immature sandstone laminae were found in the outcrops of the central and eastern parts of the study area. In some cases the thicknesses within series of consecutive foreset packages vary cyclically. A planar cross-bed set with mud-draped foreset packages was found towards the top of the formation in the eastern parts of the study area.

The mudstone-bounded foreset packages described are interpreted as tidal bundles and indicate a tidal setting for those paleogeographic parts of the formations where they occur. Bimodal paleocurrent distributions provide additional evidence for tidal deposition. West of an imaginary point near the center of the study area, no indicators of tidal sedimentation have been found and all the formations were probably deposited by fluvial processes here. However, to the east of this point, which indicates the position of a paleo-bayline, formations Rt2 and Rt3 exhibit characteristics indicative of tidal deposition. Paleocurrents, and the decrease in pebble size of the conglomerates of formation Rt2 towards the east, suggest that its sediment was derived from a western source area, most likely in response to tectonic uplift and tilting here, and transported into a tidal environment by fluvial processes. The basal disconformity of this formation is thought to be a type 2 unconformity and the result of progradation of coarse-grained gravel across a paleo-surface comprising sand.

INTRODUCTION

The vast majority of the economic gold placers of the late Archean Witwatersrand Supergroup of South Africa occur within its upper Group, the Central Rand. Sedimentological studies have shown that most of these placers are fluvial in origin and most probably were deposited in braided streams. Obviously, a downstream transition from these interpreted fluvial paleoenvironments to coastal environments can be expected, but such transitions have not been recorded within the mining areas, most of which are situated along the northeastern arcuate rim of the preserved basin (Fig. 1A). The well-known Vredefort structure is situated near the center of the preserved Witwatersrand Basin and the outcrops of this supergroup here represent distal facies, relative to those of the mining areas. The main objective of this paper is to document transitions from fluvial through tidal to shallow marine paleoenvironments for a coarse clastic sequence within the upper subgroup of the Witwatersrand Supergroup, the Turffontein, in the Vredefort structure. An additional objective is to offer hypotheses on the paleoenvironmental nature of the proposed tidal environment.

GEOGRAPHIC SETTING, REGIONAL GEOLOGY, AND STRATIGRAPHY

The well-known Vredefort structure is situated towards the northern parts of South Africa in the vicinity of the towns of Parys and Vredefort (Fig. 1B). The structure, the exposed northwestern rim of which is often portrayed as semi-circular on large-scale maps, has a diameter of about 55 km and occurs roughly near the center of the preserved oblong Witwatersrand Basin, as depicted by Pretorius (1986). The Vredefort structure comprises a central granitic hub and an arcuate rim consisting of steeply dipping sedimentary and volcanic rocks of the Dominion Group, and the Witwatersrand, Ventersdorp, and Transvaal Supergroups. With the exception of the covered southeastern parts of the structure, the beds constituting the rim are mostly overturned, dipping periclinally towards the hub. The collar of the Vredefort structure can be considered to comprise a number of approximate straight-line segments, separated by inferred faults. Within each segment, tectonic strikes remain remarkably constant, but the mean tectonic dip azimuths show a gradual increase for successive clockwise segments. The Vredefort area abounds with intrusives of various ages and has been affected by thermal metamorphism, demonstrated by metamorphic facies variations within different regions of the structure (Bisschoff, 1982). Additional structural features include shatter cones (Albat, 1988) and planar shock fractures (Albat and Mayer, 1989). After much controversy and debate, the initial hypothesis of Manton (1962) that the structure represents a meteorite impact scar now seems to be accepted by the vast majority of earth scientists.

In spite of their complex post-depositional history, the arenaceous rocks of the Vredefort structure have generally remained remarkably undeformed and unaltered. In most cases, primary sedimentary structures are readily observed in these rocks and meaningful sedimentary petrographic investigations are possible. In some instances, however, sedimentary structures are obscured by the above-mentioned deformation features.

The Vredefort area was first mapped more than 60 years ago by Nel (1927), who subdivided the Central Rand Group into a

Tidalites: Processes and Products, SEPM Special Publication No. 61

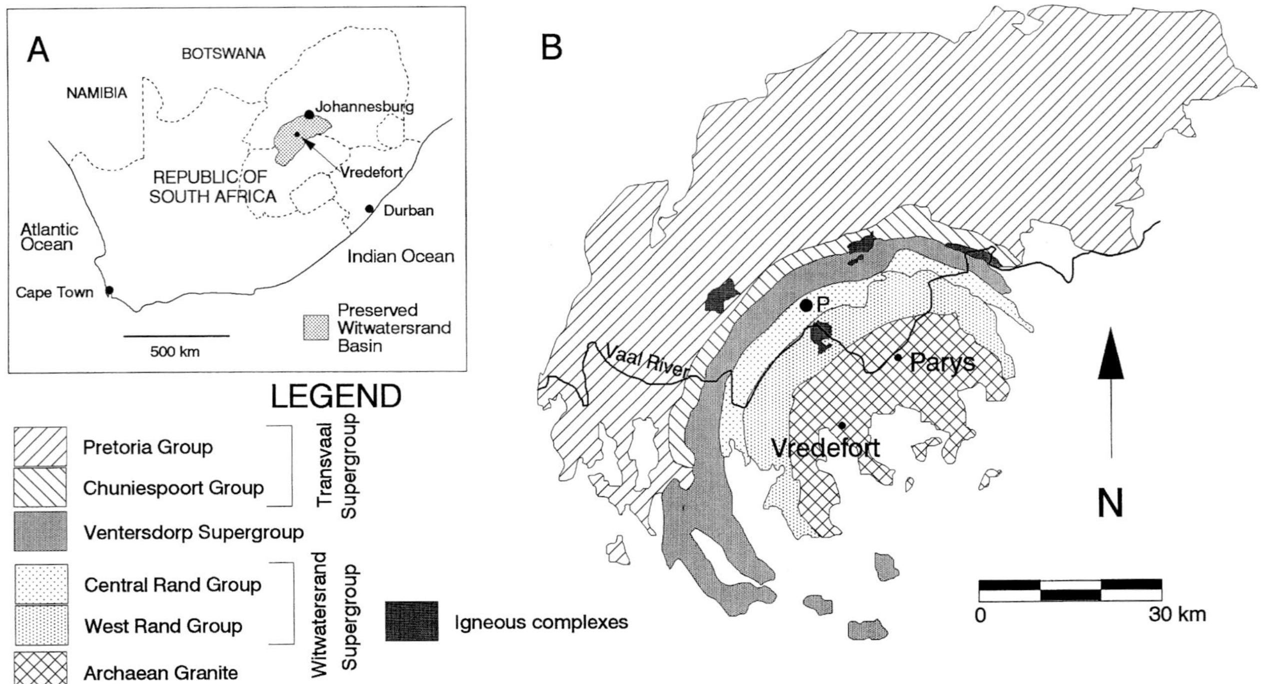

FIG. 1.—Map showing the locality of the preserved Witwatersrand Basin within South Africa (A) and the general geology of the Vredefort structure (B). The locality of the area where the stratigraphic profiles (Figs. 4 and 7) were measured is indicated by P.

lower Main Bird Series and an upper Kimberley-Elsburg Series, which are separated by a shale unit. The correlatives of this lutite can be traced over the entire preserved Witwatersrand Basin and is known as the Booysens Shale in the Johannesburg area and as the Upper Shale Marker in the Welkom goldfield. In the Vredefort structure this lutite is described as a siltstone (Mayer, 1992) and is considered an independent formation (The South African Committee for Stratigraphy, 1980), for which the stratigraphic code Rb is used in this paper.

After an initial phase of systematic mapping of the Central Rand Group, Mayer and Els (1992) proposed a lithostratigraphic subdivision of the Turffontein Subgroup (previously the Kimberley-Elsburg Series) (Fig. 2). The names proposed by Mayer and Els (1992) for the five formations were not accepted by the South African Council for Geosciences for use on a new map of the structure to be published by this Council. Instead, the codes Rt1 through Rt5 have been adopted (Fig. 2). With the exception of formation Rt2, which comprises interbedded conglomerate and sandstone, the formations are sandy in character. Some of the conglomerates and pyrite laminae of formation Rt2 are auriferous and were known as the Amazon placers (du Toit, 1953). Formations Rt1 through Rt3 constitute the subject of this study. Particular attention was paid to the latter formation, for reasons given later in this paper.

The principal marker used for correlation purposes is formation Rt4, which, because of its exceptional textural maturity, can readily be traced along the entire arcuate rim of the Vredefort structure. An additional maker horizon is the basal contact of the conglomeratic formation Rt2, which is easily recognized by an abrupt upward pebble size increase across the contact.

METHODS

Paleocurrent Directions and Distributions

For this investigation trough cross-bedding was used as the principal paleocurrent indicator from which rose diagrams, to illustrate paleocurrent distributions, were compiled. These sedimentary structures are present in all three formations studied and are especially prominent in their texturally mature beds. The steep tectonic dip of the beds in the study area afforded the opportunity to view complete (overturned) troughs, and such exposures were used to measure the plunge angles and directions of trough axes at a number of localities in the study area. For each reading, the dip direction of trough foresets was determined relative to the plunge direction of the overturned trough axis.

Planar crossbedding was used as a second paleocurrent indicator by measuring the dip angle and dip direction of a number of (overturned) foresets in each crossbed set found.

The standard stereographic correction for tectonic tilt of paleocurrents comprises simple rotation of the beds to horizontal about their tectonic strikes (Potter and Pettijohn, 1963). However, Mayer and Els (1992), have shown that this procedure yields questionable results in the Vredefort structure, and proposed an alternative correction technique for this area based upon a specific tectonic model for the diastrophism of the area. These authors consider the Vredefort structure an initially vertical, cylindrical impact-rebound diapir, which has subsequently been tilted and faulted. Paleocurrent corrections of this study were therefore carried out in successive steps to compensate for the various sequential phases of diastrophism.

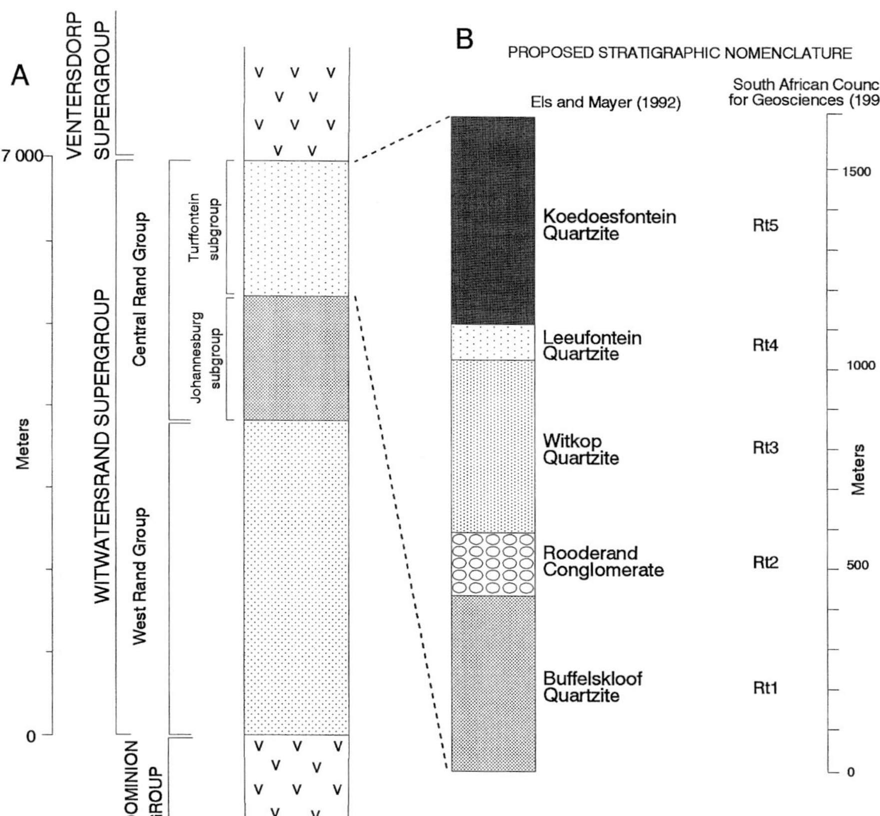

FIG. 2.—Simplified stratigraphic columns showing subdivision of the Witwatersrand Supergroup (A), and proposed lithostratigraphic subdivision of its uppermost subgroup, the Turffontein (B).

Lithofacies, Lithofacies Successions and Sedimentary Structures

These characteristics were investigated by the compilation of a detailed sedimentological profile, recorded along a traverse perpendicular to the tectonic strike, at outcrops on the easternmost parts of the farm Leeufontein 485 (Fig. 3).

For the demarcation of sedimentary units within the sedimentological profiles, maximum grain size, dominant internal structure, and textural maturity were chosen as distinguishing parameters. Mean maximum grain size was estimated by comparison with a scale of diagrammatic grain-size patterns. Paleocurrent directions were recorded throughout the profiles using trough crossbed axes and planar crossbed foresets as indicators. In addition, the preserved widths of troughs and set thicknesses of planar crossbeds were recorded and paleocurrents stereographically corrected, as outlined before.

Petrographic Investigation of Formation Rt3

Thin sections, representing 20 units of formation Rt3, were examined in order to investigate the petrographic transition from this formation, which appeared to be moderately mature throughout, to formation Rt4, which has been proved to be exceptionally mature (Els and Mayer, 1992). Standard point-counting techniques, described by Galehouse (1971), were applied to the thin-sections, counting 300 points per bed. For the determination of grain-size parameters the apparent long dimensions of 300 grains were measured, adopting a lower size limit of 30 μm to separate framework grains from matrix, as suggested by Pettijohn et al. (1972). From the grain-size measurements, statistics were calculated using Statgraphics software by Statistical Graphics Corporation.

DESCRIPTIONS AND RESULTS

Formation Rb

Although this formation does not form part of this study, it is important to mention its main characteristics for full comprehension of the environmental evolution of the Turffontein Subgroup in the Vredefort structure. In this area formation Rb has been described as a siltstone, about 70 m thick (Mayer, 1992), whereas its correlative in the Carletonville goldfield, the Booysens Shale Formation, is described by Engelbrecht et al. (1986) as a shale interbedded between a lower and an upper transition zone. Antrobus et al. (1986) consider the correlative of formation Rb in the Klerksdorp goldfield, the Modderfontein Member, to comprise fine-grained argillaceous quartzites and shales. In spite of facies differences, evident from these brief descriptions, stratigraphers are in agreement about the correlation and the widespread geographic distribution of this fine-grained unit, which has been considered the "most persistent and reliable marker of the Central Rand Group (Engelbrecht et al., 1986, p. 609)."

Formation Rt1

Because the lower part of this formation is not exposed in any part of the study area, the contact with the underlying formation Rb could not be examined. Investigation of this for-

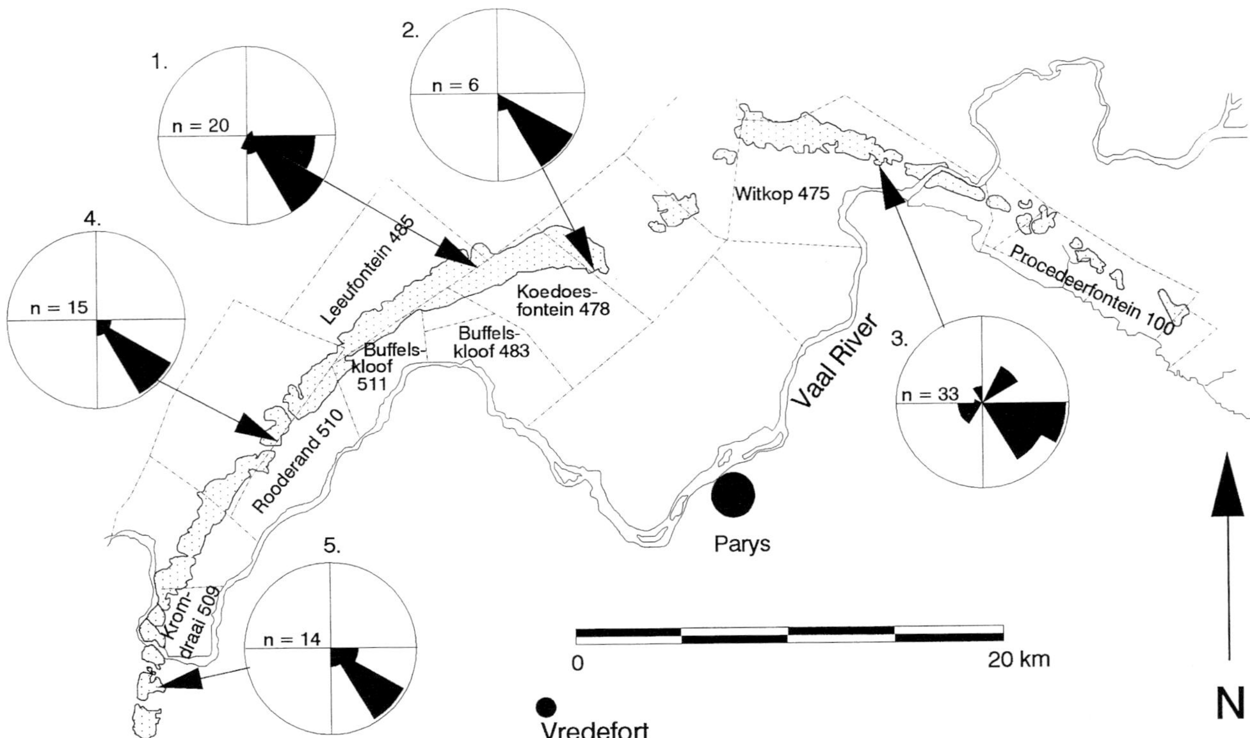

FIG. 3.—Map showing paleocurrent distributions of trough axes for the upper portion of formation Rt1 in the Vredefort structure. Note that only unimodal distributions were found in the western and central parts of the structure.

mation is further complicated by the fact that it does not crop out in the extreme eastern regions of the study area. The upper part of formation Rt1 consists predominantly of sandstones, with scattered small quartz pebbles and conglomerates comprising single-pebble layers (Fig. 4). These conglomerates are invariably laterally discontinuous, and conglomerate and sandstone are typically arranged in fining-upward sedimentary cycles. Grain size of the sandstones vary from medium to very coarse, apparently randomly. The dominant internal structure is medium- and large-scale trough cross-bedding, but planar crossbeds, with set thicknesses up to 90 cm, are also found. Paleocurrent distributions are unimodal, with a mode toward the southeast, except in the eastern outcrops, where a polymodal distribution was recorded (Figs. 3 and 4). In the distribution for this area, the secondary, northeastern mode is separated from the dominant one by only 30°. To test the statistical significance of the secondary mode, the probability of an azimuth reading of 60° (the upper azimuth limit of the secondary mode) belonging to the same population as the dominant mode was investigated. For this purpose a normal distribution of paleocurrent readings was assumed, as suggested by Potter and Pettijohn (1963). The mean (117°) and standard deviation (14°) of the dominant mode were subsequently used to test the probability in question through a tail area probability test provided in Statgraphics software. The calculated low probability (0.0036) suggests that azimuth readings of 60° would not belong to the paleocurrent population of the dominant mode and that the secondary mode is significant.

Formation Rt2

In the measured section this formation is 110 m thick and comprises quartz-pebble conglomerate, containing pebbles with long axes up to −7 phi, and interbedded, coarse-grained sandstone (Fig. 4). The conglomerates are apparently devoid of internal structure, whereas the sandstones are predominantly trough and planar cross-bedded with subordinate plane beds. Individual conglomerate beds appear to be lenticular and cannot be traced along the outcrops for more than a number of tens of meters.

Paleocurrent distributions of trough axes are unimodal in the western parts of the study area with modes towards the southeast (Figs. 4 and 5). In the remainder of the study area, east of data station Number 2 (Fig. 5), the distributions are bimodal and polymodal with variable directions of the dominant modes. Most of the distributions display modes towards the southeast.

Lateral variation of pebble size was investigated by measuring the three axes of at least 100 clasts liberated from weathered outcrops at nine localities in the western and central region. The measuring technique applied is that of Krumbein (1941). From the measurements, means for the intermediate axes were calculated for each of the nine samples and these values were used in a first-order trend surface analysis. This produced a very high correlation coefficient of 0.93 and revealed a gradual size decrease towards the southeast (Fig. 6), the same direction as that of the modes of the unimodal paleocurrent distributions.

Formation Rt3

In the measured profile formation, Rt3 is 350 m thick and consists of medium- to coarse-grained sandstone with thin small-pebble conglomerates and scattered pebbles (Fig. 7). Conglomerates are particularly common within the lower part of the formation, but are not laterally continuous. Pebble-size variation within the vertical profile appears to be random (Fig. 7).

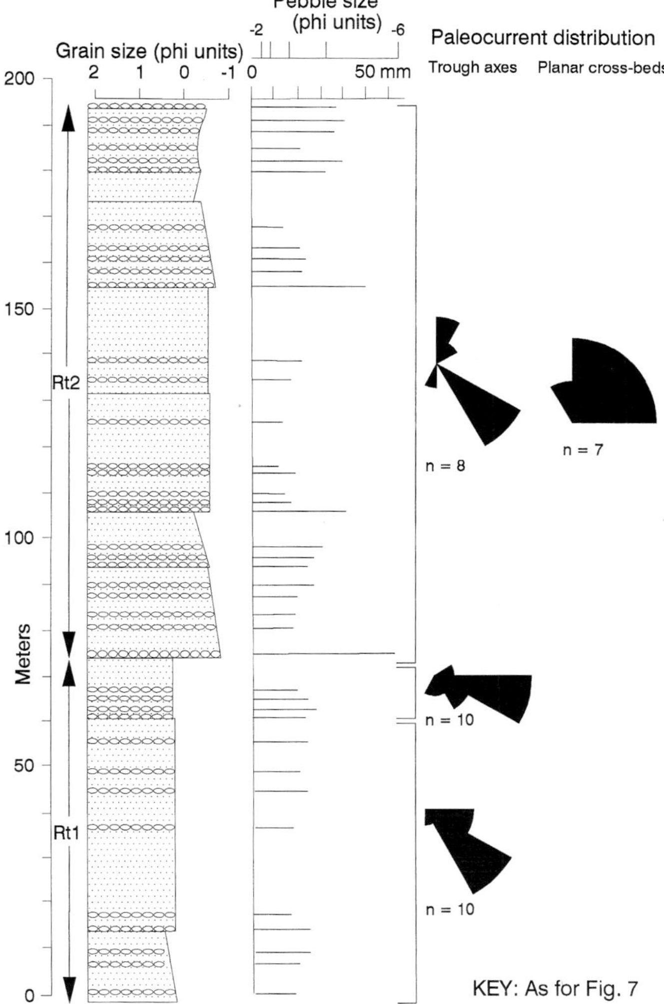

FIG. 4.—Schematic profile for formations Rt1 and Rt2, compiled from measurements along a line perpendicular to tectonic strike, on the farm Leeufontein 485 (Fig. 3).

Paleocurrent distributions of formation Rt3, too, are unimodal in the western regions and bimodal and polymodal east of data station Number 2 (Fig. 5). For the unimodal distributions the modes are, like those of formations Rt1 and Rt2, towards the southeast. Within the measured section, paleocurrent distributions for trough and planar cross-bedding are bimodal, except for one sample representing planar crossbeds in the middle of the formation (Fig. 7). However, in most cases the dominant modes for distributions of trough and planar cross-beds do not coincide. Troughs indicating opposed paleocurrents were found within the same coset. In the vertical profile the dominant paleocurrent modes of trough axes are towards the north near the base of the formation, but towards its top the modes are oriented predominantly toward the southeast. The frequency distribution of trough widths recorded within the measured section was investigated by compiling two histograms; one representing azimuth readings of the second and third quadrants (the quadrants containing the modes of unimodal distributions) and one for azimuths of the first and fourth quadrants (Fig. 8). The former histogram shows three modes, whereas the latter may be considered unimodal, if the weak tail representing trough widths of more than 160 cm is ignored.

The variation of maximum set-thickness of planar cross-beds within the vertical profile is shown on Fig. 7, which indicates no noticeable upward increase or decrease. At the locality where the section was measured, on the farm Leeufontein 485, two planar cross-bed sets, comprising foreset packages separated by laminae consisting of fine-grained material, were found (Fig. 9). In the first of these sets, the laminae separating foreset groups initially appeared to comprise mudstone, but further examination revealed that they consist of immature sandstone with grain size smaller than that of the sandstone constituting the foresets (Fig. 10). Thickness variation of consecutive foreset packages appears to be cyclic (Fig. 11A). Thickness of this set is 140 cm and the corrected mean dip of foresets is 18° in direction 025°. In the second cross-bed set found, foreset packages are separated by mudstone laminae, and, although only six packages could be measured, the thickness variations of consecutive packages also appear to be cyclic (Fig. 11B). The corrected mean dip of foresets in this 43-cm-thick cross-bed set is 15° in direction 073°. An additional example of foreset groups draped by mudstone laminae (Fig. 12) was found near the top of formation Rt3 on the farm Witkop 475 near paleocurrent data station Number 3 (Fig. 3). However, no cyclic variation of foreset package thickness could be determined for this set.

Petrographically, the sandstones of formation Rt3 consist of framework grains of monocrystalline and polycrystalline quartz and a matrix comprising predominantly sericite and fine-grained quartz. Percentage framework grains for the 19 sandstone units examined varies from 61.8% to 77.6% with a mean of 67.9% and a standard deviation of 5.56%. Applying the classification scheme of Pettijohn et al. (1972), all the units are therefore classified as quartz wackes. Within the lower 150 m of formation Rt3, the variation of mean grain-size, as measured in thin section, is apparently random. However, within the upper part of the formation there is a general upward decrease in grain size (Fig. 7).

Formation Rt4

Formation Rt4 has been studied by Els and Mayer (1992), but for full comprehension of this paper, it necessary to summarize its dominant characteristics and the environmental interpretation by Els and Mayer (1992). The most prominent characteristic of formation Rt4, which comprises exclusively sandstones, is the very high textural maturity of the arenites, which contain up to 98% quartz framework grains. In addition to trough and planar cross-beds, formation Rt4 exhibits structures such as hummocky cross-stratification and parallel plane-lamination that is laterally continuous for several meters. Except for the western parts of the study area, paleocurrent distributions are bimodal or polymodal. A sharp, planar contact separates formation Rt4 from the overlying formation Rt5, which has previously tentatively been interpreted as fluvial in origin (Els and Mayer, 1992). Formation Rt4 was interpreted to be of shallow marine origin, probably deposited in the near-shore zone.

DISCUSSION AND INTERPRETATIONS

General Paleoenvironmental Setting

The unimodal paleocurrent distributions found for all three formations in the western part of the study area constitute the

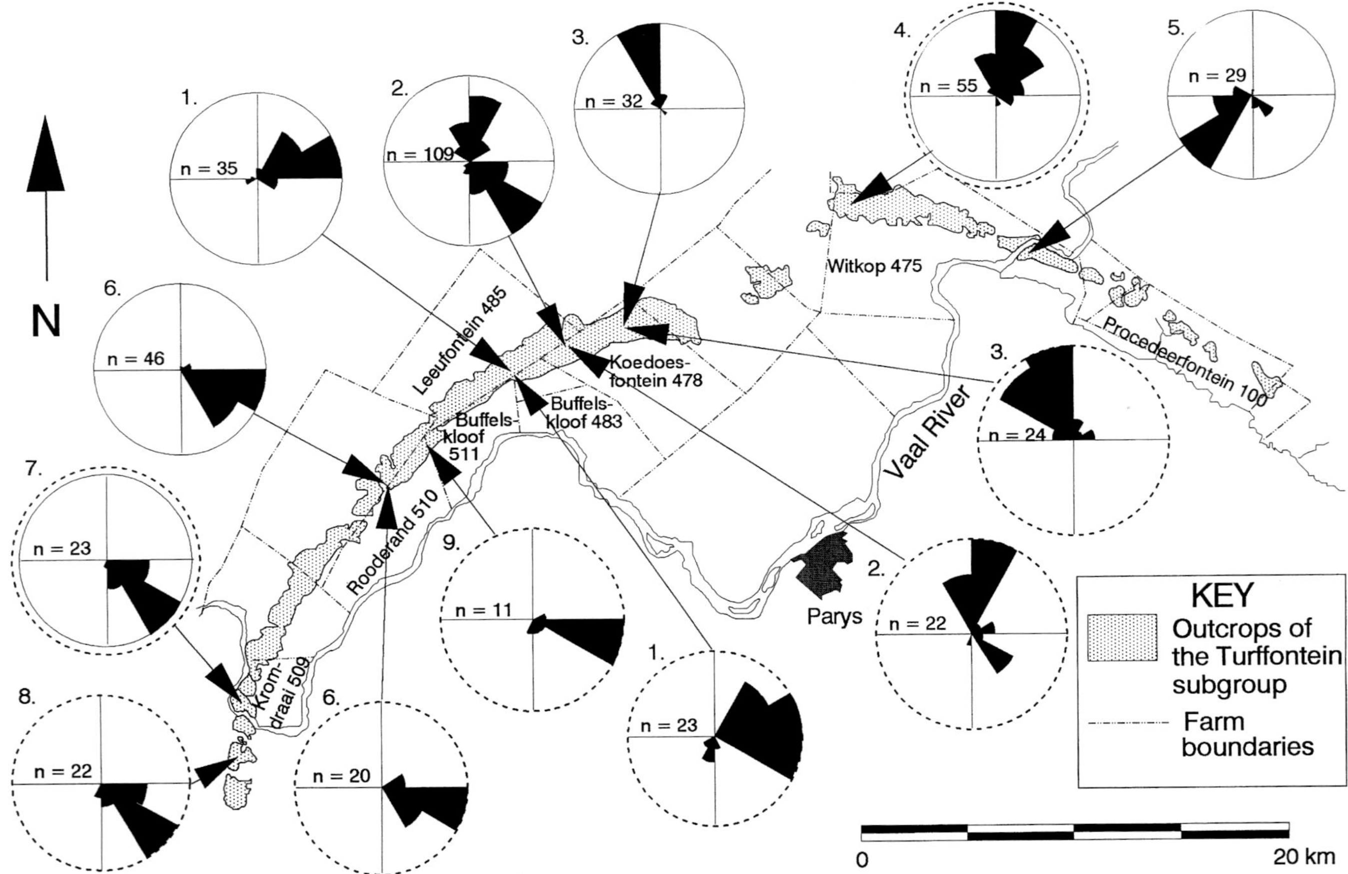

FIG. 5.—Map showing paleocurrent distributions of trough axes for formation Rt2 (compass roses in circles of dashed lines) and formation Rt3 (compass roses enclosed in circles of solid lines). Compass roses enclosed within both types of circles depict distributions of combined readings for the two formations.

most significant indication of fluvial deposition in this region (Visher, 1972; Selley, 1982; Long; 1978, Pettijohn et al., 1972). Additional supporting criteria for fluvial deposition in this area are (a) repetitive fining-upward conglomerate-sandstone sedimentary cycles (Ethridge and Westcott, 1984; Nemec and Steel, 1984), (b) the dominance of large- and medium-scale trough and planar cross-beds (Long, 1978), and (c) a gradual downstream decrease in pebble size (Minter and Loen, 1991).

The mudstone-bounded foreset packages found in formation Rt3 in the central and eastern parts of the study area are considered tidal bundles similar to those described by Visser (1980) and provide convincing evidence for tidal deposition of this formation (Nio and Yang, 1991). The cyclic thickness variation of foreset packages lend further support to the hypothesis of tidal deposition of formation Rt3 (Nio and Yang, 1991).

The bimodal and polymodal paleocurrent distributions of the central and eastern regions provide additional support for the hypothesis of tidal deposition of formation Rt3, and also formation Rt2 in this area (Nio and Yang, 1991). In some cases, for example, station 3 for formation Rt3 (Fig. 5), the second mode of the distribution is weak and represents only two readings. The principal mode, however, is towards the north-northwest, opposite to the southeastern mode of unimodal distributions of the western regions, which are thought to indicate offshore direction. The principal mode for formation Rt3 at station 3 therefore probably represents a flood-dominated tidal regime here. The polymodal paleocurrent distribution of the upper part of formation Rt1 in the eastern parts of the study area suggests tidal conditions during the late depositional stages of the formation in this area.

The reason for the change in the vertical profile of formation Rt3 of the dominant paleocurrent modes (Fig. 7) is not certain, but this phenomenon could be ascribed to a change from flood to ebb-dominated deposition.

In summary, both fluvial and tidal facies are present for all three formations studied. For formations Rt2 and Rt3 a paleobayline separating fluvial and tidal facies is situated between data stations 1 and 6 (Fig. 5). For formation Rt1, the paleobayline is located farther to the east (Fig. 3). (Bayline, rather than shoreline is used to indicate the boundary between tidal and fluvial deposits, which, according to the definition of Posamentier and Vail [1988], are deposited exclusively above sealevel.)

Nature of the Proposed Tidal System

From the data presented it can be concluded that differences between the fluvial and tidal facies of formations Rt2 and Rt3 are subtle; the tidal facies is recognized mainly by bimodal and polymodal paleocurrent distributions, and the occurrence of

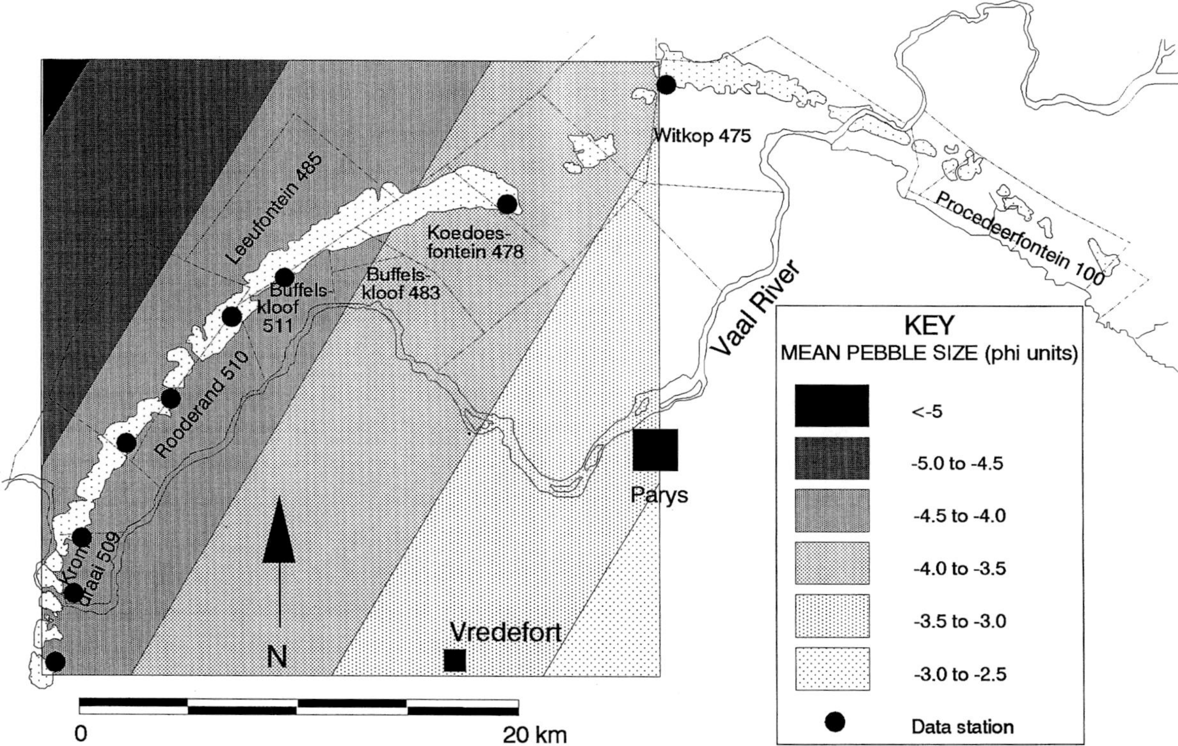

FIG. 6.—First-order trend surface map of mean pebble size (true intermediate axes) of the conglomerates of formation Rt2. Note that pebble size decreases towards the southeast, the general direction of the unimodal paleocurrent modes found for this formation in the western parts of the study area (Fig. 5).

rare tidal bundles and mud-draped cross-bed foresets. In terms of typical lithofacies and lithofacies associations, the fluvial and tidal facies are remarkably similar. For example, in the case of the conglomeratic formation Rt2, both facies are characterized by massive, clast-supported, large-pebble conglomerate, coarse-grained, medium- to large-scale trough and planar cross-bedded sandstone and occasional plane-bedded sandstone. In both facies, conglomerate and sandstone units are typically arranged in fining-upward sedimentary cycles and the conglomerates are characteristically lenticular. Mudstone and siltstone units are rare and very thin in both the fluvial and tidal facies. Because these characteristics are typical of braided stream deposits, it is concluded that both facies represent braided stream deposits. These proposed braided tidal systems are envisaged to have been similar to pure fluvial braided streams, comprising temporary islands of sediment separated by bifurcating and rejoining channels (Miall, 1977). Within such channels migrating dunes produced ebb- and flood-oriented trough cross-bedding, with trough axes oriented parallel to the main flow direction. Planar cross-bed sets probably formed in a variety of straight-crested bedforms such as bars and sandwaves (Harms et al., 1975; Miall, 1977). Because the foresets in some bars, especially the linguoid and diagonal types, do not dip in the principal flow direction (Miall, 1977), paleocurrent modes obtained from such bars would be oblique or even perpendicular to trough axis orientations. Sediment periodically deposited in the tidal zone was most probably supplied exclusively by the upstream fluvial system. This sediment was reworked by braided tidal currents.

Relative Sea-level Changes and Sedimentation Rates

The fine-grained character, considerable thickness and widespread occurrence of formation Rb and its correlatives suggest that it is probably of shallow marine origin and most likely deposited on the continental shelf after a relative sea-level rise. If this hypothesis is considered in conjunction with the conformable, transitional contacts of the correlatives of this formation in the Carletonville goldfield (Engelbrecht et al., 1986), and in the Klerksdorp goldfield (Antrobus et al., 1986), it seems likely that the lower part of the overlying formation Rt1 could represent a transition from shelf to paralic and fluvial deposition during a subsequent offlap. Because of the lack of outcrops of this lower part of the formation, this hypothesis remains unproved. Also, the reason for the proposed water-depth decrease is unknown, but a relative sea-level fall during the onset of deposition of formation Rt2 seems a likely explanation.

The large, combined thickness of the tidal facies of formations Rt2 and Rt3 indicates a state of near-equilbrium—throughout the time of their deposition—between the rate of relative sea-level rise and that at which accommodation was filled. However, towards the end of deposition of formation Rt3, the rate of relative sea-level rise exceeded that of accommodation filling, resulting in shallow marine conditions during deposition of formation Rt4.

Basal Contact of Formation Rt2

Els (1991) used the abrupt upward pebble-size increase across the erosive basal contact of the Middelvlei gold placer to infer a disconformity at this contact. A similar abrupt size

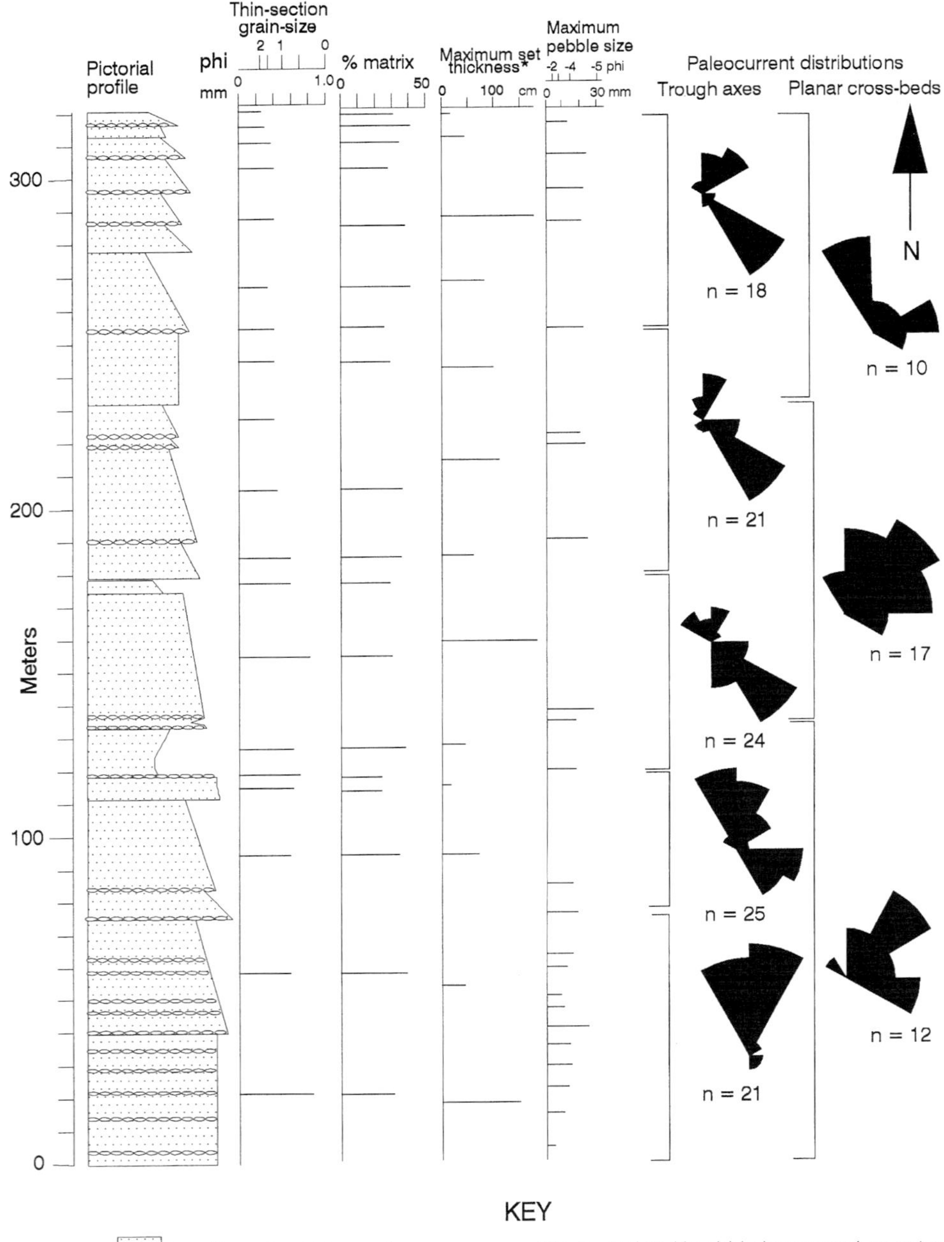

FIG. 7.—Sedimentological profile for formation Rt3, compiled from measurements along a line perpendicular to tectonic strike, on the farm Leeufontein 485 (Fig. 5). The paleocurrent directions indicated by trough axes are also represented in Fig. 5 as a combined distribution for station 2.

increase at the base of formation Rt2 is illustrated in Fig. 4, suggesting a disconformity at the base of this formation. The change in the vertical profile from the unimodal paleocurrent distributions of formation Rt1 to the bimodal distributions of formation Rt2 is abrupt, too, occurring at the bottom contact of the basal conglomerate of the latter formation (Fig. 4). This abrupt change lends further support to the existence of a disconformity at the base of formation Rt2. This disconformity probably developed due to scouring of the sands of formation Rt1 by the prograding coarse-grained gravels of formation Rt2, a process also proposed for the fluvial Middelvlei gold placer by Els (1991). The common manifestations of such progradations are upward coarsening in the vertical profile, and, in some cases, erosion of the paleo-surface (Frostick and Steel, 1993). Formations Rt1 and Rt2 clearly exhibit such characteristics. Firstly, there is convincing evidence of erosion at the base of the latter. Secondly, the general upward increase in frequency of conglomerates and in pebble size within the top part of formation Rt1 indicates upward coarsening (Fig. 4). The reason for the proposed progradation and resultant disconformity may be one of several alternatives. One mode of development of unconformities, which is not directly related to sea-level changes, is scouring by by-passing sediment, as a response to sedimentation in a basin that is almost completely filled (Frostick and Steel, 1993). Alternatively, the proposed progradation may have been induced by tectonic uplift in the source area,

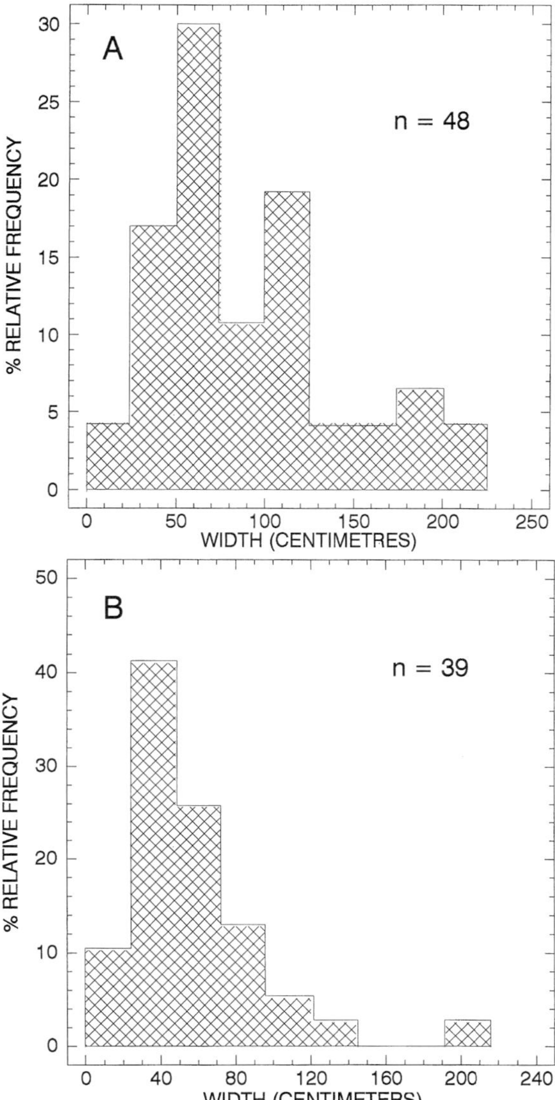

FIG. 8.—Histograms showing the frequency distributions of the widths of cross-bed troughs measured throughout the vertical profile represented on Figs. 4 and 7. Histogram (A) represents troughs that indicate transport directions of the second- and third-azimuth quadrants (in which the mode of unimodal paleocurrent distributions lies), whereas histogram (B) represents troughs of the remaining quadrants.

and could therefore be similar in origin to those associated with the progressive unconformities of Riba (1976) and the marginal unconformities of Winter and Brink (1991). However, a more plausible explanation for the postulated progradation is that it was induced by a relative sea-level fall. The lack of pronounced incision into formation Rt1 suggests a slow fall and therefore a type 2 unconformity (Posamentier and Vail, 1988) at the base of formation Rt2. This hypothesis is supported by the similarity, in terms of vertical arrangement of facies, between the model of Posamentier and Vail (1988, their Fig. 18) and the sedimentological architecture of the lower Turffontein formations. In both cases fluvial facies are disconformably overlain by coastal plain (tidal) sediments in the proximal regions, whereas deeper into the basin, coastal plain facies occurs both below and above the unconformity.

Direction of the Transgression Represented by Formation Rt4

Considering the arcuate geographic distribution of the Turffontein outcrops (Figs. 3 and 5) in conjunction with the fact that no tidal indicators have been found in the central and western outcrops of the top part of formation Rt1, a marine transgression from the southeast seems likely.

Frequency Distribution of Trough Widths

Considering the fact that during deposition of formations Rt2 and Rt3 sediment was supplied to their interpreted tidal environment in volumes sufficient to maintain equilibrium be-

FIG. 9.—Photograph showing a solitary planar cross-bed set, with foreset packages separated by laminae comprising sandstone with grain-size smaller than that of the sandy foresets, in the top part of formation Rt3. The photograph is oriented upside-down to show correct stratigraphic polarity of overturned beds (stratigraphic up toward top of photo). The scale-board, (15 cm), is oriented parallel to the lower bounding surface of the cross-bed set. (Locality: Leeufontein 485, Fig. 3).

FIG. 10.—Close-up view of the cross-beds shown in Fig. 9. The lamina (L) separating the foreset packages (F) consists of sandstone, which is finer grained than that of the packages. Diameter of the coin is 15 mm.

tween sedimentation and subsidence, it can be assumed that three types of current were active in the area: flood currents, ebb currents and currents transporting sediment from the source area. The bimodal distribution of the histogram representing troughs of the first- and fourth-azimuth quadrants (Fig. 8), therefore probably reflects mixing of two populations of troughs, namely troughs formed by ebb currents and troughs produced by streams supplying sediment to the area.

Sandstone Laminae on Cross-bed Foresets

The occurrence of fine-grained sandstone laminae between tidal bundles is apparently not common in the tidal environment, although separating siltstone laminae have been reported by Aitken and Flint (1995). The sand in laminae separating tidal bundles of this study is considered too coarse to represent suspended load and was probably deposited by traction, rather than by suspension, as are mud laminae occurring on cross-bed foresets (Visser, 1980). Considering the low corrected dip angle of the foresets of the cross-bed set in question, it seems possible that sand could have been transported up the foresets through traction by the subordinate tidal current. The apparent lack of internal structure in these laminae can possibly be ascribed to their small thickness and/or to diagenetic effects.

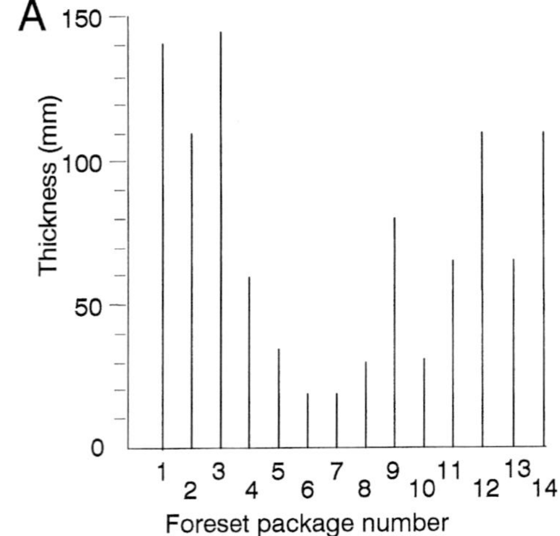

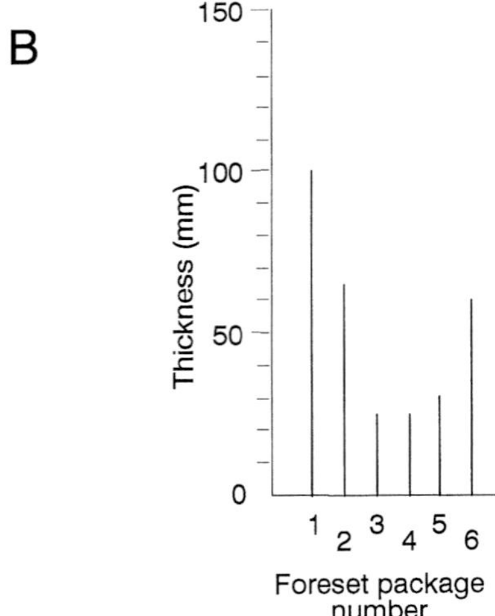

FIG. 11.—Graphical representations of the thickness variations of foreset packages of, (A) the crossbed set shown on Figure 9; and (B) an additional set, also found towards the top of formation Rt3 on the farm Leeufontein 485 (Fig. 3).

FIG. 12.—Photograph showing a low-angle, solitary, planar crossbed set, with foreset packages separated by mudstone laminae, near the base of formation Rt3. Differential erosion has caused the resistant sandstone laminae to stand out as in yardangs. The photograph is oriented upside-down to show correct stratigraphic polarity of overturned beds (stratigraphic up towards top of photo). The scale-board (15 cm) is oriented parallel to the lower bounding surface of the crossbed set. (Locality: Witkop 475, Fig. 5).

CONCLUSIONS

1. A northwest-to-southeast transition from fluvial to tidal deposition for the lower Turffontein formations is evident in the Vredefort structure. This transition is manifested mainly by bimodal and polymodal paleocurrent distributions, but tidal bundles and mud-draped cross-bed foresets have been found, too.
2. The tidal facies of the lower Turffontein formations most probably comprised gravelly and sandy braided streams. Sediment deposited in the tidal zone was supplied by an upstream, fluvial braided-stream system.
3. The large thickness of the tidal deposits suggests a state of near-equilibrium between sedimentation rate and relative sea-level rise. Tidal deposition was terminated when the rate of relative sea-level rise exceeded that of filling of accommodation space.
4. The disconformity at the base of the conglomeratic formation Rt2 probably formed as a result of a relative sea-level fall and is classified as a type 2 unconformity.

ACKNOWLEDGMENTS

This study was made possible through grants by the Bureau for Research of the Potchefstroomse Universiteit vir C.H.O. We gratefully acknowledge the assistance of the following students: Nico Denner, Sias le Roux, and Jaco van der Merwe. Thanks are due to two anonymous reviewers for their constructive criticism and useful suggestions

REFERENCES

AITKEN, J. F., AND FLINT, S. S., 1995, The application of high-resolution sequence stratigraphy to fluvial systems: A case study from the Upper Carboniferous Breathitt Group, eastern Kentucky, USA: Sedimentology, v. 42, p. 3–30.

ALBAT, H. M., 1988, Shatter cone/bedding interrelationships in the Vredefort structure: evidence for meteorite impact?: South African Journal of Geology, v. 91, p. 106–113.

ALBAT, H. M., AND MAYER, J. J. 1989, Megascopic planar shock fractures in the Vredefort structure: a potential time marker?: Tectonophysics, v. 162, p. 265–276.

ANTROBUS, E. S. A., BRINK, W. C. J., BRINK, M. C., CAULKIN, J., HUTCHISON, R. I., THOMAS, D. E., VAN GRAAN, J. A., VILJOEN, J. J., 1986, The Klerksdorp Goldfield, *in* Anhaeusser, C. R., Maske, S. eds., Mineral Deposits of Southern Africa (I): Geological Society of South Africa Publication, p. 549–598.

BISSCHOFF, A. A., 1982, Thermal metamorphism in the Vredefort Dome: Transactions of the Geological Society of South Africa, v. 85, p. 43–57.

DU TOIT, A. L., 1953, Geology of South Africa: Edinburgh, Oliver and Boyd, 611 p.

ELS, B. G., 1991, Placer formation during progradational fluvial degradation: The Late Archaean Middelvlei gold placer, Witwatersrand, South Africa: Economic Geology, v. 86, p. 29–45.

ELS, B. G., AND MAYER, J. J., 1992, Transgressive and progradational beach and nearshore facies in the Late Archaean Turffontein subgroup of the Witwatersrand supergroup, Vredefort Area, South Africa: South African Journal of Geology, v. 95, p. 60–73.

ENGELBRECHT, C. J., BAUMBACH, G. W. S., MATTHYSEN, J. L., FLETCHER, P., 1986, The West Wits Line, *in* Anhaeusser, C. R., and Maske, S., eds., Mineral Deposits of Southern Africa (I): Geological Society of South Africa Publication, p. 599–648.

ETHRIDGE, F. G., AND WESTCOTT, W. A., 1984, Tectonic setting, recognition and hydrocarbon reservoir potential of fan-delta deposits, *in* Koster, E. H., and Steel, R. J., eds., Sedimentology of gravels and conglomerates: Calgary, Canadian Society of Petroleum Geologists Memoir 10. 441 p.

FROSTICK, L. E., AND STEEL, R. J., 1993, Tectonic signatures in sedimentary basin fills: An overview, *in* Frostick, L. E., and Steel, R. J., eds., Tectonic controls and signatures in sedimentary successions: Oxford, International Association of Sedimentologists Special Publication 20, p. 1–12.

GALEHOUSE, J. S., 1971, Point counting, *in* Carver, R. E., ed., Procedures in sedimentary Petrology: New York, Wiley-Interscience, p. 385–407.

HARMS, J. C., SOUTHARD, J. B., SPEARING, D. R., AND WALKER, R. G., 1975, Depositional environments as interpreted from primary sedimentary structures and stratification sequences: Society for Sedimentary Geology Short Course No. 2, Dallas, 161 p.

KRUMBEIN, W. C., 1941, Measurements and geological significance of shape and roundness of sedimentary particles: Journal of Sedimentary Petrology, v. 11, p. 64–72.

LONG, D. G. F., 1978, Proterozoic stream deposits: Some problems of recognition and interpretation of ancient sandy fluvial systems, *in* Miall, A. D., ed., Fluvial Sedimentology: Calgary, Canadian Society of Petroleum Geologists Memoir 5, 859 p.

MANTON, W. I., 1962, The orientation and implication of shatter cones in the Vredefort Ring structure: Unpublished M.Sc. thesis, University of the Witwatersrand, Johannesburg, 167 p.

MAYER, J. J., 1992, Aspects of sedimentological, structural and stratigraphic investigations of the collar strata of the Vredefort structure: Contributions towards an understanding of the pre-history and origin of the structure: Unpublished Ph.D. Dissertation, Potchefstroomse Universiteit vir C.H.O., Potchefstroom, South Africa, 284 p.

MAYER, J. J., AND ELS, B. G., 1992, The reconstruction of palaeocurrent dispersal patterns for Turffontein strata in the Vredefort structure in relation to a tectonic model for the area: South African Journal of Geology, v. 95, p. 40–50.

MIALL, A. D., 1977, A review of the braided-river depositional environment: Earth Science Reviews, v. 13, p. 1–62.

MINTER, W. E. L. AND LOEN, J. S., 1991, Palaeocurrent dispersal patterns of Witwatersrand gold placers: South African Journal of Geology, v. 94, p. 70–85.

NEL, L. T. 1927, The geology of the country around Vredefort—an explanation of the geological map: Special Publication of the Geological Survey of South Africa, v. 6, 134 p.

NEMEC, W., AND STEEL, R. J., 1984, Alluvial and coastal conglomerates: Their significant features and some comments on gravelly mass-flow deposits, *in* Koster E. H., and Steel, R. J., eds., Sedimentology of gravels and conglomerates: Calgary, Canadian Society of Petroleum Geologists Memoir 10, 441 p.

NIO, S. D., AND YANG, C., 1991, Diagnostic criteria of clastic tidal deposits: A review, *in* Smith, D. G., Reinson, G. E., Zaitlin, B. A. and Rahmani, R. A., eds., Clastic tidal sedimentology: Calgary, Canadian Society of Petroleum Geologists, Memoir 16.

PETTIJOHN, F. J., POTTER, P. E., AND SIEVER, R., 1972, Sand and Sandstone: New York, Springer-Verlag, 618 p.

POSAMENTIER, H. W. AND VAIL, P. R., 1988, Eustatic controls on clastic deposition II—Sequence and systems tract models, *in* Wilgus, C. K., Hastings, B. S., Kendal, C. G. St. C, Posamentier, H. W., Ross, C. A., and Van Wagoner, J. C., eds., Sea-level changes: An integrated approach: Tulsa, Society for Sedimentary Geology Special Publication 42.

POTTER, P. E. AND PETTIJOHN, F. J., 1963, Paleocurrents and Basin analysis: Heidelberg, Springer-Verlag, 296 p.

PRETORIUS, D. A., 1986, The Witwatersrand Basin: Surface and subsurface geology and structure: Map (1:500 000), Geological Society of South Africa.

RIBA, O., 1976, Syntectonic unconformities of the Alt Cardener, Spanish Pyrenees: A genetic interpretation: Sedimentary Geology, v. 15, p. 213–233.

The South African Committee for Stratigraphy, 1980, Stratigraphy of South Africa. Part 1. Compiler L. E. Kent, Lithostratigraphy of the Republic of South Africa, South West Africa/Namibia, and the Transkei and Venda: Handbook 8 of the Geological Survey of South Africa, 690 p.

SELLEY, R. C., 1982, An Introduction to Sedimentology: London, Academic Press.

VISHER, G. S., 1972, Physical characteristics of fluvial deposits: Society for Sedimentary Geology Special Publication 16, p. 84–97.

VISSER, M. J., 1980, Neap-spring cycles reflected in Holocene subtidal large-scale bedform deposits: A preliminary note: Geology, v. 8, p. 543–546.

WINTER, H. DE LA R., AND BRINK, M. C., 1991, Chronostratigraphic subdivision of the Witwatersrand Basin based on a Western Transvaal composite column: South African Journal of Geology, v. 94, p. 191–203.

SUBJECT INDEX

SUBJECT INDEX

Tidalites: Processes and Products, SEPM Special Publication No. 61

T

U

V

W

Y